Cell Invasion

Medical Intelligence Unit

CRC Press
Taylor & Francis Group
6000 Broken Sound Parkway NW, Suite 300
Boca Raton, FL 33487-2742

First issued in hardback 2019

ISBN-13: 978-1-58706-073-1 (hbk)
ISBN-13: 978-1-58706-099-1 (pbk)

Visit the Taylor & Francis Web site at
http://www.taylorandfrancis.com

and the CRC Press Web site at
http://www.crcpress.com

Library of Congress Cataloging-in-Publication Data

CIP information applied for but not received at time of publication.

CONTENTS

EDITORS

Jyrki Heino
Chapter 2
Veli-Matti Kähäri
Chapter 1
Centre for Biotechnology
MediCity Research Laboratory
and Department of Medical Biochemistry
University of Turku
Turku, Finland
and Department of Biology
University of Jyväskylä, Jyväskylä, Finland

CONTRIBUTORS

Andrew H. Baker
University of Bristol and University of Glasgow
Glasgow, U.K.
Chapter 4

Anja Bosserhoff
University Hospital RWTH
Aachen, Germany
Chapter 5

Reinhard Buettner
University Hospital RWTH
Aachen, Germany
Chapter 5

Sarah J. George
University of Bristol and University of Glasgow
Glasgow, U.K.
Chapter 4

Lari Häkkinen
University of British Columbia
Vancouver, Canada
Chapter 3

Henri J. Huttunen
University of Helsinki
Helsinki, Finland
Chapter 6

Jonathan C.R. Jones
University of Texas Southwestern Medical Center
Dallas, Texas, U.S.A.
Chapter 9

Pekka Koistinen
University of Turku, Turku, Finland and University of Jyväskylä
Jyväskylä, Finland
Chapter 2

Leeni Koivisto
University of British Columbia
Vancouver, Canada
Chapter 3

Hannu Larjava
University of British Columbia
Vancouver, Canada
Chapter 3

Andrew C. Newby
University of Bristol and University of Glasgow
Glasgow, U.K.
Chapter 4

William C. Parks
University of Texas Southwestern
Medical Center
Dallas, Texas, U.S.A.
Chapter 9

Brian K. Pilcher
University of Texas Southwestern
Medical Center
Dallas, Texas, U.S.A.
Chapter 9

Heikki Rauvala
University of Helsinki
Helsinki, Finland
Chapter 6

Niina Reunanen
University of Turku
Turku, Finland
Chapter 1

M. Patrizia Stoppelli
International Institute of Genetics
and Biophysics
Napoli, Italy
Chapter 8

Bryan P. Toole
Tufts University School of Medicine
Boston, Massachusetts, U.S.A.
Chapter 7

Taina Turpeenniemi-Hujanen
University of Oulu
Oulu, Finland
Chapter 10

PREFACE

An Introduction to the Cell Invasion

Every student in cell biology is familiar with a textbook picture of a migrating cell, which generates a lamellipodium to reach extracellular matrix and form a new adhesion site. The picture indicates some critical molecular components of cell migration, namely extracellular matrix proteins and integrin-type adhesion receptors. The movement of the cell is powered by the interaction of actin microfilaments with myosin. A fourth type of molecular machinery, i.e., matrix degrading proteinases, is also essential for the penetration of a cell into adjacent tissue. Our understanding of cell locomotion has greatly increased during the past decade. Migration and invasion of cells appears to be a result of complex interplay between the numerous protein families participating in this process. Mechanisms of cell movement are of obvious interest for basic cellular and developmental biology. The same mechanisms are important in the pathogenesis of various diseases. Tumor invasion, cancer-related angiogenesis and cardiovascular diseases are just a few examples of common disorders in which cell locomotion plays a central role. Therefore, it is not surprising that pharmacological inhibitors of adhesion receptors and matrix degrading proteinases have been developed in rapidly growing numbers. The ongoing trials will show whether these new therapeutic molecules will find their way to clinical medicine.

Extracellular Matrix Forms the Platform of Cell Invasion

Extracellular matrix gives the form and the tensile strength to a tissue and serves as a platform for the cells to attach. While cells have the potency to bind to most of the structural proteins in the matrix, some proteins seem to be more important than others in terms of cell invasion. Fibronectin was found almost 30 years ago, and it has become a prototype of the matrix glycoproteins mediating cell adhesion and movement. Several studies have indicated its importance in numerous processes related to embryonic development, tissue healing, cancer progression etc. Hannu Larjava, Leeni Koivisto and Lari Häkkinen (Chapter 3) have written a chapter describing fibronectin in cell migration.

Cell Surface Integrins Mediate Adhesion to Extracellular Matrix

In the mid-1980s several research groups found that cells and platelets attach to various extracellular matrices by structurally related heterodimeric receptor complexes, later named integrins because of their ability to integrate the cytoskeleton to the matrix. The integrin family is composed of at least 24 distinct heterodimers, containing numerous receptors for fibronectin, collagens, and laminins. In their chapter Pekka Koistinen and Jyrki Heino (Chapter 2) give an introduction to the integrin family and review the changes described in the integrin expression of cancer cells.

Matrix Metalloproteinases and Serine Proteinases Make Possible the Invasion of Cells through Extracellular Matrix

Extracellular matrix forms a dense molecular meshwork, and cells cannot penetrate it without proteolytic degradation of the matrix molecules. M. Patrizia Stoppelli (Chapter 8) reviews the group of serine proteinases and the introduction to the large family of the matrix metalloproteinases and their role in cancer cell invasion can be found in the chapter written by Niina Reunanen and Veli-Matti Kähäri (Chapter 1). Despite the fact that proteinases are central proteins in the process of tumor invasion, it is often less obvious whether the cancer cells themselves produce the enzymes required for tumor invasion or whether they are released by the stromal fibroblasts. Bryan P. Toole (Chapter 7) reports about an interesting protein, named EMPPRIN, a cancer cell-related inducer of matrix metalloproteinase expression in fibroblasts. A growing number of pharmacological inhibitors of matrix metalloproteinases has made the matrix metalloproteinases even more interesting for clinical oncology. Taina Turpeenniemi-Hujanen (Chapter 10) has taken a clinical view of matrix metalloproteinases as prognostic markers and therapeutic targets in cancer.

Cell Invasion is an Interplay between Numerous Protein Families

Integrins anchor the cells to matrix, but they also mediate cellular signals regulating, for example, the expression of matrix molecules and proteinases. Inflammatory cytokines, as well as growth and differentiation factors, regulate the expression of matrix, proteinase, and adhesion receptor genes at the same time when they rule the organization of cytoskeleton. New powerful methods have made it possible to take a genome-wide approach to analyze migration-related changes in cellular gene expression. The studies have supported the idea of the important role of the above mentioned gene families, but also unveiled new concepts. In their chapter Anja Bosserhoff and Reinhard Buettner (Chapter 5) review the recent progress in this field.

Cell Invasion is an Essential Process in Many Areas of Biology and Medicine

Cell migration and invasion also play a central role during tissue repair. Brian K. Pilcher, Jonathan C.R. Jones, and William C. Parks have written a chapter focused on wound keratinocytes (Chapter 9). Researchers working on numerous other areas in biology or medicine have also realized the importance of similar cellular mechanisms. Henri J. Huttunen and Heikki Rauvala have studied cell adhesion mechanisms in the central nervous system and they review the recent progress in this field (Chapter 6). Finally, Sarah J. George, Andrew C. Newby and Andrew H. Baker have written a chapter dealing with cell invasion in cardiovascular system (Chapter 4).

Jyrki Heino and Veli-Matti Kähäri

CHAPTER 1

Matrix Metalloproteinases in Cancer Cell Invasion

Niina Reunanen and Veli-Matti Kähäri

Abstract

Controlled remodeling of extracellular matrix (ECM) is essential for growth, invasion, and metastasis of malignant tumors. Matrix metalloproteinases (MMPs) are a family of secreted zinc-dependent endopeptidases collectively capable of degrading ECM components, and there is considerable amount of evidence that they play an important role at different steps of malignant tumor growth. Recent observations also suggest that MMPs play a role in cancer cell survival. In this chapter we discuss the role of MMPs and their inhibitors in tumor cell invasion, as a basis for prognostication and targeted therapeutic intervention.

Matrix Metalloproteinases

MMPs are a family of structurally related zinc-dependent endopeptidases collectively capable of degrading essentially all components of the extracellular matrix (ECM). There is strong evidence for the role of MMPs in physiological ECM remodeling, e.g., during tissue morphogenesis, growth, uterine cycling and postpartum involution, tissue repair, and angiogenesis. In addition, MMPs play a role in pathological conditions with excessive degradation of ECM, such as rheumatoid arthritis, osteoarthritis, atherosclerotic plaque rupture, aortic aneurysms, periodontitis, autoimmune blistering disorders of the skin, dermal photoaging, tumor invasion, and tumor metastasis.[1-4]

To date, 21 human MMPs are known and they can be divided into subgroups based on their structure and substrate specificity (Fig. 1). These subgroups include collagenases, stromelysins and stromelysin-like MMPs, matrilysins, gelatinases, MMP-19-like MMPs, membrane-type MMPs (MT-MMPs), and other MMPs (Fig. 1).[1-4]

MMPs have a multi-domain structure (Fig. 1). N-terminal signal peptide directs the secretion of the proenzyme. Propeptide contains a highly conserved sequence PRCG(V/N)PD, in which the cysteine forms a covalent bond, with the catalytic zinc ion, the cysteine switch, which maintains the proMMP in latent form. The catalytic domain consists of two modules separated by a deep active site cleft with zinc ion at the bottom.[5] Three histidine residues coordinate the binding of catalytic zinc at the active site. This zinc-binding motif HExxHxxGxxH together with the zinc ion is essential for the proteolytic activity of MMPs, and is conserved among all MMPs. There is also a structural zinc, and at least one calcium ion located approximately 12 Å from the catalytic zinc. A proline-rich hinge region links the catalytic domain to the C-terminal hemopexin domain, which is highly conserved and contains four repeats showing sequence similarity to hemopexin, a plasma protein. A disulfide bridge connects the ends of the domain, which plays a functional role in substrate binding and in interactions with the tissue inhibitors of metalloproteinases (TIMPs). In addition to these domains, some MMPs possess additional domains described here.

Cell Invasion, edited by Jyrki Heino and Veli-Matti Kähäri. ©2002 Eurekah.com.

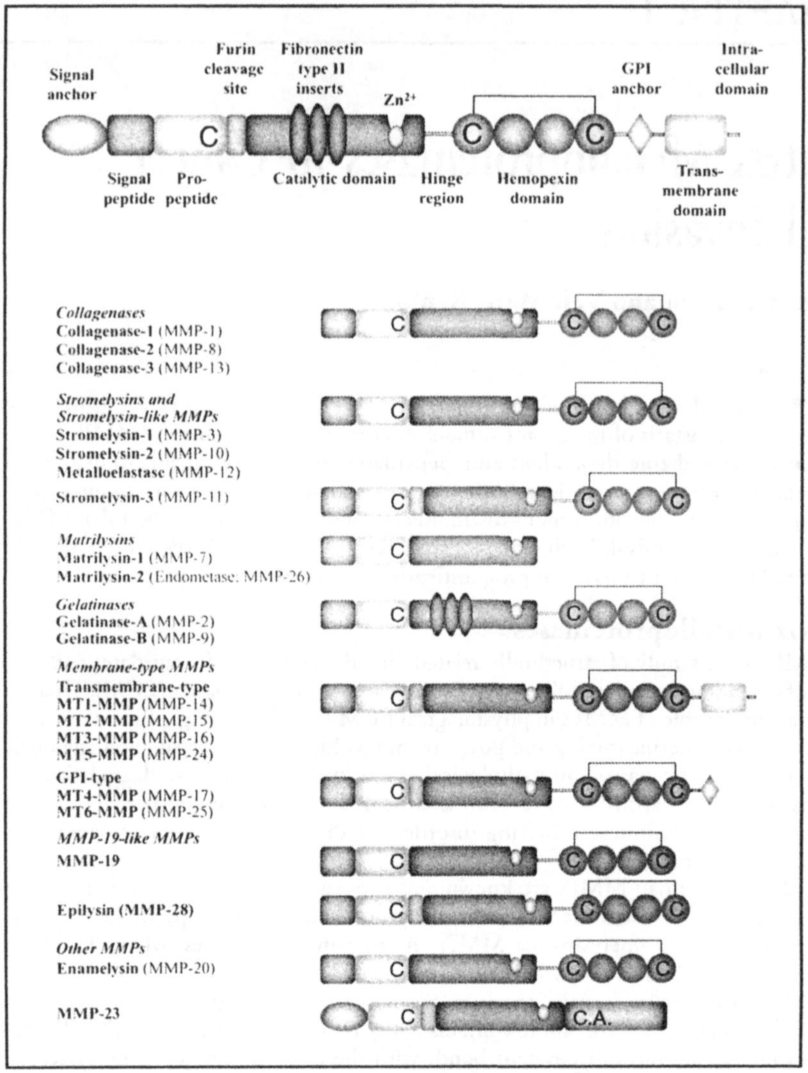

Fig. 1. Domain structures of human MMPs. C, Cysteine; GPI, Glycosylphosphatidylinositol; Cysteine Array.

Collagenases

Collagenase-1 (MMP-1), collagenase-2 (MMP-8), and collagenase-3 (MMP-13), cleave the triple helix of fibrillar collagens of types I, II, III and V. They all cleave native type I collagen between Gly^{775}-Ile^{776} of $\alpha 1$ chain or Gly^{775}-Leu^{776} of $\alpha 2$ chain results in 3/4 N-terminal and 1/4 C-terminal triple-helical fragments, which then denature spontaneously in 37°C into gelatin and are further degraded by other MMPs, such as gelatinases.[1-4] In addition, MMP-13 cleaves type I collagen at N-terminal nonhelical telopeptide.[6] MMP-1 cleaves preferentially type III collagen over other fibrillar collagens, and MMP-8 cleaves type I collagen most efficiently.[7-9] MMP-13 preferentially cleaves type II collagen, and also gelatin 40-fold more effectively than MMP-1 and MMP-8.[7-9]

Human MMP-1 is secreted as major 52 kDa and minor glycosylated 57 kDa proenzymes, and cleavage of propeptide produces active forms of 42 kDa and 47 kDa, respectively.[10] The nine residues RWTNNFREY(183-191) in the catalytic domain together with the C-terminal hemopexin domain are essential for collagenolytic activity, but additional structural elements in the catalytic domain are also required.[11] MMP-1 substrates include type I, II, III, VII, VIII, and X collagens, aggrecan, serine proteinase inhibitors, and α2 macroglobulin. In contrast to many other MMPs, MMP-1 can not cleave BM components. MMP-1 expression is detected in vivo in many physiological situations, such as embryonal development and tissue repair, but also in pathological conditions, including chronic cutaneous ulcers and malignant tumors.[12,13] Production of MMP-1 is induced by growth factors and cytokines. In incision wounds and chronic ulcers, MMP-1 is expressed by basal keratinocytes bordering the sites of active re-epithelialization[14] and is needed for keratinocyte migration on type I collagen.[15]

Collagenase-2 (MMP-8) is synthesized by polymorphonuclear leukocytes maturing in bone marrow, stored in intracellular granules, and released in response to extracellular stimuli.[16] It is also expressed by human articular cartilage chondrocytes in vivo and by mononuclear fibroblast-like cells in rheumatoid synovium.[17,18]

Collagenase-3 (MMP-13) has a wide substrate specificity.[7-9,19-22] Physiological expression of MMP-13 is limited to fetal bone development, postnatal bone remodeling, gingival wound repair, and fetal cutaneous wound repair,[23-26] suggesting a role for MMP-13 in rapid and effective remodeling of collagenous ECM in these situations. MMP-13 expression in vivo is detected in inflammatory conditions, e.g., osteoarthritis,[9] rheumatoid arthritis,[25] chronic cutaneous ulcers,[27] and chronic periodontitis,[28] and in invasive malignant tumors, such as breast carcinomas,[7] squamous cell carcinomas (SCCs) of the head and neck[29] and vulva,[30] primary and metastatic melanomas,[31,32] and transitional cell carcinoma of the urinary bladder.[33] Accordingly, recent observations show, that expression of MMP-13 enhances the invasion capacity of HT-1080 fibrosarcoma cells.[34]

Stromelysins and Stromelysin-Like MMPs

Stromelysin-1 (MMP-3) and stromelysin-2 (MMP-10), have similar structure and substrate specificity. As stromelysin-3 (MMP-11) and macrophage metalloelastase (MMP-12) differ in their structure from stromelysins, they are included in this subfamily as a subgroup of stromelysin-like MMPs.[1-4] MMP-3 and MMP-10 are expressed by keratinocytes and fibroblasts in culture and in vivo. MMP-3 is expressed by stromal cells during mammary gland development and is strongly up-regulated during post-lactational mammary involution when considerable ECM remodeling and alveolar apoptosis occur.[35,36] MMP-3 can induce apoptosis or promote proliferation depending on the differentiation status of the target cell.[37] It also triggers angiogenesis and can act as a natural tumor promoter.[36,37] MMP-3 is a potent activator of latent MMP-1.[38] Deletion of MMP-3 impairs early dermal wound contraction, suggesting a role for MMP-3 in the organization of a multicellular actin network.[39]

Stromelysin-3 (MMP-11) is expressed in many invasive human tumors,[40] and high expression levels correlate with poor clinical outcome in breast cancers patients.[40] MMP-11 is expressed by stromal fibroblasts adjacent to tumor cells,[40] but also by breast carcinoma cells that have undergone a degree of epithelial-to-mesenchymal transition.[41] MMP-11 is important in the early stages of tumorigenesis by favoring cancer cell survival in a tissue environment initially not permissive for tumor growth.[42] MMP-11-deficient mice have reduced chemical-induced tumorigenesis.[43] MMP-11 is expressed by fibroblasts in basal cell carcinomas, squamous cell carcinomas, and benign dermatofibromas,[44,45] MMP-11 prodomain contains a furin cleavage site, and the proenzyme is processed intracellularly and released as a mature enzyme.[46] MMP-11 degrades α1-proteinase inhibitor (α1PI),[47] but the ECM substrates have not been identified, although the existence of such substrates has been suggested.[48]

Macrophage metalloelastase (MMP-12) is constantly expressed by macrophages[49] and in spindle-shaped stromal cells of placenta.[49] MMP-12 is expressed by macrophages in atherosclerotic lesions,[50] abdominal aortic aneurysms,[51] and intestinal ulcerations.[52] In skin, MMP-12 is expressed by tumor cells in cutaneous SCCs,[53] and by macrophages in areas devoid of normal elastic fibers or with disrupted basement membrane.[54] Deletion of MMP-12 in mice leads to impaired macrophage recruitment and protects from cigarette smoke-induced emphysema.[55]

Matrilysins

Matrilysin (MMP-7) and matrilysin-2 (endometase, MMP-26) are the smallest MMPs, both lacking the hinge region and the hemopexin domain.[56,57] MMP-7 has a wide substrate specificity.[50,58] Unlike most MMPs, it is constitutively expressed by many epithelial cell types, often ductal epithelium of adult exocrine glands in skin, salivary glands, pancreas, liver, and breast, and by glandular epithelium of the intestine and reproductive organs.[59,60] MMP-7 is also expressed in the lumenal surface of dysplastic glands in the early-stage human colorectal tumors.[61] As MMP-7 activates antibacterial peptides, defensins, in intestinal mucosa, it may function similarly also in the other epithelial sites of expression.[62] Mice deficient of MMP-7 have reduced intestinal mucosal defence, but also reduced intestinal tumorigenesis.[61,62] MMP-7 is required for repair of airway epithelial injuries.[63]

The pro-domain of MMP-26 has a unique cysteine switch sequence, PHCGVPDGSD.[56] MMP-26 is most homologous to MMP-12. MMP-26 is expressed in human uterus and placenta and in endometrial, lung, and prostate adenocarcinomas.[56,57,64] MMP-26 may be involved in tissue-remodeling events associated with tumor progression or reproductive processes including implantation and menstruation.[57]

Gelatinases

Gelatinase-A (72-kDa gelatinase, MMP-2) and gelatinase-B (92-kDa gelatinase, MMP-9) have three tandem repeats of 58 amino acid residue long fibronectin type II-like modules in the catalytic domain. As gelatinases degrade components of basement membranes, they are believed to play a crucial role in processes requiring basement membrane disruption, such as tumor invasion and tissue infiltration of T lymphocytes.[65-67] MMP-2 is also thought to be important in malignancies, as its activation correlates with tumor spread and poor prognosis.[68] MMP-2-deficient mice show reduced angiogenesis and tumor progression,[69] and MMP-9-deficient mice show impaired metastasis formation and tumor growth.[70]

Membrane-Type MMPs

To date, six MT-MMPs have been described: MT1-MMP (MMP-14), MT2-MMP (MMP-15), MT3-MMP (MMP-16), MT4-MMP (MMP-17), MT5-MMP (MMP-24), and MT6-MMP (MMP-25). They have a RxKR motif between the propeptide and the catalytic domain, which can be cleaved intracellularly in the trans-Golgi network by members of the proprotein convertase family, such as furin resulting in activation of MT-MMPs.[71,72] They are bound to cell membrane with a stem/transmembrane/cytosolic domain of approximately 25 amino acids at the C-terminal end, except for MT4-MMP and MT6-MMP, which are glycosylphosphatidylinositol (GPI)-anchored. Localization of MT-MMPs at the cell surface implies that they play a role in cell-matrix interactions and in cell invasion.[73,74] MT1-, MT3-, MT5-, and MT6-MMP activate proMMP2, and MT1- and MT2-MMP also activate proMMP-13.[75,76] The activation of proMMP-13 is enhanced in the presence of a latent form of MMP-2.[77] Thus, in the degradation of ECM, a proprotein convertase/MT-MMP/MMP cascade may play an important role at the level of zymogen activation.[78]

MT1-MMP (MMP-14) is widely expressed in normal tissues.[79,80] In addition, MT1-MMP expression has been detected in tumor cells and adjacent stromal cells in a large variety of tumors.[80,81] MT1-MMP expression in stromal cells is thought to represent a tumor-induced host response similar to that in wound healing.[82] MT1-MMP can cleave several ECM components (Table I).[22,73,83] Expression of MT1-MMP in human fetal membranes and in early human placenta suggests a role for MT1-MMP in trophoblast invasion.[84,85] MT1-MMP-deficient

Table I. Human MMPs, their expression profile, and substrates

Enzyme	Expression or function	Substrates
Collagenases		
Collagenase-1 (MMP-1)	Development, tissue repair, malignant tumors	Col I, II, III, VII, VIII, X, aggrecan, MBP, serpins, α2M
Collagenase-2 (MMP-8)	Leukocytes, cartilage	Col I, II, III, aggrecan,, serpins, α2M
Collagenase-3 (MMP-13)	Bone development, invasive tumors	Col I, II, III, IV, IX, X, XIV, gelatin, fibronectin, laminin, large tenascin C, aggrecan, fibrillin, osteonectin,serpins
Stromelysins		
Stromelysin-1 (MMP-3)	Keratinocytes and fibroblasts	Col IV, V, VII, IX, X, XIV, fibronectin, elastin, gelatin, laminin, aggrecan, nidogen, fibrillin, osteonectin, α1PI, MBP, decorin
Stromelysin-2 (MMP-10)	Keratinocytes and fibroblasts	Col IV, V, IX, X, XIV, fibronectin, elastin, gelatin, laminin, aggrecan, nidogen
Stromelysin-like MMPs		
Stromelysin-3 (MMP-11)	Invasive human carcinomas	α1PI
Metalloelastase (MMP-12)	Constant expression in macrophages	Col IV, gelatin, fibronectin, laminin, MBP, elastin, vitronectin, nidogen, α1PI, fibrillin, plasminogen, apolipoprotein A, proteoglycans
Matrilysins		
Matrilysin-1 (MMP-7)	Constant expression in ductal epithelial cells of exocrine glands	Col IV, elastin, fibronectin, laminin, nidogen, tenascin, osteonectin, MBP, decorin, versican, α1PI
Matrilysin-2 (MMP-26)	Uterus, placenta, reproductive processes	Col IV, gelatin, fibronectin, fibrin, fibrinogen, type I gelatin, α1PI, β-casein, TACE-substrate
Gelatinases		
Gelatinase A (MMP-2)	Degradation of BM and of fibrillar collagens after initial cleavage by	Col I, IV, V, VII, X, gelatin, fibronectin, tenascin, fibrillin, osteonectin, MBP, decorin, α2M
Gelatinase B (MMP-9)	collagenases, invasion of malignant tumors	Col IV, V, VII, XI, XIV, XVII, gelatin, elastin, fibrillin, osteonectin, fibronectin, MBP, α2M
Membrane-type MMPs		
Transmembrane type		
MT1-MMP (MMP-14)	Skeletal development, angiogenesis, trophoblast invasion, tumors	Col I, II, III, gelatin, fibronectin, laminin, vitronectin, aggrecan, tenascin, nidogen, perlecan, fibrin, fibrillin, α1PI, α2M
MT2-MMP (MMP-15)	Liver, brain, placenta, heart	fibronectin, laminin, aggrecan, tenascin, nidogen, perlecan

continued on next page

Table I. Human MMPs, their expression profile, and substrates (cont.)

Enzyme	Expression or function	Substrates
Membrane-type MMPs		
Transmembrane type		
MT3-MMP (MMP-16)	Brain, heart, placenta, carcinomas	Col III, gelatin, casein, fibronectin, laminin, aggrecan, vitronectin, α1PI, α2M
MT5-MMP (MMP-24)	Brain development and tumors	ND
GPI type		
MT4-MMP (MMP-17)	Brain, leucocytes, colon, ovary, testis, breast carcinoma	gelatin, TNF-α precursor, fibrin, fibronectin
MT6-MMP (MMP-25)	Leukocytes, lung, spleen	Col IV, gelatin, fibronectin, fibrin
MMP-19-like MMPs		
MMP-19	Capillary endothelial cells, inflamed synovium	Col IV, gelatin, laminin, nidogen, tenascin, fibronectin, aggrecan, COMP
MMP-28	Testis, lung, keratinocytes	ND
Other MMPs		
Enamelysin (MMP-20)	Tooth development	Amelogenin, aggrecan, COMP
MMP-23	Reproductive processes	Synthetic MMP substrates

Modified from Murphy and Knäuper, 1997. α1-proteinase inhibitor, α1PI; α2-macroglobulin, α2M; collagen type, Col; cartilage oligomeric matrix protein, COMP; myelin basic protein, MBP; not determined, ND.

mice have severe defects in skeletal development and angiogenesis[86,87] and activation of proMMP-2 is also deficient in these mice,[87] emphasizing the importance of MMP-2 activation by MT1-MMP in tumor growth and metastasis,[88] and in the cartilage destruction of rheumatoid arthritis.[89]

MT2-MMP is expressed in vivo in liver, placenta, testis, colon, intestine, pancreas, kidney, lung, heart, and skeletal muscle.[90] Human cell-associated, but not soluble, MT2-MMP is defective in proMMP-2 activation due to substitution of seven amino acids within the catalytic domain, as compared to mouse MT2-MMP capable of activating proMMP-2.[91]

MT3-MMP (MMP-16) is expressed in brain, heart, and placenta, in oral malignant melanoma,[80] and brain microglial cells.[92] MT3-MMP is also expressed as alternatively spliced soluble variant that activates proMMP-2 and cleaves type III collagen and fibronectin (Table I).[93]

MT4-MMP (MMP-17) has the least degree of sequence identity to other MT-MMPs.[94] Its lacks the cytoplasmic tail, and instead of a transmembrane domain, it is attached to the cell membrane with a GPI anchor.[95] MT4-MMP can be shed from the cell membrane by TIMP-insensitive metalloproteinases, possibly by certain ADAM family members.[95] MT4-MMP does not activate proMMP-2, but sheds proform of tumor necrosis factor-α (TNF-α).[96] MT4-MMP mRNA is expressed in vivo in brain, leukocytes, colon, ovary, testis, and in breast carcinomas.[94]

MT5-MMP (MMP-24) is predominantly expressed in kidney, pancreas, lung, and brain tissues, especially in brain tumors.[78,97] MT5-MMP can activate MMP-2 in a TIMP-2-sensitive fashion. MT5-MMP may have a role in synaptic plasticity during nervous system development,[98] whereas activation of proMMP-2 in tumor tissues may facilitate tumor progression.[97]

MT6-MMP (MMP-25, leucolysin), is specifically expressed by peripheral blood leukocytes,[99] and in lung and spleen tissue.[100] It has also been found in anaplastic astrocytomas and glioblastomas.[100] It shares sequence similarity to MT4-MMP, and is also GPI-anchored.[100] MT6-MMP may serve as a potent proteolytic tool for leukocytes during inflammatory responses[99,101] and may also facilitate tumor progression through its ability to activate proMMP-2 at the colon carcinoma and brain tumor cell membrane.[100]

Matrix Metalloproteinase-19-Like MMPs

Human MMP-19 is expressed in injured and acutely inflamed synovium, especially in capillary endothelial cells.[102-104] The substrate specificity of MMP-19 (Table I) also suggests a role for MMP-19 in angiogenesis.[104,105]

Epilysin (MMP-28) is most closely related to MMP-19, with which it shares 46% amino acid identity in the catalytic domain.[106,107] MMP-28 gene contains 8 exons in contrast to other MMPs, which usually have 10 exons. Exon 4 is alternatively spliced to a transcript that does not encode the N-terminal half of the catalytic domain. MMP-28 has a furin activation sequence (RRKKR) but has no transmembrane sequence. MMP-28 is expressed in testis and lung and at lower levels in heart, colon, intestine, and brain.[106,107] MMP-28 can be detected in the basal and suprabasal epidermis of intact skin and in wounded skin; MMP-28 is seen in basal keratinocytes both at and some distance from the wound edge.[107]

Other MMPs

Enamelysin (MMP-20) has an expression pattern restricted to ameloblasts and odontoblasts of developing teeth. It is able to degrade amelogenin, the major protein component of the enamel matrix, aggrecan, and cartilage oligomeric matrix protein (COMP).[108] Enamelysin has been suggested to play a central role in matrix remodeling during tooth development and enamel maturation.

MMP-23 lacks the signal sequence, the cysteine switch, and the C-terminal domain lacks any similarity with hemopexin. N-terminal signal-anchor localizes MMP-23 to the cell membrane. A single proteolytic cleavage on the RRRR motif activates MMP-23 prior to secretion.[109] Although MMP-23 mRNA is found in heart, intestine, colon, placenta, lung, and pancreas, the predominant expression in ovary, testis, and prostate suggests a specialized role in reproductive processes, e.g., during ovarian follicle development.[110]

Regulation of MMP Activity

In general, MMP production and activity is strictly regulated in vivo. Cells within intact tissues usually do not store MMPs, and constitutive expression is minimal. Neutrophils are an exception, as they store MMP-8 and MMP-9 in secretory granules for rapid release. In addition, expression of MMP-7 is constitutive in ductal epithelium of adult exocrine glands. Activity of MMPs is regulated at multiple levels including transcription, modulation of mRNA half-life, secretion, localization, activation, and inhibition. The natural inhibitors include tissue inhibitors of metalloproteinases (TIMPs 1-4) and nonspecific proteinase inhibitors.

Regulation at the Transcriptional Level

The expression of MMPs in general is regulated by growth factors, cytokines, chemical agents like phorbol esters, physical stress, oncogenic transformation, cell-cell and cell-ECM interactions.[1-4] MMP genes responsive to extracellular stimuli (MMP-1, MMP-13, MMP-3, MMP-10, MMP-7, MMP-12, MMP-9, and MMP-19) contain an AP-1 (activator protein-1) binding site in the proximal promoter approximately at position -70 with respect to the transcription initiation site. Another distal AP-1 or related element is found in the promoters of MMP-1, MMP-3, and MMP-9. Jun and Fos bind to the AP-1 cis-element and activate the transcription of the MMP gene. The proximal or distal AP-1 site is often accompanied by another cis-element, PEA3 (polyomavirus enhancer A-binding protein-3) site, that binds ETS transcription factors. AP-1 and PEA3 together confer responsiveness to a variety of growth factors, oncogene products, and tumor promoters.[111] Transforming growth factor-β (TGF-β) inhibitory element is found in MMP-1, MMP-7, and MMP-13 promoters.

Promoter regions of MMP-2, MMP-11, and MT1-MMP genes do not contain a conserved AP-1 element.[4,112,113] The promoter region of human MT1-MMP has an Sp-1 site crucial for maintaining MT1-MMP transcription, four CCAAT-boxes, and additional unidentified positive and negative regulatory sequences.[113] It lacks the typical MMP promoter regulatory sites, TATA-box and AP-1 and TGF-β-responsive elements, as does its murine counterpart.[114]

Zymogen Activation

Most MMPs are secreted as inactive proenzymes, and their proteolytic activity is regulated by zymogen activation and enzyme inhibition. MT-MMPs, MMP-11, and MMP-28 are activated intracellularly by Golgi-associated furin-like proteases. For MMP-23, a single cleavage both activates it and releases it from the cell surface, where it is anchored.[109] The latency of proMMPs is maintained by the so-called cysteine switch,[115] i.e., a covalent bond between the cysteine residue in the prodomain and the Zn^{2+} in the catalytic domain. This interaction is disrupted by activation of the proMMP by proteinases such as plasmin, trypsin, kallikrein, chymase, and mast cell tryptase.[1-4] ProMMPs can also be activated by mercurial compounds (aminophenyl mercuric acetate), SH-reactive agents, reactive oxygen, and detergents.[3,115] Many MMPs can activate other MMPs forming a complex network regulating the tissue proteolysis. TIMP-2 N-terminal domain and the active site of MT1-MMP can associate to form a proMMP-2 receptor at the cell surface, leaving the C-terminus of TIMP-2 free to bind proMMP-2. This allows efficient activation of proMMP-2 by adjacent TIMP-2-free, active MT1-MMP and may be a common mechanism for proMMP-2 activation by MT-MMPs.[76,116,117] Thus, at low concentrations TIMP-2 enhances the activation process by concentrating MMP-2 to the site where the activator is available and in high concentrations, TIMP-2 inhibits MMP-2 activation.[116,117] In tissues, physiological MMP activators are likely to include tissue or plasma proteinases or opportunistic bacterial proteinases. The plasminogen activator/plasmin system is an important activator of proMMPs in pathological situations.[118]

Localization and Trafficking

An important aspect of the regulation of MMP activity is localizing the proteolytic activity to the pericellular space.[119] Anchoring MMPs to the cell surface prevents them from rapidly diffusing away and also keeps them under close regulatory control. Binding to cell surface also allows positioning of MMPs for activation, their interaction with cell surface adhesion molecules or receptors, regulation of their turnover, and focused pericellular proteolysis, as with MMP-14 localized in invadopodia.[120] Accordingly, heparan sulfate proteoglycan has been shown to anchor MMP-7 on cell surface and possibly in the basement membrane in vivo.[121] Similarly, MMP-2 interacts with integrin αvβ3,[122] MMP-9 binds to cell surface hyaluronan receptor CD44[123], and integrin αvβ6[124], and proMMP-1 interacts with collagen receptor α2β1.[125]

Inhibition of MMP Activity

MMP activity can be inhibited by tissue inhibitors of metalloproteinases (TIMPs), by serine proteinase inhibitors (serpins), and by nonspecific serum proteinase inhibitors, such as α2-macroglobulin, which is important in blocking MMP activity in the synovial fluid, serum, and other body fluids.[1-4] Serpins are glycoproteins of 50–100 kDa, abundant in all human tissues, and involved in controlling the general proteolytic activity in several tissues. Serpins include α1-antitrypsin (α1-proteinase inhibitor) and plasminogen activator inhibitor (PAI)-1 and PAI-2.

Tissue Inhibitors of Metalloproteinases (TIMPs)

Tissue inhibitors of metalloproteinases (TIMPs) -1, -2, -3, and -4 are important endogenous regulators of MMP activity in tissue (see George et al, in this book). TIMPs inhibit the MMP activity through non-covalent binding of the active zinc-binding sites of MMPs at molar equivalence.[126] Although the primary amino acid sequence identity between the TIMPs is low, about 30%, their tertiary structure is remarkably similar. They have 12 conserved cysteine residues required for the formation of six disulfide bonds, which hold the two domains in a rigid conformation.[127] N-terminal domain, that contains the TIMP consensus sequence VIRAK, is necessary for MMP inhibition.[128-130] The C-terminal domains are more divergent and appear to be important in forming differences in specificity to each TIMP family member.[131-133]

In general, TIMPs can inhibit the activity of all MMPs in vitro, except for MT1-MMP and MT3-MMP, which are not inhibited by TIMP-1.[76,134] Due to structural similarities in the active site of MMPs and ADAMs, some ADAMs are also inhibited by TIMPs. TIMP-1 and TIMP-3 inhibit ADAM-10, whereas TNF-α convertase (TACE, ADAM-17) is only inhibited by TIMP-3.[135,136] TIMP-1 and TIMP-3 inhibit aggrecanase-1 (ADAM-TS4) (aggrecanase-1) and TIMP-3 also inhibits aggrecanase-2 (ADAM-TS5).[137,138] TIMP-1 has a role in tissue remodeling during embryonal growth and tumor progression, in gonadal steroidogenesis, ovulation, pregnancy, and parturition, and in inhibiting migration of vascular smooth muscle cells and endothelial cells, angiogenesis, tumor cell invasion and metastasis.[127,128,139-141] It also possesses growth factor activity[142] and inhibits shedding of heparin-binding epidermal growth factor and its receptor HER-2.[143,144]

TIMP-2 is expressed in a constitutive manner by cells in culture. TIMP-2 can inhibit the shedding of TNF-α receptors (TNF-αRI and II).[145] Mice with an inactivating mutation of TIMP-2 gene have normal phenotypes, in spite of severe impairment of proMMP-2 activation in vivo.[146] In human tumors, high levels of TIMP-2 at the interface between malignant cells and stromal cells correlate with poor prognosis, suggesting that overexpression of TIMP-2 reflects a host reaction against highly invasive cells.[147]

Whereas other TIMPs are present in soluble form, TIMP-3 is insoluble, bound to the ECM.[148] TIMP-3 promotes the detachment of transformed cells from the ECM and accelerates morphological changes associated with cell transformation.[141] In addition, up-regulation of TIMP-3 has been associated with a block in the G_1 phase of the cell cycle during differentiation of HL-60 leukemia cells.[149] TIMP-3 can block the cellular shedding of proTNF-α, L-selectin and IL-6 receptor.[135,136,150,151] Adenovirus-mediated gene delivery of TIMP-3 inhibits invasion and induces apoptosis in various normal and malignant cells.[141,152,153] TIMP-4 is mostly expressed in the adult human heart, though very low levels of mRNA and protein are found in many tissues.[154]

MMPs in Tumor Growth and Invasion

MMPs have a dual role in tumor growth and metastasis processes. They promote tumor growth by degrading matrix barriers and by enhancing angiogenesis.[4] On the other hand, MMPs can limit tumor neovascularization. Angiostatin is a specific inhibitor of endothelial cell proliferation and one the most effective and specific natural inhibitors of angiogenesis. It is cleaved from plasminogen by the action of plasmin and plasmin reductase,[155] and by MMPs, the most efficient being MMP-12, followed by MMP-9, MMP-3, and MMP-7, with collagenases exhibiting practically no activity.[156,157] In α1 integrin knockout mice, which lack the α1β1 collagen receptor, the synthesis of MMP-7 and MMP-9 is markedly increased. This increased the plasma levels of angiostatin, leading to markedly decreased vascularization of implanted tumors.[158] Endostatin, another natural angiogenesis inhibitor, is a fragment of type XVIII collagen.[159] Although MMPs are not required for generation of endostatin, they are involved in the processing of collagen XVIII.[160]

MMPs have also other functions.[161] They can release active growth factors and angiogenic factors from the cell surface and ECM.[162] They can cleave growth factor binding proteins[163] and cell surface growth factor receptors.[164] They can generate an α1-antitrypsin cleavage product that promotes tumor growth and invasion.[165] They may alter cell cycle checkpoint control and promote genomic instability by affecting cell adhesion.[166] MMPs can also induce programmed cell death in anchorage-dependent cells.[36] This can either inhibit tumor progression or promote it by enhancing selection of anchorage-independent and apoptosis-resistant subpopulations.[37]

Tumor growth involves alterations in the stromal ECM and malignant tumors often induce a fibroproliferative response in the adjacent stroma, characterized by increased expression of type I and III procollagens.[167] During metastasis formation, malignant cells detach from the primary tumor, invade through stromal tissue, enter the circulation, arrest at the peripheral vascular bed, and extravasate, invade the target organ, and form a metastatic colony.[4,163] Tumor cells must escape the host immune surveillance and only a fraction of circulating tumor cells establish metastatic colonies.[168] Tumor induced angiogenesis is essential for growth of the primary tumor and metastases, and new blood vessels are sites for entry of tumor cells into the circulation. It is conceivable that proteolytic degradation of ECM plays a crucial role in all the above mentioned aspects of tumor development.

A considerable body of evidence is available implicating MMPs in cancer spread. A number of studies have demonstrated a positive correlation between MMP expression and invasive and metastatic potential of malignant tumors including, colon, lung, head and neck, basal cell, breast, thyroid, prostate, ovarian, and gastric carcinomas.[4] For example, expression of MMP-1 is correlates with poor prognosis in colorectal cancer and esophageal cancer[169,170] and MMP-2 and MMP-3 expression is associated with lymph node metastasis and vascular invasion in SCC of esophagus.[171] Similarly, high expression of MMP-13 in SCCs of the head and neck and vulva is associated with their metastasis capacity.[29,30] MMP-11 expression correlates with increased local invasiveness in head and neck SCCs[172] and the level of MMP-2 expression with poor prognosis of cervical SCCs.[173] In general, all MMPs, the expression of which has been documented in malignant tumors, can also be expressed by non-neoplastic cells. However, MMP-13, MMP-7, MMP-12, and MMP-14 are expressed by malignantly transformed keratinocytes in SCCs, but not in normal keratinocytes, indicating that their expression serves as a marker for transformation.[29,30] In addition, MMP-2 expression serves as a marker for malignant transformation of cervical epithelial cells.[173,174]

MMPs are mainly produced by nonmalignant stromal cells in malignant tumors. Tumor cells also secrete factors, such as extracellular MMP inducer (EMMPRIN), which enhance the expression of MMPs by stromal fibroblasts (see Toole, in this book). In addition, growth factors and cytokines secreted by tumor infiltrating inflammatory cells as well as by tumor or stromal cells modulate MMP expression. Tumor invasion involves interaction between tumor cells, adjacent stromal cells, and infiltrating inflammatory cells and it is likely that all these cells express distinct MMPs, which may complement each other's substrate specificity and form a network of MMP cascades, in which one MMP cleaves a particular native or partially degraded ECM component and activates other MMPs.

TIMPs in Tumor Growth and Invasion

A number of studies have demonstrated the expression of TIMPs in tumor stroma and tumor tissue. In general, there is convincing evidence that overexpression of TIMPs by cancer cells or by the host reduces invasive and metastatic capacity of tumor cells. In cutaneous and oral SCCs expression of TIMP-1, TIMP-2, and TIMP-3 is detected in stromal cells adjacent to the tumor,[175-177] suggesting that their expression represents a host attempt to limit tumor invasion and tumor-induced angiogenesis. This notion is supported by observations indicating that the presence of TIMP-1 and TIMP-2 in SCCs correlates with less aggressive growth.[178] However, in breast cancer TIMP-2 expression correlates with tumor recurrence[179] and in cervical carcinomas TIMP-2 expression correlates with poor prognosis.[180] Similarly, in malignant breast cancer TIMP-1 expression is enhanced, as compared to nonmalignant breast tumor.[181] However, the MMP:TIMP ratio is elevated in cervival carcinomas with poor prognosis, indicating that evaluation of either MMP or TIMP expression alone is not sufficient for prognostication of malignancies.[180]

Cancer cell invasion can be inhibited by recombinant TIMPs or by overexpression of either TIMPs using a variety of gene delivery vehicles. TIMP-2 inhibited the invasion of HT-1080 fibrosarcoma cells in vitro[182,183] but had no effect on tumor cell growth. Overexpression of

TIMP-1 reduced metastasis of gastric carcinoma cells.[184] TIMP-1 also reduced the growth rate and invasion of astrocytoma and mammary carcinoma cells[185,186] and prevented metastasis of gastric carcinoma cells.[187] The ability of TIMP-1 to inhibit tumor development at different stages has been demonstrated by transgenic mouse models. Constitutive overexpression of TIMP-1 in the liver suppressed tumor initiation, growth and angiogenesis in transgenic mice, which develop hepatocellular carcinomas as a result of SV40 T antigen expression.[188] These observations were also supported by a recent study, in which TIMP-1 overexpression in the brain prevented tumor formation.[189]

Overexpression of TIMP-2 reduced the MMP activity and suppressed growth of melanomas in the skin of immunodeficient mice[190] and melanoma cells overexpressing TIMP-2 showed a reduced metastatic capacity.[191] Melanoma cells overexpressing TIMP-1 were shown to have reduced metastasis capacity due to inhibition of tumor growth following extravasation.[192] TIMP-4 overexpression in breast carcinoma cells also inhibits invasion in vitro and tumor growth in vivo and results in reduction in lymph node and lung metastasis.[193] Together these studies highlight both similar and diverse effects of overexpression of individual TIMPs on tumor cell phenotype in vivo. Adenoviral delivery of TIMP-2 has been shown to inhibit growth of liver metastases.[194] In constrast, systemic delivery of TIMP-4 enhances mammary tumorigenesis but inhibits growth of Wilms' tumor in vivo.[195,196]

Further studies have demonstrated distinct effects of individual TIMPs on cell survival. Overexpression of TIMP-2 reduced invasion and angiogenesis but also protected the melanoma cells from apoptosis although it increased necrosis.[197] TIMP-1 has also been shown to promote survival of B-cells through modulation of CD40 levels.[198] However, we and others have shown that adenoviral-mediated gene delivery of TIMP-3 promotes apoptosis of a number of malignant cell types associated with reduced capacity of TIMP-3 transduced cells to bind to ECM components.[141,152,153] TIMP-3 expressing stable colon carcinoma cell lines display reduced tumor growth,[199] and in these cells TIMP-3 overexpression resulted in apoptosis through stabilization of TNF-α receptors on the cell surface.[200] This suggests that individual TIMPs may modulate the levels of death proteins from the cell surface, as demonstrated by findings that shedding of FAS from the cell surface is mediated by MMP-induced cleavage and is inhibited by synthetic MMP inhibitors.[201] Furthermore, TIMP-3 inhibits activity of TNF-α convertase (ADAM-17) providing further evidence for complex regulation of death ligands and receptors by MMPs and TIMPs,[202] which may play an important role in survival, growth, and invasion of malignant cells.

At present, several synthetic MMP inhibitors are in clinical trials evaluating their ability to inhibit growth and invasion of malignant tumors in vivo (see Turpeenniemi-Hujanen, in this book). Gene delivery of TIMP-1, -2, -3, and -4 into malignant cells may also be a potent way of inhibiting tumor growth and invasion.[145,152,153,196] Furthermore, an effective way of inhibiting MMP expression may be blocking signaling pathways mediating activation of MMP gene expression.[4] The on-going clinical trials are expected to show whether synthetic MMP inhibitors have a place in the therapeutic arsenal aimed at inhibiting growth, invasion, and metastasis of malignant tumors.

Acknowledgments

The original work of authors has been supported by grants from the Academy of Finland, Sigrid Jusélius Foundation, the Cancer Foundation of Finland, and Turku University Central Hospital, and by research contract with Finnish Life and Pension Insurance Companies.

References

1. Cawston T. Matrix metalloproteinases and TIMPs: properties and implications for the rheumatic diseases. Mol Med Today 1998; 4:130-137.
2. Nagase H, Woessner JF, Jr. Matrix metalloproteinases. J Biol Chem 1999; 274:21491-21494.
3. Murphy G, Knäuper V. Relating matrix metalloproteinase structure to function: why the "hemopexin" domain? Matrix Biol 1997; 15:511-518.

4. Westermarck J, Kähäri V-M. Regulation of matrix metalloproteinase expression in tumor invasion. FASEB J 1999; 13:781-792.
5. Borkakoti N. Structural studies of matrix metalloproteinases. J Mol Med 2000; 78:261-268.
6. Krane SM, Byrne MH, Lemaître V et al. Different collagenase gene products have different roles in degradation of type I collagen. J Biol Chem 1996; 271:28509-28515.
7. Freije JM, Diez-Itza I, Balbín M et al. Molecular cloning and expression of collagenase-3, a novel human matrix metalloproteinase produced by breast carcinomas. J Biol Chem 1994; 269:16766-16773.
8. Knäuper V, López-Otín C, Smith B et al. Biochemical characterization of human collagenase-3. J Biol Chem 1996; 271:1544-1550.
9. Mitchell PG, Magna HA, Reeves LM et al. Cloning, expression, and type II collagenolytic activity of matrix metalloproteinase-13 from human osteoarthritic cartilage. J Clin Invest 1996; 97:761-768.
10. Wilhelm SM, Eisen AZ, Teter M et al. Human fibroblast collagenase: glycosylation and tissue-specific levels of enzyme synthesis. Proc Natl Acad Sci USA 1986; 83:3756-3760.
11. Chung L, Shimokawa K-i, Dinakarpandian D et al. Identification of the 183RWTNNFREY191 region as a critical segment of matrix metalloproteinase 1 for the expression of collagenolytic activity. J Biol Chem 2000; 275:29610-29617.
12. McGowan KA, Bauer EA, Smith LT. Localization of type I human skin collagenase in developing embryonic and fetal skin. J Invest Dermatol 1994; 102:951-957.
13. Saarialho-Kere UK. Patterns of matrix metalloproteinase and TIMP expression in chronic ulcers. Arch Dermatol Res 1998; 290 Suppl:S47-54.
14. Saarialho-Kere UK, Chang ES, Welgus HG et al. Distinct localization of collagenase and tissue inhibitor of metalloproteinases expression in wound healing associated with ulcerative pyogenic granuloma. J Clin Invest 1992; 90:1952-1957.
15. Pilcher BK, Dumin JA, Sudbeck BD et al. The activity of collagenase-1 is required for keratinocyte migration on a type I collagen matrix. J Cell Biol 1997; 137:1445-1457.
16. Hasty KA, Pourmotabbed TF, Goldberg GI et al. Human neutrophil collagenase. A distinct gene product with homology to other matrix metalloproteinases. J Biol Chem 1990; 265:11421-11424.
17. Cole AA, Chubinskaya S, Schumacher B et al. Chondrocyte matrix metalloproteinase-8. Human articular chondrocytes express neutrophil collagenase. J Biol Chem 1996; 271:11023-11026.
18. Hanemaaijer R, Sorsa T, Konttinen YT et al. Matrix metalloproteinase-8 is expressed in rheumatoid synovial fibroblasts and endothelial cells. J Biol Chem 1997; 272:31504-31509.
19. Fosang AJ, Last K, Knäuper V et al. Degradation of cartilage aggrecan by collagenase-3 (MMP-13). FEBS Lett 1996; 380:17-20.
20. Knäuper V, Cowell S, Smith B et al. The role of the C-terminal domain of human collagenase-3 (MMP-13) in the activation of procollagenase-3, substrate specificity, and tissue inhibitor of metalloproteinase interaction. J Biol Chem 1997; 272:7608-7016.
21. Sasaki T, Gohring W, Mann K et al. Limited cleavage of extracellular matrix protein BM-40 by matrix metalloproteinases increases its affinity for collagens. J Biol Chem 1997; 272:9237-9243.
22. Ashworth JL, Murphy G, Rock MJ et al. Fibrillin degradation by matrix metalloproteinases: implications for connective tissue remodelling. Biochem J 1999; 340:171-181.
23. Ravanti L, Häkkinen L, Larjava H et al. Transforming growth factor-β induces collagenase-3 expression by human gingival fibroblasts via p38 mitogen-activated protein kinase. J Biol Chem 1999; 274:37292-37300.
24. Johansson N, Saarialho-Kere U, Airola K et al. Collagenase-3 (MMP-13) is expressed by hypertrophic chondrocytes, periosteal cells, and osteoblasts during human fetal bone development. Dev Dyn 1997; 208:387-395.
25. Ståhle-Bäckdahl M, Sandstedt B, Bruce K et al. Collagenase-3 (MMP-13) is expressed during human fetal ossification and re-expressed in postnatal bone remodeling and in rheumatoid arthritis. Lab Invest 1997; 76:717-728.
26. Ravanti L, Toriseva M, Penttinen R et al. Expression of human collagenase-3 (MMP-13) by fetal skin fibroblasts is induced by transforming growth factor b via p38 mitogen-activated protein kinase. FASEB J 2001; 15:1098-1100.
27. Vaalamo M, Mattila L, Johansson N et al. Distinct populations of stromal cells express collagenase-3 (MMP-13) and collagenase-1 (MMP-1) in chronic ulcers but not in normally healing wounds. J Invest Dermatol 1997; 109:96-101.
28. Uitto V-J, Airola K, Vaalamo M et al. Collagenase-3 (matrix metalloproteinase-13) expression is induced in oral mucosal epithelium during chronic inflammation. Am J Pathol 1998; 152:1489-1499.
29. Johansson N, Airola K, Grénman R et al. Expression of collagenase-3 (matrix metalloproteinase-13) in squamous cell carcinomas of the head and neck. Am J Pathol 1997; 151:499-508.
30. Johansson N, Vaalamo M, Grénman S et al. Collagenase-3 (MMP-13) is expressed by tumor cells in invasive vulvar squamous cell carcinomas. Am J Pathol 1999; 154:469-480.

31. Airola K, Karonen T, Vaalamo M et al. Expression of collagenases-1 and -3 and their inhibitors TIMP-1 and -3 correlates with the level of invasion in malignant melanomas. Br J Cancer 1999; 80:733-743.
32. Nikkola J, Vihinen P, Vlaykova T et al. High collagenase-1 expression correlates to favourable chemoimmunotherapy response in human metastatic melanoma. Melanoma Res 2001; 11:157-166
33. Boström PJ, Ravanti L, Reunanen N et al. Expression of collagenase-3 (matrix metalloproteinase-13) in transitional-cell carcinoma of the urinary bladder. Int J Cancer 2000; 88:417-423.
34. Ala-aho R, Johansson N, Baker AH et al. Expression of collagenase-3 (MMP-13) enhances invasion of human fibrosarcoma HT-1080 cells. Int J Cancer 2001; (In press)
35. Lund LR, Rømer J, Thomasset N et al. Two distinct phases of apoptosis in mammary gland involution: proteinase-independent and -dependent pathways. Development 1996; 122:181-193.
36. Thomasset N, Lochter A, Sympson CJ et al. Expression of autoactivated stromelysin-1 in mammary glands of transgenic mice leads to a reactive stroma during early development. Am J Pathol 1998; 153:457-467.
37. Sternlicht MD, Bissell MJ, Werb Z. The matrix metalloproteinase stromelysin-1 acts as a natural mammary tumor promoter. Oncogene 2000; 19:1102-1113.
38. Unemori EN, Bair MJ, Bauer EA et al. Stromelysin expression regulates collagenase activation in human fibroblasts. Dissociable control of two metalloproteinases by interferon-γ. J Biol Chem 1991; 266:23477-23482.
39. Bullard KM, Lund L, Mudgett JS et al. Impaired wound contraction in stromelysin-1-deficient mice. Ann Surg 1999; 230:260-265.
40. Rouyer N, Wolf C, Chenard MP et al. Stromelysin-3 gene expression in human cancer: an overview. Invasion Metastasis 1994; 14:269-275.
41. Ahmad A, Hanby A, Dublin E et al. Stromelysin 3: an independent prognostic factor for relapse-free survival in node-positive breast cancer and demonstration of novel breast carcinoma cell expression. Am J Pathol 1998; 152:721-728.
42. Noël AC, Lefebvre O, Maquoi E et al. Stromelysin-3 expression promotes tumor take in nude mice. J Clin Invest 1996; 97:1924-1930.
43. Masson R, Lefebvre O, Noël A et al. In vivo evidence that the stromelysin-3 metalloproteinase contributes in a paracrine manner to epithelial cell malignancy. J Cell Biol 1998; 140:1535-1541.
44. Unden AB, Sandstedt B, Bruce K et al. Stromelysin-3 mRNA associated with myofibroblasts is overexpressed in aggressive basal cell carcinoma and in dermatofibroma but not in dermatofibrosarcoma. J Invest Dermatol 1996; 107:147-153.
45. Thewes M, Worret WI, Engst R et al. Stromelysin-3 (ST-3): immunohistochemical characterization of the matrix metalloproteinase (MMP)-11 in benign and malignant skin tumours and other skin disorders. Clin Exp Dermatol 1999; 24:122-126.
46. Pei D, Weiss SJ. Furin-dependent intracellular activation of the human stromelysin-3 zymogen. Nature 1995; 375:244-247.
47. Pei D, Majmudar G, Weiss SJ. Hydrolytic inactivation of a breast carcinoma cell-derived serpin by human stromelysin-3. J Biol Chem 1994; 269:25849-25855.
48. Noël A, Boulay A, Kebers F et al. Demonstration in vivo that stromelysin-3 functions through its proteolytic activity. Oncogene 2000; 19:1605-1612.
49. Belaaouaj A, Shipley JM, Kobayashi DK et al. Human macrophage metalloelastase. Genomic organization, chromosomal location, gene linkage, and tissue-specific expression. J Biol Chem 1995; 270:14568-14575.
50. Halpert I, Sires UI, Roby JD et al. Matrilysin is expressed by lipid-laden macrophages at sites of potential rupture in atherosclerotic lesions and localizes to areas of versican deposition, a proteoglycan substrate for the enzyme. Proc Natl Acad Sci USA 1996; 93:9748-9753.
51. Curci JA, Liao S, Huffman MD et al. Expression and localization of macrophage elastase (matrix metalloproteinase-12) in abdominal aortic aneurysms. J Clin Invest 1998; 102:1900-1910.
52. Vaalamo M, Karjalainen-Lindsberg ML, Puolakkainen P et al. Distinct expression profiles of stromelysin-2 (MMP-10), collagenase-3 (MMP-13), macrophagemetalloelastase (MMP-12), and tissue inhibitor of metalloproteinases-3 (TIMP-3) in intestinal ulcerations. Am J Pathol 1998; 152:1005-1014.
53. Kerkelä E, Ala-aho R, Jeskanen L et al. Expression of human macrophage metalloelastase (MMP-12) by tumor cells in skin cancer. J Invest Dermatol 2000; 114:1113-1119.
54. Vaalamo M, Kariniemi AL, Shapiro SD et al. Enhanced expression of human metalloelastase (MMP-12) in cutaneous granulomas and macrophage migration. J Invest Dermatol 1999; 112:499-505.
55. Hautamaki RD, Kobayashi DK, Senior RM et al. Requirement for macrophage elastase for cigarette smoke-induced emphysema in mice. Science 1997; 277:2002-2004.

56. Park HI, Ni J, Gerkema FE et al. Identification and characterization of human endometase (matrix metalloproteinase-26) from endometrial tumor. J Biol Chem 2000; 275:20540-20544.
57. Uría JA, López-Otín C. Matrilysin-2, a new matrix metalloproteinase expressed in human tumors and showing the minimal domain organization required for secretion, latency, and activity. Cancer Res 2000; 60:4745-4751.
58. Sires UI, Murphy G, Baragi VM et al. Matrilysin is much more efficient than other matrix metalloproteinases in the proteolytic inactivation of a1-antitrypsin. Biochem Biophys Res Commun 1994; 204:613-620.
59. Saarialho-Kere U, Crouch EC, Parks WC. Matrix metalloproteinase matrilysin is constitutively expressed in adult human exocrine epithelium. J Invest Dermatol 1995; 105:190-196.
60. Wilson CL, Heppner KJ, Rudolph LA et al. The metalloproteinase matrilysin is preferentially expressed by epithelial cells in a tissue-restricted pattern in the mouse. Mol Biol Cell 1995; 6:851-869.
61. Wilson CL, Heppner KJ, Labosky PA et al. Intestinal tumorigenesis is suppressed in mice lacking the metalloproteinase matrilysin. Proc Natl Acad Sci USA 1997; 94:1402-1407.
62. Wilson CL, Ouellette AJ, Satchell DP et al. Regulation of intestinal a-defensin activation by the metalloproteinase matrilysin in innate host defense. Science 1999; 286:113-117.
63. Dunsmore SE, Saarialho-Kere UK, Roby JD et al. Matrilysin expression and function in airway epithelium. J Clin Invest 1998; 102:1321-1331.
64. Benoit De Coignac A, Elson G et al. Cloning of MMP-26. A novel matrilysin-like proteinase. Eur J Biochem 2000; 267:3323-3329.
65. Aimes RT, Quigley JP. Matrix metalloproteinase-2 is an interstitial collagenase. Inhibitor-free enzyme catalyzes the cleavage of collagen fibrils and soluble native type I collagen generating the specific 3/4- and 1/4-length fragments. J Biol Chem 1995; 270:5872-5876.
66. Fridman R, Toth M, Pena D et al. Activation of progelatinase B (MMP-9) by gelatinase A (MMP-2). Cancer Res 1995; 55:2548-2555.
67. Esparza J, Vilardell C, Calvo J et al. Fibronectin upregulates gelatinase B (MMP-9) and induces coordinated expression of gelatinase A (MMP-2) and its activator MT1-MMP (MMP-14) by human T lymphocyte cell lines. A process repressed through RAS/MAP kinase signaling pathways. Blood 1999; 94:2754-2766.
68. Yu AE, Hewitt RE, Kleiner DE et al. Molecular regulation of cellular invasion—role of gelatinase A and TIMP-2. Biochem Cell Biol 1996; 74:823-831.
69. Itoh T, Tanioka M, Yoshida H et al. Reduced angiogenesis and tumor progression in gelatinase A-deficient mice. Cancer Res 1998; 58:1048-1051.
70. Itoh T, Tanioka M, Yoshida H et al. Reduced angiogenesis and tumor progression in gelatinase A-deficient mice. Cancer Res 1998; 58:1048-1051.
71. Sato H, Kinoshita T, Takino T et al. Activation of a recombinant membrane type 1-matrix metalloproteinase (MT1-MMP) by furin and its interaction with tissue inhibitor of metalloproteinases (TIMP)-2. FEBS Lett 1996; 393:101-104.
72. Steiner DF. The proprotein convertases. Curr Opin Chem Biol 1998; 2:31-39.
73. d'Ortho MP, Will H, Atkinson S et al. Membrane-type matrix metalloproteinases 1 and 2 exhibit broad-spectrum proteolytic capacities comparable to many matrix metalloproteinases. Eur J Biochem 1997; 250:751-757.
74. Hotary K, Allen E, Punturieri A et al. Regulation of cell invasion and morphogenesis in a three-dimensional type I collagen matrix by membrane-type matrix metalloproteinases 1, 2, and 3. J Cell Biol 2000; 149:1309-1323.
75. Knäuper V, Will H, López-Otín C et al. Cellular mechanisms for human procollagenase-3 (MMP-13) activation. Evidence that MT1-MMP (MMP-14) and gelatinase A (MMP-2) are able to generate active enzyme. J Biol Chem 1996; 271:17124-17131.
76. Will H, Atkinson SJ, Butler GS et al. The soluble catalytic domain of membrane type 1 matrix metalloproteinase cleaves the propeptide of progelatinase A and initiates autoproteolytic activation. Regulation by TIMP-2 and TIMP-3. J Biol Chem 1996; 271:17119-17123.
77. Cowell S, Knäuper V, Stewart M L et al. Induction of matrix metalloproteinase activation cascades based on membrane-type 1 matrix metalloproteinase: associated activation of gelatinase A, gelatinase B and collagenase 3. Biochem J 1998; 331:453-458.
78. Pei D. Identification and characterization of the fifth membrane-type matrix metalloproteinase MT5-MMP. J Biol Chem 1999; 274:8925-8932.
79. Sato H, Takino T, Okada Y et al. A matrix metalloproteinase expressed on the surface of invasive tumour cells. Nature 1994; 370:61-65.
80. Takino T, Sato H, Shinagawa A et al. Identification of the second membrane-type matrix metalloproteinase (MT-MMP-2) gene from a human placenta cDNA library. MT-MMPs form a unique membrane-type subclass in the MMP family. J Biol Chem 1995; 270:23013-23020.

81. Okada A, Bellocq JP, Rouyer N et al. Membrane-type matrix metalloproteinase (MT-MMP) gene is expressed in stromal cells of human colon, breast, and head and neck carcinomas. Proc Natl Acad Sci USA 1995; 92:2730-2734.
82. Okada A, Tomasetto C, Lutz Y et al. Expression of matrix metalloproteinases during rat skin wound healing: evidence that membrane type-1 matrix metalloproteinase is a stromal activator of pro-gelatinase A. J Cell Biol 1997; 137:67-77.
83. Ohuchi E, Imai K, Fujii Y et al. Membrane type 1 matrix metalloproteinase digests interstitial collagens and other extracellular matrix macromolecules. J Biol Chem 1997; 272:2446-2451.
84. Fortunato SJ, Menon R, Lombardi SJ. Expression of a progelatinase activator (MT1-MMP) in human fetal membranes. Am J Reprod Immunol 1998; 39:316-322.
85. Hurskainen T, Seiki M, Apte SS et al. Production of membrane-type matrix metalloproteinase-1 (MT-MMP-1) in early human placenta: a possible role in placental implantation? J Histochem Cytochem 1998; 46:221-229.
86. Holmbeck K, Bianco P, Caterina J et al. MT1-MMP-deficient mice develop dwarfism, osteopenia, arthritis, and connective tissue disease due to inadequate collagen turnover. Cell 1999; 99:81-92.
87. Zhou Z, Apte SS, Soininen R et al. Impaired endochondral ossification and angiogenesis in mice deficient in membrane-type matrix metalloproteinase I. Proc Natl Acad Sci USA 2000; 97:4052-4057.
88. Nabeshima K, Inoue T, Shimao Y et al. Front-cell-specific expression of membrane-type 1 matrix metalloproteinase and gelatinase A during cohort migration of colon carcinoma cells induced by hepatocyte growth factor/scatter factor. Cancer Res 2000; 60:3364-3369.
89. Yamanaka H, Makino K, Takizawa M et al. Expression and tissue localization of membrane-types 1, 2, and 3 matrix metalloproteinases in rheumatoid synovium. Lab Invest 2000; 80:677-687.
90. Therét N, Musso O, L'Helgoualc'h A et al. Differential expression and origin of membrane-type 1 and 2 matrix metalloproteinases (MT-MMPs) in association with MMP2 activation in injured human livers. Am J Pathol 1998; 153:945-954.
91. Miyamori H, Takino T, Seiki M et al. Human membrane type-2 matrix metalloproteinase is defective in cell-associated activation of progelatinase A. Biochem Biophys Res Commun 2000; 267:796-800.
92. Yoshiyama Y, Sato H, Seiki M et al. Expression of the membrane-type 3 matrix metalloproteinase (MT3-MMP) in human brain tissues. Acta Neuropathol (Berl) 1998; 96:347-350.
93. Matsumoto S, Katoh M, Saito S et al. Identification of soluble type of membrane-type matrix metalloproteinase-3 formed by alternatively spliced mRNA. Biochim Biophys Acta 1997; 1354:159-170.
94. Puente XS, Pendás AM, Llano E et al. Molecular cloning of a novel membrane-type matrix metalloproteinase from a human breast carcinoma. Cancer Res 1996; 56:944-949.
95. Itoh Y, Kajita M, Kinoh H et al. Membrane type 4 matrix metalloproteinase (MT4-MMP, MMP-17) is a glycosylphosphatidylinositol-anchored proteinase. J Biol Chem 1999; 274:34260-34266.
96. English WR, Puente XS, Freije JM et al. Membrane type 4 matrix metalloproteinase (MMP17) has tumor necrosis factor-α convertase activity but does not activate pro-MMP2. J Biol Chem 2000; 275:14046-14055.
97. Llano E, Pendás AM, Freije JP et al. Identification and characterization of human MT5-MMP, a new membrane- bound activator of progelatinase A overexpressed in brain tumors. Cancer Res 1999; 59:2570-2576.
98. Jaworski DM. Developmental regulation of membrane type-5 matrix metalloproteinase (MT5-MMP) expression in the rat nervous system. Brain Res 2000; 860:174-177.
99. Pei D. Leukolysin/MMP25/MT6-MMP: a novel matrix metalloproteinase specifically expressed in the leukocyte lineage. Cell Res 1999; 9:291-303.
100. Velasco G, Cal S, Merlos-Suarez A et al. Human MT6-matrix metalloproteinase: identification, progelatinase A activation, and expression in brain tumors. Cancer Res 2000; 60:877-882.
101. English WR, Velasco G, Stracke JO et al. Catalytic activities of membrane-type 6 matrix metalloproteinase (MMP25). FEBS Lett 2001; 491:137-142.
102. Pendás AM, Knäuper V, Puente XS et al. Identification and characterization of a novel human matrix metalloproteinase with unique structural characteristics, chromosomal location, and tissue distribution. J Biol Chem 1997; 272:4281-4286.
103. Sedlacek R, Mauch S, Kolb B et al. Matrix metalloproteinase MMP-19 (RASI-1) is expressed on the surface of activated peripheral blood mononuclear cells and is detected as an autoantigen in rheumatoid arthritis. Immunobiol 1998; 198:408-423.
104. Kolb C, Mauch S, Krawinkel U et al. Matrix metalloproteinase-19 in capillary endothelial cells: expression in acutely, but not in chronically, inflamed synovium. Exp Cell Res 1999; 250:122-130.
105. Stracke JO, Hutton M, Stewart M et al. Biochemical characterization of the catalytic domain of human matrix metalloproteinase 19. Evidence for a role as a potent basement membrane degrading enzyme. J Biol Chem 2000; 275:14809-14816.

106. Lohi JL, Wilson CL, Roby JD et al. Epilysin: A novel human matrix metalloproteinase (MMP-28) expressed in testis and keratinocytes and in response to injury. J Biol Chem 2000; 276:10134-10144.
107. Marchenko GN, Strongin AY. MMP-28, a new human matrix metalloproteinase with an unusual cysteine-switch sequence is widely expressed in tumors. Gene 2001; 265:87-93.
108. Llano E, Pendás AM, Knäuper V et al. Identification and structural and functional characterization of human enamelysin (MMP-20). Biochemistry 1997; 36:15101-15108.
109. Pei D, Kang T, Qi H. Cysteine array matrix metalloproteinase (CA-MMP)/MMP-23 is a type II transmembrane matrix metalloproteinase regulated by a single cleavage for both secretion and activation. J Biol Chem 2000; 275:33988-97.
110. Velasco G, Pendás AM, Fueyo A et al. Cloning and characterization of human MMP-23, a new matrix metalloproteinase predominantly expressed in reproductive tissues and lacking conserved domains in other family members. J Biol Chem 1999; 274:4570-4576.
111. Gutman A, Wasylyk B. The collagenase gene promoter contains a TPA and oncogene-responsive unit encompassing the PEA3 and AP-1 binding sites. EMBO J 1990; 9:2241-2246.
112. Benbow U, Brinckerhoff CE. The AP-1 site and MMP gene regulation: What is all the fuss about? Matrix Biol 1997; 15:519-526.
113. Lohi J, Lehti K, Valtanen H et al. Structural analysis and promoter characterization of the human membrane-type matrix metalloproteinase-1 (MT1-MMP) gene. Gene 2000; 242:75-86.
114. Haas TL, Stitelman D, Davis SJ et al. Egr-1 mediates extracellular matrix-driven transcription of membrane type 1 matrix metalloproteinase in endothelium. J Biol Chem 1999; 274:22679-22685.
115. Springman EB, Angleton EL, Birkedal-Hansen H et al. Multiple modes of activation of latent human fibroblast collagenase: evidence for the role of a Cys73 active-site zinc complex in latency and a "cysteine switch" mechanism for activation. Proc Natl Acad Sci USA 1990; 87:364-368.
116. Butler GS, Butler MJ, Atkinson SJ et al. The TIMP-2 membrane type 1 metalloproteinase "receptor" regulates the concentration and efficient activation of progelatinase A. J Biol Chem 1998; 273:871-880.
117. Zucker S, Drews M, Conner C et al. Tissue inhibitor of metalloproteinase-2 (TIMP-2) binds to the catalytic domain of the cell surface receptor, membrane type 1-matrix metalloproteinase 1 (MT1-MMP). J Biol Chem 1998; 273:1216-1222.
118. Carmeliet P, Moons L, Lijnen R et al. Urokinase-generated plasmin activates matrix metalloproteinases during aneurysm formation. Nat Genet 1997; 17:439-444.
119. Murphy G, Gavrilovic J. Proteolysis and cell migration: creating a path? Curr Opin Cell Biol 1999 Oct;11(5):614-21.
120. Nakahara H, Howard L, Thompson EW et al. Transmembrane/cytoplasmic domain-mediated membrane type 1-matrix metalloprotease docking to invadopodia is required for cell invasion. Proc Natl Acad Sci USA 1997; 94:7959-7964.
121. Yu WH, Woessner JF, Jr. Heparan sulfate proteoglycans as extracellular docking molecules for matrilysin (matrix metalloproteinase 7). J Biol Chem 2000; 275:4183-4191.
122. Brooks PC, Stromblad S, Sanders LC et al. Localization of matrix metalloproteinase MMP-2 to the surface of invasive cells by interaction with integrin $\alpha v\beta 3$. Cell 1996; 85(5):683-93.
123. Yu Q, Stamenkovic I. Cell surface-localized matrix metalloproteinase-9 proteolytically activates TGF-β and promotes tumor invasion and angiogenesis. Genes Dev 2000; 14:163-76.
124. Thomas GJ, Lewis MP, Hart IR et al. $\alpha v\beta 6$ integrin promotes invasion of squamous carcinoma cells through up-regulation of matrix metalloproteinase-9. Int J Cancer 2001; 92:641-50.
125. Dumin JA, Dickeson SK, Stricker TP et al. Pro-collagenase-1 (Matrix metalloproteinase-1) binds the $\alpha 2\beta 1$ integrin upon release from keratinocytes migrating on type I collagen. J Biol Chem 2001; 276:29368-29374.
126. Gomis-Ruth FX, Maskos K, Betz M et al. Mechanism of inhibition of the human matrix metalloproteinase stromelysin-1 by TIMP-1. Nature 1997; 389:77-81.
127. Gomez DE, Alonso DF, Yoshiji H et al. Tissue inhibitors of metalloproteinases: structure, regulation and biological functions. European J Cell Biol 1997; 74:111-122.
128. Murphy G, Houbrechts A, Cockett MI et al. The N-terminal domain of tissue inhibitor of metalloproteinases retains metalloproteinase inhibitory activity. Biochemistry 1991; 30:8097-8102.
129. DeClerck YA, Yean TD, Lee Y et al. Characterization of the functional domain of tissue inhibitor of metalloproteinases-2 (TIMP-2). Biochem J 1993; 289:65-69.
130. Langton KP, Barker MD, McKie N. Localization of the functional domains of human tissue inhibitor of metalloproteinases-3 and the effects of a Sorsby's fundus dystrophy mutation. J Biol Chem 1998; 273:16778-16781.
131. Goldberg GI, Strongin A, Collier IE et al. Interaction of 92-kDa type IV collagenase with the tissue inhibitor of metalloproteinases prevents dimerization, complex formation with interstitial collagenase, and activation of the proenzyme with stromelysin. J Biol Chem 1992; 267:4583-4591.

132. Ogata Y, Itoh Y, Nagase H. Steps involved in activation of the pro-matrix metalloproteinase 9 (progelatinase B)-tissue inhibitor of metalloproteinases-1 complex by 4-aminophenylmercuric acetate and proteinases. J Biol Chem 1995; 270:18506-18511.
133. Bigg HF, Shi YE, Liu YE et al. Specific, high affinity binding of tissue inhibitor of metalloproteinases-4 (TIMP-4) to the COOH-terminal hemopexin-like domain of human gelatinase A. TIMP-4 binds progelatinase A and the COOH- terminal domain in a similar manner to TIMP-2. J Biol Chem 1997; 272:15496-15500.
134. Shimada T, Nakamura H, Ohuchi E et al. Characterization of a truncated recombinant form of human membrane type 3 matrix metalloproteinase. Eur J Biochem 1999; 262:907-914.
135. Amour A, Slocombe PM, Webster A et al. TNF-α converting enzyme (TACE) is inhibited by TIMP-3. FEBS Lett 1998; 435:39-44.
136. Amour A, Knight CG, Webster A et al. The in vitro activity of ADAM-10 is inhibited by TIMP-1 and TIMP-3. FEBS Lett 2000; 473:275-279.
137. Tortorella MD, Burn TC, Pratta MA et al. Purification and cloning of aggrecanase-1: a member of the ADAMTS family of proteins. Science 1999; 284:1664-1666.
138. Kashiwagi M, Tortorella M, Nagase H et al. TIMP-3 is a potent inhibitor of aggrecanase-1 (ADAM-TS4) and aggrecanase-2 (ADAM-TS5). J Biol Chem 2001; 276:12501-12504
139. George SJ, Johnson JL, Angelini GD et al. Adenovirus-mediated gene transfer of the human TIMP-1 gene inhibits smooth muscle cell migration and neointimal formation in human saphenous vein. Hum Gene Ther 1998; 9:867-877.
140. Fernandez HA, Kallenbach K, Seghezzi G et al. Inhibition of endothelial cell migration by gene transfer of tissue inhibitor of metalloproteinases-1. J Surg Res 1999; 82:156-162.
141. Ahonen M, Baker AH, Kähäri V-M. Adenovirus-mediated gene delivery of tissue inhibitor of metalloproteinases-3 inhibits invasion and induces apoptosis in melanoma cells. Cancer Res 1998; 58:2310-2315.
142. Hayakawa T, Yamashita K, Tanzawa K et al. Growth-promoting activity of tissue inhibitor of metalloproteinases-1 (TIMP-1) for a wide range of cells. A possible new growth factor in serum. FEBS Lett 1992; 298:29-32.
143. Dethlefsen SM, Raab G, Moses MA et al. Extracellular calcium influx stimulates metalloproteinase cleavage and secretion of heparin-binding EGF-like growth factor independently of protein kinase C. J Cell Biochem 1998; 69:143-153.
144. Codony-Servat J, Albanell J, Lopez-Talavera JC et al. Cleavage of the HER2 ectodomain is a pervanadate-activable process that is inhibited by the tissue inhibitor of metalloproteases-1 in breast cancer cells. Cancer Res 1999, 59:1196-1201.
145. Lombard MA, Wallace TL, Kubicek MF et al. Synthetic matrix metalloproteinase inhibitors and tissue inhibitor of metalloproteinase (TIMP)-2, but not TIMP-1, inhibit shedding of tumor necrosis factor-a receptors in a human colon adenocarcinoma (Colo 205) cell line. Cancer Res 1998; 57:4001-4007.
146. Caterina JJ, Yamada S, Caterina NC et al. Inactivating mutation of the mouse tissue inhibitor of metalloproteinases-2(TIMP-2) gene alters proMMP-2 activation. J Biol Chem 2000; 275:26416-26422.
147. Henriet P, Blavier L, Declerck YA. Tissue inhibitors of metalloproteinases (TIMP) in invasion and proliferation. APMIS 1999; 107:111-119.
148. Leco KJ, Khokha R, Pavloff N et al. Tissue inhibitor of metalloproteinases-3 (TIMP-3) is an extracellular matrix-associated protein with a distinctive pattern of expression in mouse cells and tissues. J Biol Chem 1994; 269:9352-9360.
149. Wick M, Burger C, Brusselbach S et al. A novel member of human tissue inhibitor of metalloproteinases (TIMP) gene family is regulated during G1 progression, mitogenic stimulation, differentiation, and senescence. J Biol Chem 1994; 269:18953-18960.
150. Hargreaves PG, Wang F, Antcliff J et al. Human myeloma cells shed the interleukin-6 receptor: inhibition by tissue inhibitor of metalloproteinase-3 and a hydroxamate-based metalloproteinase inhibitor. Br J Haematol 1998; 101:694-702.
151. Borland G, Murphy G, Ager A. Tissue inhibitor of metalloproteinases-3 inhibits shedding of L-selectin from leukocytes. J Biol Chem 1999; 274:2810-2815.
152. Baker AH, Zaltsman AB, George SJ et al. Divergent effects of tissue inhibitor of metalloproteinase-1, -2, or -3 overexpression on rat vascular smooth muscle cell invasion, proliferation, and death in vitro. TIMP-3 promotes apoptosis. J Clin Invest 1998; 101:1478-1487.
153. Baker AH, George SJ, Zaltsman AB et al. Inhibition of invasion and induction of apoptotic cell death of cancer cell lines by overexpression of TIMP-3. Br J Cancer 1999; 79:1347-1355.
154. Greene J, Wang M, Liu YE et al. Molecular cloning and characterization of human tissue inhibitor of metalloproteinase 4. J Biol Chem 1996; 271:30375-30380.

155. Gately S, Twardowski P, Stack MS et al. The mechanism of cancer-mediated conversion of plasminogen to the angiogenesis inhibitor angiostatin. Proc Natl Acad Sci USA 1997; 94:10868-10872.
156. Cornelius LA, Nehring LC, Harding E et al. Matrix metalloproteinases generate angiostatin: effects on neovascularization. J Immunol 1998; 161:6845-6852.
157. Patterson BC, Sang QA. Angiostatin-converting enzyme activities of human matrilysin (MMP-7) and gelatinase B/type IV collagenase (MMP-9). J Biol Chem 1997; 272:28823-5.
158. Pozzi A, Moberg PE, Miles LA et al. Elevated matrix metalloprotease and angiostatin levels in integrin a1 knockout mice cause reduced tumor vascularization. Proc Natl Acad Sci USA 2000; 97:2202-2207.
159. O'Reilly MS, Boehm T, Shing Y et al. Endostatin: an endogenous inhibitor of angiogenesis and tumor growth. Cell 1997; 88:277-285.
160. Felbor U, Dreier L, Bryant RA et al. Secreted cathepsin L generates endostatin from collagen XVIII. EMBO J 2000; 19:1187-1194.
161. Lukashev ME, Werb Z. ECM signalling: orchestrating cell behaviour and misbehaviour. Trends Cell Biol 1998; 8:437-441.
162. Suzuki M, Raab G, Moses MA et al. Matrix metalloproteinase-3 releases active heparin-binding EGF-like growth factor by cleavage at a specific juxtamembrane site. J Biol Chem 1997; 272:31730-31737.
163. Fowlkes JL, Enghild JJ, Suzuki K et al. Matrix metalloproteinases degrade insulin-like growth factor-binding protein-3 in dermal fibroblast cultures. J Biol Chem 1994; 269:25742-25746.
164. Levi E, Fridman R, Miao HQ et al. Matrix metalloproteinase 2 releases active soluble ectodomain of fibroblast growth factor receptor 1. Proc Natl Acad Sci USA 1996; 93:7069-7074.
165. Kataoka H, Uchino H, Iwamura T et al. Enhanced tumor growth and invasiveness in vivo by a carboxyl-terminal fragment of a1-proteinase inhibitor generated by matrix metalloproteinases: a possible modulatory role in natural killer cytotoxicity. Am J Pathol 1999; 154:457-468.
166. Tlsty TD. Cell-adhesion-dependent influences on genomic instability and carcinogenesis. Curr Opin Cell Biol 1998; 10:647-653.
167. Kauppila S, Saarela J, Stenbäck F et al. Expression of mRNAs for type I and type III procollagens in serous ovarian cystadenomas and cystadenocarcinomas. Am J Pathol 1996; 148:539-548.
168. Stetler-Stevenson WG, Aznavoorian S, Liotta LA. Tumor cell interactions with the extracellular matrix during invasion and metastasis. Annu Rev Cell Biol 1993; 9:541-573.
169. Murray GI, Duncan ME, O'Neil P et al. Matrix metalloproteinase-1 is associated with poor prognosis in colorectal cancer. Nat Med 1996; 2:461-462.
170. Murray GI, Duncan ME, O'Neil P et al. Matrix metalloproteinase-1 is associated with poor prognosis in oesophageal cancer. J Pathol 1998; 185:256-261.
171. Shima I, Sasapuri Y, Kusukawa J et al. Production of matrix metalloproteinase-2 and metalloproteinase-3 related to malignant behavior of esophageal carcinoma. A clinicopathological study. Cancer 1992; 70:2747-2753.
172. Muller D, Wolf C, Abecassis J et al. Increased stromelysin 3 gene expression is associated with increased local invasiveness in head and neck squamous cell carcinomas. Cancer Res 1993 53:165-169.
173. Davidson B, Goldberg I, Kopolovic J et al. MMP-2 and TIMP-2 expression correlates with poor prognosis in cervical carcinoma- a clinicopathologic study using immunohistochemistry and mRNA in situ hybridization. Gynecol Oncol 1999; 73:372-382.
174. Talvensaari-Mattila A, Apaja-Sarkkinen M, Höyhtyä M et al. Matrix metalloproteinase 2 immunoreactive protein appears early in cervical epithelial dedifferentiation. Gynecol Oncol 1999; 72:306-311.
175. Wagner SN, Ockenfels HM, Wagner C et al. Differential expression of tissue inhibitor of metalloproteinases-2 by cutaneous squamous and basal cell carcinomas. J Invest Dermatol 1996; 106:321-326.
176. Airola K, Ahonen M, Johansson N et al. Human TIMP-3 is expressed during fetal development, hair growth cycle, and cancer progression. J Histochem Cytochem 1998; 46:437-447.
177. Sutinen M, Kainulainen T, Hurskainen T et al. Expression of matrix metalloproteinases (MMP-1 and -2) and their inhibitors (TIMP-1, -2 and -3) in oral lichen planus, dysplasia, squamous cell carcinoma and lymph node metastasis. Br J Cancer 1998; 77:2239-2245.
178. Polette M, Clavel C, Muller D et al. Detection of mRNAs encoding collagenase I and stromelysin 2 in carcinomas of the head and neck by in situ hybridization. Invasion Metastasis 1991; 11:76-83.
179. Visscher DW, Höyhtyä M, Ottosen SK et al. Enhanced expression of tissue inhibitor of metalloproteinase-2 (TIMP-2) in the stroma of breast carcinomas correlates with tumor recurrence. Int J Cancer 1994; 59:339-344.
180. Nuovo GJ, MacDonnell PB, Simsir A et al. Correlation of the in situ detection of polymerase chain reaction-amplified metalloproteinase complementary DNAs and their inhibitors with prognosis in cervical carcinoma. Cancer Res 1995; 55:267-275.

181. Yoshiji H, Gomez DE, Thorgeirsson UP. Enhanced RNA expression of tissue inhibitor of metalloproteinases-1 (TIMP-1) in human breast cancer. Int J Cancer 1996; 69:131-134.
182. Albini A, Melchiori A, Santi L et al. Tumor cell invasion inhibited by TIMP-2. J Natl Cancer Inst 1991; 83:775-779.
183. Declerck YA, Yean TD, Chan D et al. Inhibition of tumor invasion of smooth-muscle cell layers by recombinant human metalloproteinase inhibitor. Cancer Res 1991; 51:2151-2157.
184. Tsuchiya Y, Sato H, Endo Y et al. Tissue inhibitor of metalloproteinase-1 is a negative regulator of the metastatic ability of a human gastric-cancer cell-line, Kkls, in the chick-embryo. Cancer Res 1993; 53:1397-1402.
185. Matsuzawa K, Fukuyama K, Hubbard SL et al. Transfection of an invasive human astrocytoma cell line with a TIMP-1 cDNA: Modulation of astrocytoma invasive potential. J Neuropath Exp Neurol 1996; 55:88-96.
186. Alonso DF, Skilton G, DeLorenzo MS et al. Histopathological findings in a highly invasive mouse mammary carcinoma transfected with human tissue inhibitor of metalloproteinases-1. Oncol Reports 1998; 5:1083-1087.
187. Watanabe M, Takahashi Y, Ohta T et al Inhibition of metastasis in human gastric cancer cells transfected with Tissue inhibitor of metalloproteinase 1 gene in nude mice. Cancer 1996; 77:1676-1680.
188. Martin, DC, Ruther U, Sanchez-Sweatman OH et al. Inhibition of SV40 T antigen-induced hepatocellular carcinoma in TIMP-1 transgenic mice. Oncogene 1996; 13:569-576.
189. Kruger A, Sanchez-Sweatman OH, Martin DC et al. Host TIMP-1 overexpression confers resistance to experimental brain metastasis of a fibrosarcoma cell line. Oncogene 1998; 16:2419-2423.
190. Montgomery AM, Mueller BM, Reisfeld RA et al. Effect of tissue inhibitor of the matrix metalloproteinases-2 expression on the growth and spontaneous metastasis of a human melanoma cell line. Cancer Res 1994; 54:5467-5473.
191. Oku, T, Ata N, Yonezawa K et al. Antimetastatic and antitumor effect of a recombinant human tissue inhibitor of metalloproteinases-2 in murine melanoma models. Biol Pharm Bull 1997; 20:843-849.
192. Koop S, Khokha R, Schimdt EE et al. Overexpression of metalloproteinase inhibitor in B16F10 cells does not affect extravasation but reduces tumor growth. Cancer Res 1994; 54:4791-4797.
193. Wang, M, Liu YE, Greene J et al. Inhibition of tumor growth and metastasis of human breast cancer cells transfected with tissue inhibitor of metalloproteinase 4. Oncogene 1997; 14:2767-74.
194. Brand K, Baker AH, Perez-Canto A et al. Treatment of colorectal liver metastases by adenoviral transfer of tissue inhibitor of metalloproteinases-2 into the liver tissue. Cancer Res 2000; 60:5723-5730.
195. Jiang Y, Wang M, Celiker MY et al. Stimulation of mammary tumorigenesis by systemic tissue inhibitor of matrix metalloproteinase 4 gene delivery. Cancer Res 2001; 61:2365-2370.
196. Celiker MY, Wang M, Atsidaftos E et al. Inhibition of Wilms' tumor growth by intramuscular administration of tissue inhibitor of metalloproteinases-4 plasmid DNA. Oncogene 2001; 20:4337-4343.
197. Valente P, Fassina G, Melchiori A et al. TIMP-2 over-expression reduces invasion and angiogenesis and protects B16F10 melanoma cells from apoptosis. Int J Cancer 1998; 75:246-253.
198. Guedez L, Courtemanch L, Stetler-Stevenson M. Tissue inhibitor of metalloproteinase (TIMP)-1 induces differentiation and an antiapoptotic phenotype in germinal center B cells. Blood 1998; 92:1342-1349.
199. Bian J, Wang Y, Smith MR et al. Suppression of in vivo tumor growth and induction of suspension cell death by tissue inhibitor of metalloproteinases (TIMP)-3. Carcinogenesis 1996; 17:1805-11.
200. Smith MR, Kung HF, Durum SK et al. TIMP-3 induces cell death by stabilizing TNF-α receptors on the surface of human colon carcinoma cells. Cytokine 1997; 9:770-780.
201. Kayagaki N, Kawasaki A, Ebata T et al. Metalloproteinase-mediated release of human Fas ligand. J Exp Med 1995; 182:1777-1783.
202. Amour A, Slocombe PM, Webster A et al. TNF-α converting enzyme (TACE) is inhibited by TIMP-3. FEBS Lett 1998; 435:39-44.

CHAPTER 2

Integrins in Cancer Cell Invasion

Pekka Koistinen and Jyrki Heino

Integrins

Integrins are a family of transmembrane glycoprotein receptors that mediate cell-matrix and cell-cell interactions.[1,2] Integrins consist of an α and a β subunit. To date 24 distinct integrin heterodimers have been described, consisting of 18 α and 8 β subunits. The extracellular environment is known to have a pivotal role in cellular behavior. As receptors, integrins mediate anchorage and migration of cells via recognition of variable extracellular matrix molecules. Moreover, intracellular signals generated by integrins often influence gene expression, affecting the regulation of cell survival, differentiation and proliferation. Apart from their role in physiological events, integrins are also involved in many pathological conditions such as inflammation and tumor progression.

Integrins as Extracellular Matrix Receptors

There are 12 integrin heterodimers containing β1 subunits, which are receptors for various extracellular matrix (ECM) molecules such as collagens, laminins, fibronectin and tenascin. Integrins α1β1 and α2β1 are known to be primarily collagen receptors. The α1β1 integrin seems to bind principally to type IV collagen, but it can also recognize types I-VI and type XIII collagen,[3,4,5] while integrin α2β1 is a receptor for types I-VIII collagens.[4,6-8] Furthermore, both α1β1 and α2β1 can bind to laminins 1, 2 and 4, and α2β1 will bind to tenascin, as well.[9-14] Integrin α3β1 can interact with some collagen subtypes, but it may not be primarily a collagen receptor.[15] Instead, laminins 1 and 5 seem to be the main ligands for α3β1.[16] Similarly, laminins are the ligands for integrins α6β1 and α7β1. The α6 subunit can also form a heterodimer with β4. This combination (α6β4) serves as a laminin receptor in hemidesmosomes, special adhesion sites linking epithelial cells to basement membranes. Recently, two novel collagen-binding integrin subunits, α10 and α11, have been discovered. Each of these form heterodimers with β1 integrin. Integrin α10β1 was originally purified due to its binding to type II collagen,[17] and integrin α11β1 has been shown to bind to at least type I collagen.[18,19]

Three different β1 integrins are known to mediate cell-fibronectin interactions. Although integrins α4β1 and α5β1 are both fibronectin receptors, their binding occurs through discrete interactions: α5β1 binds to an RGD motif, whereas α4β1 recognizes another short binding sequence, LDV. Moreover, integrin α8β1 is also capable of binding fibronectin in RGD-dependent manner.[20]

Another large subfamily of integrins is composed of heterodimers sharing the αv subunit. Integrins αvβ1, αvβ3, αvβ5, αvβ6 and αvβ8 have been found to bind to fibronectin through the RGD motif, and most of them function as vitronectin receptors as well. Integrin αvβ5 is a preferential vitronectin receptor while αvβ6 has a higher affinity for fibronectin. In addition to the αv integrins, the integrin αIIbβ3 can bind to the RGD motif in both vitronectin and fibronectin, although its main function is to mediate platelet binding to fibrinogen.

Cell Invasion, edited by Jyrki Heino and Veli-Matti Kähäri. ©2002 Eurekah.com.

Table 1. Integrin receptors for extracellular matrix

Ligands	β1 integrins	αV integrins	Others
Collagens	α1β1, α2β1, α3β1, α10β1, α11β1		
Laminins	α1β1, α2β1, α3β1, α6β1, α7β1		α6β4
Fibronectin	α4β1, α5β1, α8β1	αVβ1, αVβ3, αVβ6, αVβ8	α4β7, αIIbβ3
Vitronectin		αVβ1, αVβ3, αVβ5, αVβ8	αIIbβ3
Fibrinogen			αIIbβ3
Tenascin	α2β1, α8β1, α9β1		

Integrin Structure

Each integrin subunit possesses a large extracellular domain and a transmembrane stretch. Additionally, each has a C-terminal, intracellular domain which is variable length but is usually 40-50 amino acids long. The shortest of these is in the α1 subunit, while the β4 subunit has an exceptionally long cytoplasmic tail composed of 1018 amino acids.

Integrin α subunits vary in size from 150 to 200 kDa. All α subunits contain seven N-terminal repeats of about 60 amino acids which form an extracellular structure known as the β-propeller. Together with the β subunit, this structure forms the ligand binding site. In most cases the β-propeller contains divalent cation binding sequences in repeats 4-6. Integrin α subunits may be composed of a heavy and a light chain which are bridged by a disulfide bond. The collagen binding subunits (α1, α2, α10, and α11) as well as certain cell–cell adhesion integrins (αD, αE, αL, αM and αX subunits) have an extra "inserted" domain (I domain) about 200 amino acids in length. This feature, located between the second and third cation-binding domains on top of the β-propeller structure,[21] participates in ligand binding. X-ray crystallography has revealed that the I domain has both a Rosmann fold[22] and a Mg^{2+} coordination site. The presence of Mg^{2+} has been shown to be necessary for ligand binding.[23,24]

Integrin β subunits are highly homologous to each other. Most are 90-110 kDa in size, with the exception of β4 which is considerably larger (210 kDa). Several conserved, cysteine-rich repeats are present in the C-terminus of each β subunit.[1,25-29] A highly conserved, 200-amino acid region of the β subunit, the βI-like domain, contains a sequence of 12 amino acids that binds divalent cations. This structure, which is important for the function of the β subunit, has been reported to possess ligand-binding capability.[30-33] However, the βI-like domain may not have autonomous folding or function which are characteristic for α-subunit I domains; the presence of a neighboring β-propeller may be required to stabilize the active conformation.[34]

Integrins in Human Cancer

Since the first observations describing transformation-related changes in the integrin expression pattern,[35] several in vitro and in vivo studies have demonstrated the association between the regulation of integrin expression and cancer. Changes in the integrin pattern during malignant transformation are highly dependent on the type of the cancer. An altered integrin pattern allows the cancer cells to recognize variable matrices, but it may also lead to altered signalling and changes in gene expression.

Altered Integrin Expression in Three Common Types of Human Cancer

Melanoma

The action of integrins has been widely studied in malignant melanoma. In particular, integrin αvβ3 has been identified as a tumor progression marker. The increase in its expression correlates with the conversion of melanoma growth from radial to vertical.[36] However, there are metastatic melanoma cells that express little or no αvβ3 integrin on their surface,[37,38] suggesting that no single adhesion receptor is irreplaceable in melanoma progression. The binding of αvβ3 integrin is known to be RGD-dependent. Nevertheless, there may be an additional ligand binding site in αvβ3, which activates a cellular signalling cascade independently from RGD binding.[39] Moreover, other molecules are also involved in αvβ3-related melanoma cell invasion. The function of a cell adhesion receptor, E-cadherin, is apparently related to melanoma cell growth control. Restored expression of E-cadherin was found to inhibit melanoma cell invasion by decreasing the expression of β3 integrin and MelCAM/MUC18, both receptors typical for invasive melanoma cells. In addition, expression of E-cadherin induced apoptosis in these cells.[40]

Integrins α2β1 and α3β1, but not α6 integrins, have been found to play a role in melanoma cell migration on type IV collagen and laminin. In several studies, the expression of α2β1 and α3β1 was increased in metastatic cells when compared to cells in the primary tumor.[41-46]

Roles for integrins as prognostic markers in primary and metastatic human melanomas have been suggested. The expression of β1 integrin in primary melanomas has been proposed to augur the emergence of regional lymph node metastases.[47] In contrast, the expression of β1 integrins in metastasis of melanoma has been shown to predict a longer disease-free period.[48] Expression of integrin β3 has been reported to correlate positively to lung metastases. Furthermore, β3-positive melanomas are associated with poorer survival rates.[49] Although these results are partly conflicting, they may be explained by differences in integrin function in primary sites as compared to metastatic sites.

It has also been proposed that laminin receptors, especially integrin α6β1 are involved in metastatic processes, possibly during extravasation. Multiple binding epitopes on α6β1 may contribute to the migratory and adhesive properties mediated by this integrin.[45,50] Intravenously injected anti-α6 integrin antibody inhibited the formation of lung metastases of a highly invasive mouse melanoma cell line.[51] Recently, in addition to receptor-ECM interactions, reciprocal association of different cell surface proteins has been suggested to play an important role in cancer cells. In particular, it has been found that α3β1 and α6β1 integrins can interact with CD36, a protein located in membrane rafts which participates in integrin related signal transduction. This interaction has been found to facilitate haptotactic cell migration.[52] In contrast to α6 integrins, the expression of laminin receptor α7β1 seems to inhibit malignant features such as cell growth and metastasis in mouse melanoma cell lines.[53]

Fibronectin-binding integrin α5β1 has been described as having a stimulatory effect on the growth of quiescent human melanoma cells. This effect may be mediated in an RGD-dependent manner.[54] Another fibronectin receptor, integrin α4β1, has been found to facilitate melanoma cell migration.[55] Moreover, there is evidence that to achieve a proper α4β1-mediated binding to fibronectin, melanoma cells need direct contact with a cell surface chondroitin sulfate glycosaminoglycan.[56] This interaction may modulate tumor invasion capacity via activation of a Rho-family GTPase, Cdc42, a Cdc42-associated protein, Ack-1 and the adapter protein p130cas followed by appropriate cytoskeletal reorganization.[57] However, the mouse homologue to human melanoma proteoglycan (NG2) has been reported to have a negative effect on α4β1-mediated binding of mouse melanoma cells. Nevertheless, the expression of NG2 increased the lung metastasis formation in an experimental model.[58] Like Cdc42, another GTPase, RhoC, has been proposed to enhance invasion of human and mouse melanoma, probably by modulating cytoskeletal assembly. In the same study, the expression of fibronectin and thymosin β4, a regulator of actin polymerization, was increased in both human and mouse melanoma cells with high metastatic capacity.[59] These findings stress the role of intra- and extracellular regulators of cytoskeletal organization in melanoma cell invasion. Fibronectin-binding integrins may act as important mediators in this process. Furthermore, αIIbβ3 which is typically found in platelets, may also occur on metastatic murine melanoma cells as well as on human melanoma cells and may contribute to invasion.[60,61]

Breast Cancer

Changes in the expression pattern of integrins in breast cancer, the most common cancer among females, have been reported in several studies. Immunohistochemical analyses of poorly differentiated breast adenocarcinomas have shown marked decreases of integrin α2β1. Expression of α5β1 and αVβ3 integrins has also been reduced, although to a lesser degree.[62] In situ hybridization of poorly differentiated breast adenocarcinoma has shown a reduction in the mRNA levels of α2, α5, and β1 integrins as compared to differentiated mammary epithelium.[63] These studies emphasize the importance of α2β1 integrin in normal mammary epithelium. Further, by reexpressing integrin α2β1, the ability of poorly differentiated breast cancer cells to differentiate into gland-like structures has been restored, and the in vivo tumorigenicity has been clearly reduced.[64] Some of the phenotypic changes introduced by reexpression of α2β1 integrin can be attributed to concomitantly increased expression of α6β4 integrin.[65] Thus, it is possible that during malignant transformation, changes in the expression of a key integrin lead to altered expression of other integrins.

Links between the integrins and other important factors contributing to the progression of breast tumors have also been traced. It has been postulated that hepatocyte growth factor/scatter factor (HGF/SF) is able to increase the adhesion of breast cancer cells to laminins 1 and 5, fibronectin, and vitronectin through a PI3 kinase-related mechanism. Several integrins, such as β1, β3, β4 and β5 may be affected by HGF-mediated regulation of integrin avidity.[66] The interactions between bone matrix proteins and invading breast cancer cells are pivotal because bone is a typical site of breast cancer metastasis. Integrins αvβ3 and αvβ5 are involved in bone sialoprotein-induced cancer cell adhesion, proliferation and migration. However, the function of these integrins differs so that αvβ3, whose expression is suggested to be in relation to metastatic potential, participates in cell migration, whereas αvβ5 is likely to facilitate cell adhesion and proliferation.[67,68] In highly invasive mammary epithelial cells, the process of osteopontin-induced migration, which is dependent mainly on αvβ3 integrin, involves the activation of the HGF receptor (Met).[69] Likewise, insulin-like growth factor I (IGF-I) stimulates an increase in the activity of integrin αvβ5 and MMP-9, thereby increasing cell migration through vitronectin-coated filters. Additionally, IGF-I is reportedly able to stimulate cell migration on type IV collagen and vitronectin coated membranes. These effects are presumably mediated by αvβ5 and α2β1 integrins.[70,71]

TGF-β has also been reported to participate in the migration of mammary epithelial cells. TGF-β-related responses have been linked to the PI3-kinase/Akt-signaling route, and they have been proposed to result in the delocalization of E-cadherin and β1 integrin from cell junctions.[72] Additionally, interaction between E-cadherin and αv integrins has been postulated. The introduction of dominant-negative E-cadherin into cells has led to increased migration on vitronectin as a result of increased activity of the αvβ1 and αvβ5 integrins.[73] Furthermore, epidermal growth factor and its receptor (EGFR), as well as the breast cancer-related protein ERBB-2, have been reported to mediate upregulation of β1 integrin function and breast cancer progression. Activation of PI3-kinase plays a role in the signalling of both of these receptors.[74] It has also been suggested that the overexpression of the melanoma and breast cancer-related protein ERBB-2 may induce fibronectin-dependent invasion with concomitant down-regulation of α4 integrin cell surface expression.[75]

Table 2. Changes in the integrin expression pattern of histological sections in three types of cancer. As a result of cancer progression the expression of integrins is either increased (↑) or decreased (↓).

Cancer	Integrins	Type of cancer	Reference
Melanoma	β3↑	primary and metastatic human melanomas	Albelda et al, 1990[192]
	α2↑, α3↑, α4↑, β3↑, α6β1↓	primary and metastatic human melanomas	Moretti et al, 1993[193]
	α4↑, α5β↑, αvβ3↑, α6↓, β4↓	primary and metastatic human melanomas	Danen et al, 1994[194]
	αvβ3↑, αvβ5↓	primary and metastatic human melanomas	Danen et al, 1995[37]
	α3β1↑	primary and metastatic human melanomas	Natali et al, 1993[42]
	β1↑	primary human melanomas	Hieken et al, 1996[47]
	β1↓	metastasis of human melanomas	Vihinen et al, 2000[48]
	β3↑	primary human melanomas	Hieken et al, 1999[49]
Breast cancer	α2β1↓	breast adenocarcinomas	Zutter et al, 1990[62]
	α1β1↓, α2β1↓, α3β1↓, α6β1↓, αvβ1↓, αvβ5↓	sections from primary human breast cancers	Gui et al, 1995[82]
Prostate cancer	β4↓	sections from human prostatic intraepithelial neoplasia	Davis et al, 2001[102]
	α6β4↓	human prostatic intraepithelial neoplasia	Nagle et al, 1995[105]
	α2β1↑, α6↑	lymph node metastases of human prostate cancer	Bonkhoff et al, 1993[93]
	α5↑, β1↑	rat prostatic adenocarcinoma	MacCalman et al, 1994[112]

Integrin α3β1 is involved in breast cancer cell migration and invasion by regulating the production of MMP-2. It may also participate in the rearrangement of cytoskeleton. Both phenomena are PI3-kinase dependent.[76] Cdc42 and Racl may be the signaling factors between integrins and PI3-kinase.[77] Moreover, the expression of α3β1 integrin is reported to be related to the metastatic capacity of breast cancer cells by increasing the activity of MMP-9.[78] The expression of a variety of MMPs may be modulated by integrin-associated cell–cell contacts and cell–matrix contacts. Coculturing of invasive breast carcinoma cells with bone marrow fibroblasts has been shown to increase the production of MMP-1 and MMP-2 in culture supernatant. Elevated levels of MMPs have contributed to cell migration through fibroblast monolayers, and migration has been inhibited by antibodies against variable integrins.[79]

Lung, liver and brains are typical loci for breast cancer metastasis. Increased expression or, in contrast, loss of multiple integrins has been found to assist the formation of invasive phenotypes. In addition to changes in the integrin expression pattern, general features of malignant cells are changes in cell polarization, adherens junctions and cytoskeletal organization.[80] In an experimental model the migratory potential of breast cancer cells has been improved by coexpression of the intermediate filaments keratin and vimentin. This may be due to the decrease in α2 and α3 integrins and the increase in α6 integrin expression.[81]

Immunostaining of breast cancer tissue sections has indicated that the loss of α1β1, α2β1, α3β1, α6β1, αvβ1 and αvβ5 integrins is associated with the formation of axillary metastases.[82] Somewhat controversially, it has been reported that increased levels of α2, α4, α6 and αv integrins may increase the malignant capacity of the breast cancer cells. The expression of these integrins is inhibited by the tumor suppressor gene maspin which, in contrast, induces the expression of α5 integrin.[83,84] The expression of α6β4 integrin may also inhibit malignant properties of breast cancer cells by inducing apoptosis via activation of p53. However, the expression of this integrin may facilitate carcinoma progression if p53 is in a mutated, inactive form.[85,86] It has also been suggested that integrin α6β4-mediated invasion occurs through PI3-kinase signalling.[87]

In human breast cancer, integrin αvβ3 has been found in both active and inactive forms. It has been proposed that the active form promotes metastatic capability via interaction with platelets.[88] In one experimental mouse model the metastatic capacity of human breast cancer cells has been reduced by using the snake venom disintegrin, contortrostatin, which can bind to αvβ3 integrin.[89] Another vitronectin receptor, αvβ5, has also been found to have a role in the invasion process. Some results indicate that αvβ5-dependent breast cancer cell migration may be partly regulated by urokinase. Urokinase receptor (uPAR) can interact with αvβ5 and αvβ1 integrins, and αvβ5-mediated cytoskeletal rearrangement and activation of protein kinase C occurs in response to urokinase. uPAR-triggered, integrin-dependent cell migration probably occurs via activation of Ras, MEK, ERK and myosin light chain kinase.[90,91] The pineal gland hormone, melatonin, may have an inhibitory effect on invasion via induction of β1 integrin and E-cadherin expression.[92]

Prostate Carcinoma

Only limited observations targeted to integrin expression patterns in prostate cancer have been made. In one report all primary and metastatic carcinomas expressed α2β1 and α6 integrins. The expression level of α2β1 was downregulated in grade I and II tumors. In grade III tumors the expression was heterogenous, but α2β1 expression was again upregulated in lymph node metastases.[93] Some integrins that have been implicated in melanoma and breast cancer progression may also be responsible for the development of prostate carcinoma. An example of this is the vitronectin receptor αvβ3 which has been found to increase metastatic capability. This may be partly due to its interaction with fibronectin and vitronectin bound on the surfaces of endothelial cells.[94,95] Such binding may help cancer cells to extravasate. In prostate cancer cells, as well as in other types of cancer cells studied, the ligand binding of αvβ3 integrin can lead to phosphophorylation of focal adhesion kinase (FAK) and trigger the PI3-kinase/Akt

Table 3. Integrins that either contribute (↑) or decrease (↓) malignant properties in cell lines of three types of cancer

Cancer	Integrins	Cell types	Reference
Melanoma	αvβ3↑	WM1552C and WM1341D melanoma cells	Hsu et al, 1998[36]
	αvβ3↑, αvβ5↓, α7β1↓	K1735 murine melanoma cells	Li et al, 1998[195]; Ziober et al, 1999[53]
	α2β1↑	MV3 and BLM human melanoma cells	Klein et al, 1991[41]
	α2β1↑	four human melanoma cell lines	Etoh et al, 1993[44]
	α2β1↑, α6β1↑	six human melanoma cell lines	Danen et al, 1993[45]
	α3β1↑	Me665/2 human melanoma cells	Melchiori et al, 1995[46]
	α6↑	B16/129 melanoma cells	Ruiz et al, 1993[51]
	α5β1↑	human melanoma cell lines	Mortarini et al, 1992[54]
	α4β1↑	A375-SM human melanoma cells	Mould et al, 1994[55]
	αIIbβ3↑	WM-983B human melanoma cells	Trikha et al, 1997[61]
Breast cancer	α6β4↓	mm5MT murine breast carcinoma cells	Sun et al, 1998[65]
	αvβ3↑, αvβ5↑, α3β1↑, αβ4↓	MDA-MB-231 human breast carcinoma cells	Jones et al, 1997[85]; Sung et al 1998[67]; Sugiura et al, 1999[76]; Morini et al, 2000[78]
	αvβ3↑	MDA-MB-435 human breast carcinoma cells and metastatic human breast carcinoma cells	Tuck et al, 2000[69]; Felding-Haberman et al, 2001[88]
	αvβ5↑, α2β1↑	MCF-7 human breast cancer cells	Doerr et al, 1996[70]; Carriero et al, 1999[90]; Mira et al, 1999[71]

continued on next page

Table 3. Cont.

	αVβ5↑, αVβ1↑	ZR75-1 human breast cancer cells	von Schlippe et al, 2000[73]
Prostate cancer	αVβ3↑, α3β1↑, α5β1↑, α6↑, αIIbβ3↑	DU145 prostate cancer cells	Rabinovitz et al, 1995[101]; Trikha et al. 1996[123]; Romanov et al, 1999[94]
	α3β1↓, α6β4↑, α2β1↑, α3β1↑	PC-3 human prostate cancer cells	Dedhar et al, 1993[107]; Kostenuik et al, 1997[119]; Festuccia et al, 1999[120]; Bonaccorsi et al, 2000[108]
	α4β1↓	four different human prostate cancer cell lines	Rokhling et al, 1995[110]

pathway. However, this activation seems to be ligand specific.[96,97] Mutation in a tumor suppressor gene PTEN, a negative regulator of Akt, and activation of integrin-linked kinase (ILK) may also play a role in malignant transformation.[98,99] Akt may also inhibit the expression of p27(Kip1), an inhibitor of certain cyclin/cyclin-dependent kinase complexes and cell cycle progression.[100]

Partially conflicting reports exist about the role of laminin receptor integrins in the progression of prostate cancer. It has been proposed that the increased expression of α6 integrin may contribute to the invasive capacity of prostate cancer cells.[95,101] On the other hand, the loss of β4 integrin, the counterpart of the α6 subunit, has been reported to occur in prostate cancer progression concomitantly with the loss of its ligand, laminin-5.[102] Furthermore, malignant prostate cancer cells may lose their ability to polarize and regulate a normal acinar morphogenesis because of decreases or changes in the distribution of α6β1 integrin.[103,104] The reduction of integrin α6β4 as a tumor progression-promoting factor may be explained by the fact that α6β4 participates in the formation of hemidesmosomes, which are pivotal for the attachment of differentiated epithelial cells to the basal lamina.[105,106] The decrease in α6β4 integrin expression may occur due to androgen regulation in androgen-sensitive cancer cells. In contrast, androgen-independent prostate cancer cells may maintain their expression of α6β4, which supports their high invasion capacity, whereas α6β1 and α3β1 expression may be related to less invasive phenotypes.[107,108] The differences in the functions of α6β1 and α6β4 integrins may be partially distinguised from melanoma by their distinct signal transduction pathways.[109]

Taken together, laminin-receptor integrins may maintain differentiated phenotypes when they are localized in a polarized manner. Alternatively, if the tumor cell has been able to break the polarized alignment of the laminin receptors, these integrins may promote the malignant phenotype. Androgens may be involved in this regulation of integrin distribution.

Several fibronectin-binding integrins have been proposed as having distinct roles in the progression of prostate cancer. When comparing the integrin expression patterns between tumorigenic and nontumorigenic cell lines, α4 integrin was found to be expressed only on nontumorigenic cells. However, RGD-binding integrins α5β1 and α3β1 were present in tumor cells.[110,111] Likewise, in rat prostate adenocarcinoma cell lines mRNA levels of α5 and β1 integrin subunits were increased as compared to normal cells of the rat prostate gland, accompanied by the loss of E-cadherin.[112] Moreover, plasma fibronectin was necessary to stimulate invasion of human and rat prostate carcinoma cells through in vitro basement membrane analog.[113]

Much attention has been drawn to the role of β1 integrin splice variants in prostate carcinoma. It has been postulated that the β1 integrin splice variant, β1C, inhibits proliferation of prostate cells by regulating cell cycle inhibitor, p27(Kip1). As seen at both at mRNA and protein level, β1C is downregulated in prostate cancer cells. Interestingly, the total β1 integrin is similarly downregulated at the mRNA level, but at the protein level there is no difference in total β1 expression between malignant and nonmalignant prostate cells.[114-117] The activation of another cell cycle inhibitor, cyclin-depended kinase inhibitor, p21WAF1, by interferon-α (IFN-α) leads to a phenotype in which the expression of α3 integrin is increased. This may be characteristic for the nontumorigenic state.[118]

Bone is the main target of metastatic prostate cancer cells. Bone cells may produce proteins that facilitate the migratory and invasive potentials of prostate cancer cells. One of these mediators might be TGF-β1, which is secreted by osteoblasts. TGF-β1 increases the cell adhesion to type I collagen and invasion through reconstituted basement membranes. In this process, increases in α2β1 and α3β1 expression levels have been reported.[119,120] Similarly, in studies on the adhesion of prostate epithelial cells or human prostate carcinoma cells to type I collagen or to the stroma of human bone marrow, it has been evident that the interactions are predominantly mediated by α2β1 integrin.[119,121,122]

Integrin αIIbβ3 has also been suggested as having a role in the metastasis of prostate cancer. Cell surface expression of αIIbβ3 integrin is reported to be regulated by protein kinase C.[123,124]

Integrins and Invasion

As the main link between a cell and the ECM, integrins have an essential role in the invasion process. The cellular integrin expression pattern is highly variable between cancer types, in individual tumors and even in separate tumor cells inside a single tumor. Thus, it is difficult to estimate the involvement of an individual adhesion receptor. The data that have accumulated with respect to integrin expression in various types of human cancer allow some conclusions to be drawn. Some integrins, especially αvβ3 seem to promote tumor progression and metastasis. The fact that some aggressive cells are negative for this integrin indicates that none of the adhesion receptors are irreplaceable.[37] It is also obvious that some integrins have distinct functions depending on cell type. For example, α2β1 integrin may participate in the maintenance of differentiated cell phenotype in breast epithelial cells, and therefore it is often downregulated in breast cancer.[63] However, in melanoma,[41] prostate[93] and gastric carcinoma, [125] α2β1 expression is associated to tumor progression and invasion. In addition, many experimental models support the idea that α2β1 integrin is essential for cancer cell migration, invasion and metastasis formation.[126-128]

It is also important to remember that downregulation of an integrin subunit in cancer cells does not necessarily mean that the receptor is unimportant for malignant phenotype or that it has a tumor suppressor function. Maximum cell migration speed is dependent on optimal ligand concentration, integrin expression, and ligand-integrin affinity.[129] Therefore, a decrease in integrin expression may actually promote cell migration. The role of the integrins is not limited to their function as a mechanical bond in cell–matrix contact sites, but, after binding extracellular ligands, integrins are also capable of sending signals into the cell. The cancer-related genomic instability leads to variable changes in the expression of cellular signalling molecules and most probably affect also integrin-linked pathways. Thus, the function of an integrin may change during tumor progression. The integrin may be downregulated during transformation because it supports a normal phenotype or inhibits cell growth. However, it may also play an essential role in invasive cancer.

Integrins and MMPs

Integrins are needed in cell movement, but they might also have other roles in cancer invasion. Importantly, they induce the expression of different ECM degrading proteases, especially members of the MMP family. The interplay between integrins and the matrix metalloproteinases may be one of the key phenomena in the invasion process.

αVβ3-Activated MMP Expression

Interaction of αvβ3 with fibronectin and vitronectin induces the expression of MMP-2.[130-132] In melanoma cells, the modulation of proMMP-2 to its invasion-contributing active form is regulated by membrane type-1 metalloproteinase (MT1-MMP), whose activation may be inhibited by tissue inhibitor of metalloproteinases-2 (TIMP-2).[133]

In our own experiments with human osteosarcoma cells we have used an intracellular, single-chain anti-αv-integrin antibody to prevent the transport of maturing αv integrin from the endoplasmic reticulum to the cell surface.[134] In antibody-transfected clones, the cell surface expression of αv was reduced by approximately 70-100%. When cells were plated on fibronectin or vitronectin matrices we could see a marked reduction in the RNA levels of MMP-2. Furthermore, cell spreading and adhesion decreased significantly in comparison with control cells.

Collagen Receptor-Activated MMP Expression

We have hypothesized[135] that the invasion of some cells through fibrillar collagen can be a stepwise process in which α2β1 integrin binds to type I collagen in three dimensional matrices. The binding upregulates expression of MMP-1,[136,137] which is able to degrade fibrillar collagens. Collagen degradation by MMP-1 is followed by denaturation, uncovering RGD sequences. The RGD motif acts as a ligand for αvβ3 integrin, which in turn may induce the production of MMP-2. Finally, this enzyme completes the degradation of collagen.

Fibronectin Receptor-Activated MMP Expression

Fibronectin receptors α4β1 and α5β1 have been found to have different influences on the expression of three MMPs in rabbit synoviocytes. As already described above, these integrins bind to separate regions of the fibronectin molecule. Thus different splice variants of fibronectin may have distinct effects on cellular gene expression. integrin α5β1 increases the expression of MMP-1, MMP-3 and MMP-9,[138] whereas the production of these MMPs is reduced after fibronectin binding via α4β1 integrin.[139] However, in human intestinal mesenchymal cells, the binding of activating, monoclonal anti-α4 antibody leads to upregulation of MMP-2-activating membrane type-1-matrix metalloproteinase (MT1-MMP), and thus increases the activated form of MMP-2.[140]

Cellular Signals Between Integrins and MMP Genes

Formation of focal adhesion sites and clustering of integrins are needed for ligand binding and signal transduction to the cell nucleus.[141-143] Signalling cascades may be activated when the cytoplasmic tails of integrin subunits bind to specific proteins inside the cell. The activation of focal adhesion kinase (FAK) within focal adhesion sites requires both ligand binding to integrins and intact cytoskeleton. FAK has binding capacity to β1, β2 and β3 subunits,[144] as well as to cytoskeletal proteins such as paxillin, talin and potentially, vinculin.[145-147] Interestingly, paxillin has been reported to bind also α4 integrin cytoplasmic tail.[148] Moreover, integrin-ligand binding results in FAK autophosphorylation. Phosphorylated tyrosine 397 acts in turns as a binding site for kinases like Csk, Fyn, and Src.[149] These events may lead to induction of MAPK/ERK/JNK pathway and among other things, MMP production. In many cases cytokines participate in MMP induction in an activating or synergistic manner.

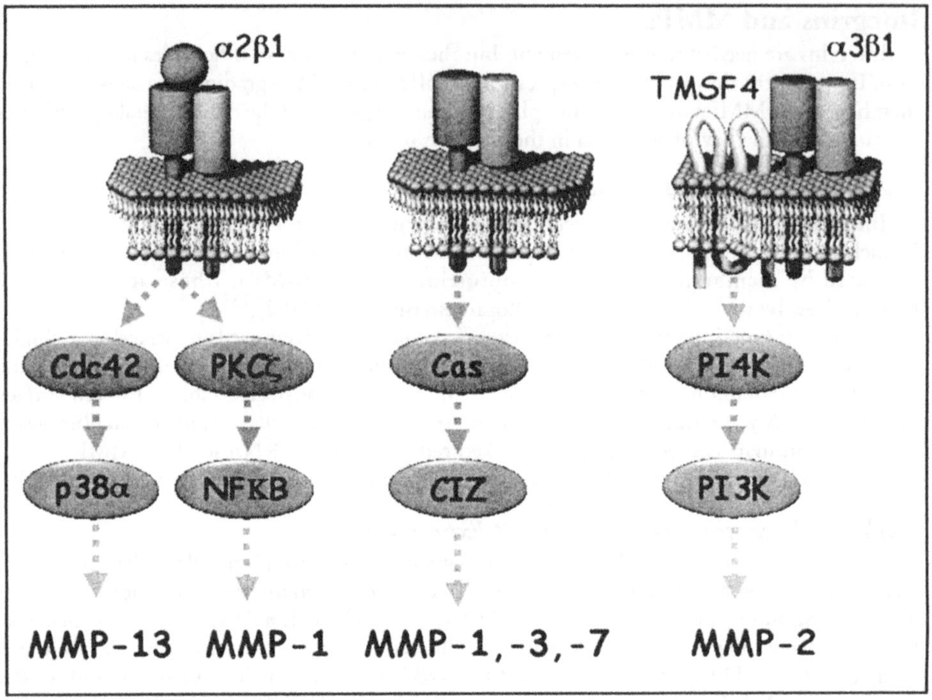

Fig. 1. Integrin related signalling leads to production of MMPs essential for cancer progression. (A) α2β1 integrin regulates MMP-1 and MMP-13 through distinct pathways. Ligand binding leads to activation of protein kinase Cζ (PKCζ) and tetraspanin, nuclear factor κB (NFκB), which upregulates MMP-1 expression. MMP-13 is upregulated via Cdc42, a GTPase of the Rho family, and the α isoform of p38 (p38α). (B) Ligand binding by multiple integrins can trigger a signalling cascade through p130cas, a tyrosine-phosphorylated, SH3 domain-containing docking protein, which is located in focal adhesions. p130cas activates metalloproteinase-activating zinc finger protein, CIZ, which leads to activation of several MMPs. (C) integrin α3β1 forms a cell surface complex with tetraspannin protein TMSF4. PI4-kinase is activated by an integrin-TMFS4 complex. PI4-kinase can then activate the PI3-kinase-dependent signalling route that results in the expression of MMP-2.[136,154,156,157,178,183,186,187]

Integrin-Related Signals and the Expression of Gelatinases (MMP-2 and MMP-9)

When ovarian carcinoma cells are cultured inside a three-dimensional collagen gel, MMP-2 is activated. The activation has also been seen when human breast cancer cells have been cultured in collagen gels. Furthermore, the induction of pro-MMP-2 and TIMP-2 have taken place after treatment of cells with soluble anti-β1 integrin antibody. When cells have come into contact with beads coated with an integrin aggregation-promoting, anti-β1 antibody, the pro-MMP-2 activator, MT-1-MMP, has been accumulated, suggesting that integrin aggregation may have a part in metalloproteinase activation processes. The activation of MMP-2 by β1 integrins has been suggested as being dependent on tyrosine kinases.[150,151]

Treatment of human glioblastoma cell lines with anti-α3β1 integrin antibody has been suggested to reduce the cell surface expression of α3β1 and to increase both MMP-2 activity and the cells' ability to invade the synthetic basement membrane, Matrigel.[152] Similar results have been obtained with rhabdomyosarcoma cells, although the cell surface expression of α3β1 after antibody treatment was not studied.[153] Other studies have given more evidence about the connection between α3β1 integrin and MMP-2. It has been suggested that α3β1 integrin and tetraspanin protein complexes regulate the production of MMP-2 and cell invasion via PI3K-dependent signalling routes. Moreover, actin cytoskeleton rearrangement may also take place in the invasion process. At least talin and MARCKS but not vinculin are reported to codistribute with tetraspanins.[154,155] In addition to PI3K, PI4K is also thought to be activated by certain integrin α3β1/tetraspannin complexes. This signalling route may have an effect on cell motility.[156,157] Tetraspanins are known to be of the family of transmembrane proteins, which form membrane complexes with at least integrins α3β1 and α6β1; several of them have been suggested as tumor-specific antigens.[155,158-160] Besides tetraspanins, integrin α3β1 has been described as forming a complex with the EMMPRIN/basigin/OX47/M6, a transmembrane protein, which contains two immunoglubulin domains.[161] So far it is not known whether this protein complex has an effect on integrin signalling.

In colon adenocarcinoma cells another laminin binding receptor, α6β4 integrin, may also facilitate signals that lead to increased expression of MMP-2 and tumor invasion.[162] In addition to regulation of MMP-2, it has been proposed that α3β1 integrin participates in the regulation of MMP-9 expression. Anti-α3 or anti-β1, but not other anti-integrin antibodies, have been found to induce the expression of MMP-9 in human keratinocytes.[163] The association between α3β1 integrin and MMP-9 expression has been established with immortalized mouse keratinocytes, as well.[164] Metastatic cells from mammary carcinomas showed increased expression of α3β1 integrin when compared to primary carcinomas. Function-blocking anti-α3 antibody could inhibit in vitro migration and invasion of these cells, and the antibody treatment could also reduce the activity of MMP-9.[78]

MMP-9 expression and invasion of ovarian cancer cells have been found to be induced by the peritoneal expression of fibronectin. These effects could be blocked by both an anti-α5 integrin antibody and RGD polypeptides.[165] It has also been shown that in ovarian carcinoma, the cell–cell interaction between carcinoma cells and fibroblasts causes induction of proMMP-2 release from fibroblasts. Type I collagen and anti-β1 integrin antibody could then induce the activation of proMMP-2 produced by tumor derived fibroblasts.[166] Collagen receptor integrins (at least α2β1 but not α3β1) seem to have a role in the activation of MMP-2, which is reported to happen via MT1-MMP.[167] Both MMP-9 activity and the retinoic acid-dependent invasion of squamous cell carcinoma lines have been downregulated by applying anti-β1 integrin antibodies or mitogen-activated protein kinase inhibitors.[168]

The mechanisms underlying the integrin-associated induction of MMP-9 have been elucitated in a model where MMP-9 gene expression was studied during macrophage differentiation with human myeloid leukemia cells. In accordance with previous studies, monoclonal anti-fibronectin or anti-α5β1 integrin antibodies could inhibit fibronectin-induced cell adhesion and MMP-9 expression. The macrophage-differentiation and cell-adhesion promoting agent, phorbol 12-myristate 13-acetate (PMA), could induce MMP-9 expression but there was no induction in PKC-β deficient variants. The disruption of cytoskeletal integrity by the inhibitors of actin polymerization, cytochalasins B and D, inhibited the production of MMP-9.[169] Similar findings have been made by studying human glioma cells.[170] On the other hand, cytochalasin D treatment could markedly increase fibronectin-induced MMP-9 and MMP-2 expression in human T lymphocytes.[132] These findings indicate the importance of the cytoskeletal arrangement in proteinase regulation in addition to protein kinases.

Increasing evidence indicates a correlation between the expressions of α5β1 integrin and MMP-9. PMA was found to increase mRNA and protein levels of both fibronectin and fibronectin-binding α5β1 integrin in human myeloid leukemia cells. Furthermore, anti-TNF-α antibody could inhibit the expression of α5 and β1 integrins resulting in decreased cell adhesion, spreading and MMP-9, but not MMP-2, expression.[171] Fibronectin-dependent production of MMP-2 and MMP-9 has been reduced in human T lymphocyte culture by applying monoclonal anti-α4, -α5 and -αv antibodies.[132] Both platelet-derived growth factor and basic fibroblast growth factor have synergistic roles in upregulation of MMP-9. These functions occur in combination with either TNF-α or interleukin-α in rabbit and human dermal fibroblasts.[172] One possible signalling cascade leading to MMP-9 expression may go through Raf-1 and mitogen-activated protein kinases.[173] On the other hand, it has been suggested that in human T-cells, ligation with fibronectin induces both activating and inhibiting signals for MMP-2 and MMP-9 expression. The activation occurs through Src-type tyrosine kinase while inhibition takes place through a Ras/MAP kinase pathway.[132]

Nuclear binding sites of nuclear factor-κB (NF-KB), Sp-1 and activator protein-1 (AP-1) may be involved at least in TNF-α-related induction of MMP-9 expression.[172,174] In human breast cancer cells, overexpression of bcl-2 was found to increase NF-κB-dependent transcriptional MMP-9 activity.[175] In a murine model, spontaneous metastasis could be prevented by retroviral delivery of dominant negative NF-κB which also resulted in downregulation in MMP-9 expression. Moreover, TIMP-1 and -2 were upregulated.[176] However, v-Src-mediated activation of the AP-1 binding site and of a GT box located downstream from the AP-1 site has been described as being involved in an inflammatory cytokine-independent pathway by which the expression of MMP-9 is activated.[177] These kinds of signalling mechanisms may correspondingly be involved in MMP-9-dependent invasion of other types of cancer cells.

Integrin-Activated Signals and the Expression of Collagenases (MMP-1 and MMP-13)

Integrin α2β1 has been shown to upregulate MMP-1 expression in osteosarcoma cells when the cells are in contact with type I collagen.[136] In dermal fibroblasts, the α2β1-dependent upregulation of MMP-1 inside a three-dimensional collagen gel has been reportedly mediated by increased activity of PKC-ζ and NF-κB, a transcription factor downstream of PKC-ζ.[178] Additionally, α2β1-mediated induction of MMP-1 can be abrogated by using tyrosine kinase inhibitors.[179,180] integrin αv may also have a regulatory role in MMP-1 gene expression with the dermatan sulfate proteoglycan, decorin. When rabbit synovial fibroblasts were plated on matrices of either decorin and vitronectin or decorin and a 120-kDa fragment of fibronectin, the expression of MMP-1 was induced.[181] Integrin α5β1-induced collagenase expression has been reported to be mediated via PEA3- and AP1-binding sites in a collagenase promoter.[182]

Recently, a novel metalloproteinase-activating zinc finger protein, CIZ, has been described. It can function together with p130cas, which is a tyrosine-phosphorylated, SH3 domain-containing docking protein located in focal adhesions. p130cas is activated as a result of integrin ligand binding, and it can further activate CIZ. CIZ is suggested to be a nucleocytoplasmic shuttling protein that can bind to certain consensus sequences in matrix metalloproteinase genes. CIZ overexpression has been found to upregulate the transcription of MMP-1, MMP-3 and MMP-7.[183] The activation of p130cas via tyrosine phosphorylation after integrin-mediated cell adhesion onto fibronectin, vitronectin, laminin and collagen is in close relation to FAK. Depolymerization of actin networks with cytochalasin D could hinder phosphorylation.[184,185] So far it is not known how a specific integrin receptor can cause the upregulation of a specific MMP by using this signaling route.

Beyond the modulation of MMP-1 and MT-1-MMP/MMP-2 expression, collagen-binding integrins play a role in the regulation of MMP-13 (collagenase-3). Culturing of human dermal fibroblasts inside a three-dimensional collagen lattice induced the expression and proteolytic activation of MMP-13. This induction could be inhibited by function-blocking anti-α2 integrin antibodies. Similarly, activating anti-β1 integrin antibody could enhance the MMP-13 induction. Thus, ligand binding by α2β1 integrin stimulated the production of MMP-13. Furthermore, it was found that activation of p38 is required for MMP-13 induction and that the induction was repressed by active ERK.[186] Integrin α2β1 has been shown to activate the α isoform of p38 via Cdc42 in a process that requires the cytoplasmic domain of α2 subunit.[187]

Integrins as MMP Receptors

There is an increasing body of evidence that some integrins can modulate the action of MMPs by binding directly to them. This kind of interaction is understandable in terms of cell movement because it enables the cell to direct proteolytic activity in the desired manner at its migrating edge. After cleavage of the ECM components, the migrating cell can use the same integrin for attachment. The first integrin–MMP interaction that was identified was between αvβ3 and MMP-2 on angiogenic blood vessels and melanoma cells.[188] There is some evidence that vascular invasion is inhibited when the interaction of integrin αvβ3 and MMP-2 is abrogated.[189] Additionally, αvβ3 integrin may have some modulatory properties on MMP-2 activity. It has been suggested that integrin αvβ3 could inhibit the cleavage of pro-MMP-2 to the active form by receptor clustering without any change in MT1-MMP expression.[190] MT1-MMP and MMP inhibitor TIMP-2 have been localized with caveolin-1 in the same domain of cell membranes with MMP-2 and αvβ3 integrin by using immunofluorescence and confocal microscopy, although direct interaction with αvβ3 integrin has not been shown.[191] The binding of MMPs to cell surface integrins may be a more general phenomenon. Thus, it is possible that pericellular proteolysis is both activated and targeted by integrins.

Acknowledgments

We wish to thank Dr. Wendy Connors for critical reading of the manuscript, and Mr. Rolf Sara for computer graphics.

References

1. Ruoslahti E. Integrins. J Clin Invest 1991; 87:1-5.
2. Hynes RO. Integrins: Versatility, modulation, and signaling in cell adhesion. Cell 1992; 69:11-25.
3. Kern A, Eble J, Golbik R, Kuhn K. Interaction of type IV collagen with the isolated integrins alpha 1 beta 1 and alpha 2 beta 1. Eur J Biochem 1993; 215:151-159.
4. Ruggiero F, Comte J, Cabanas C, Garrone R. Structural requirements for alpha 1 beta 1 and alpha 2 beta 1 integrin mediated cell adhesion to collagen V. J Cell Sci 1996; 109:1865-1874.
5. Nykvist P, Tu H, Ivaska J, Kapyla J, Pihlajaniemi T et al. Distinct recognition of collagen subtypes by alpha(1)beta(1) and alpha(2)beta(1) integrins. Alpha(1)beta(1) mediates cell adhesion to type XIII collagen. J Biol Chem 2000; 275:8255-8261
6. Staatz WD, Fok KF, Zutter MM, Adams SP, Rodriguez BA et al. Identification of a tetrapeptide recognition sequence for the alpha 2 beta 1 integrin in collagen. J Biol Chem 1991; 266:7363-7367.
7. Saelman EU, Nieuwenhuis HK, Hese KM, de Groot PG, Heijnen HF et al. Platelet adhesion to collagen types I through VIII under conditions of stasis and flow is mediated by GPIa/IIa (alpha 2 beta 1-integrin). Blood 1994; 83:1244-1250.
8. Chen M, O'Toole EA, Li YY, Woodley DT. Alpha 2 beta 1 integrin mediates dermal fibroblast attachment to type VII collagen via a 158-amino-acid segment of the NC1 domain. Exp Cell Res 1999; 249:231-239.
9. Colognato H, MacCarrick M, O'Rear JJ, Yurchenco PD.The laminin alpha2-chain short arm mediates cell adhesion through both the alpha1beta1 and alpha2beta1 integrins. J Biol Chem 1997; 272:29330-29336.
10. Ettner N, Gohring W, Sasaki T, Mann K, Timpl R. The N-terminal globular domain of the laminin alpha1 chain binds to alpha1beta1 and alpha2beta1 integrins and to the heparan sulfate-containing domains of perlecan. FEBS Lett 1998; 430:217-221.
11. Elices MJ, Hemler ME. The human integrin VLA-2 is a collagen receptor on some cells and a collagen/laminin receptor on others. Proc Natl Acad Sci USA 1989; 86:9906-9910.
12. Languino LR, Gehlsen KR, Wayner E, Carter WG, Engvall E et al. Endothelial cells use alpha 2 beta 1 integrin as a laminin receptor. J Cell Biol. 1989; 109:2455-2462
13. O'Connell PJ, Faull R, Russ GR, D'Apice AJ. VLA-2 is a collagen receptor on endothelial cells. Immunol Cell Biol 1991; 69:103-110.
14. Sriramarao P, Mendler M, Bourdon MA. Endothelial cell attachment and spreading on human tenascin is mediated by alpha 2 beta 1 and alpha v beta 3 integrins. J Cell Sci 1993; 105:1001-1012.
15. Krishnamurti U, Chen Y, Michael A, Kim Y, Fan WW et al. integrin-mediated interactions between primary/T-sv40 immortalized human glomerular epithelial cells and type IV collagen. Lab Invest 1996; 74:650-657.

16. Tashiro K, Monji A, Yoshida I, Hayashi Y, Matsuda K et al. An IKLLI-containing peptide derived from the laminin alpha1 chain mediating heparin-binding, cell adhesion, neurite outgrowth and proliferation, represents a binding site for integrin alpha3beta1 and heparan sulphate proteoglycan. Biochem J 1999; 340:119-126.
17. Camper L, Hellman U, Lundgren-Akerlund E. Isolation, cloning, and sequence analysis of the integrin subunit alpha10, a beta1-associated collagen binding integrin expressed on chondrocytes. J Biol Chem 1998; 273:20383-20389.
18. Lehnert K, Ni J, Leung E, Gough SM, Weaver A et al. Cloning, sequence analysis, and chromosomal localization of the novel human integrin alpha11 subunit (ITGA11). Genomics 1999; 60:179-187
19. Velling T, Kusche-Gullberg M, Sejersen T, Gullberg D. cDNA cloning and chromosomal localization of human alpha(11) integrin. A collagen-binding, I domain-containing, beta(1)-associated integrin alpha-chain present in muscle tissues. J Biol Chem 1999; 274:25735-25742.
20. Müller U, Bossy B, Venstrom K, Reichardt LF. integrin alpha 8 beta 1 promotes attachment, cell spreading, and neurite outgrowth on fibronectin. Mol Biol Cell 1995; 6:433-448.
21. Stewart M, Thiel M, Hogg N. Leukocyte integrins. Curr Opin Cell Biol 1995; 7:690-696.
22. Lee JO, Rieu P, Arnaout MA, Liddington R. Crystal structure of the A domain from the alpha subunit of integrin CR3 (CD11b/CD18). Cell 1995; 80:631-638.
23. Tuckwell DS, Reid KB, Barnes MJ, Humphries MJ. The A-domain of integrin alpha 2 binds specifically to a range of collagens but is not a general receptor for the collagenous motif. Eur J Biochem 1996; 241:732-739.
24. Michishita M, Videm V, Arnaout MA. A novel divalent cation-binding site in the A domain of the beta 2 integrin CR3 (CD11b/CD18) is essential for ligand binding. Cell 1993; 72:857-867.
25. Hynes RO. Integrins: A family of cell surface receptors. Cell 1987; 48:549-554.
26. Albelda SM, Buck CA. Integrins and other cell adhesion molecules. FASEB J 1990; 4:2868-2880.
27. Larson RS, Springer TA. Structure and function of leukocyte integrins. Immunol Rev 1990; 114:181-217.
28. Hemler ME, Elices MJ, Parker C, Takada Y. Structure of the integrin VLA-4 and its cell-cell and cell-matrix adhesion functions. Immunol Rev 1990; 114:45-65.
29. Springer TA. Folding of the N-terminal, ligand-binding region of integrin alpha-subunits into a beta-propeller domain. Proc Natl Acad Sci USA 1997; 94:65-72.
30. Takada Y, Puzon W. Identification of a regulatory region of integrin beta 1 subunit using activating and inhibiting antibodies. J Biol Chem 1993; 268:17597-17601.
31. Shih DT, Edelman JM, Horwitz AF, Grunwald GB, Buck CA. Structure/function analysis of the integrin beta 1 subunit by epitope mapping. J Cell Biol 1993; 122:1361-1371.
32. Steiner B, Trzeciak A, Pfenninger G, Kouns WC. Peptides derived from a sequence within beta 3 integrin bind to platelet alpha IIb beta 3 (GPIIb-IIIa) and inhibit ligand binding. J Biol Chem 1993; 268:6870-6873.
33. D'Souza SE, Haas TA, Piotrowicz RS, Byers-Ward V, McGrath DE et al. Ligand and cation binding are dual functions of a discrete segment of the integrin beta 3 subunit: Cation displacement is involved in ligand binding. Cell 1994; 79:659-667
34. Huang C, Lu C, Springer TA. Folding of the conserved domain but not of flanking regions in the integrin beta2 subunit requires association with the alpha subunit. Proc Natl Acad Sci USA 1997; 94:3156-3161.
35. Plantefaber LC, Hynes RO. Changes in integrin receptors on oncogenically transformed cells. Cell 1989; 56:281-290.
36. Hsu MY, Shih DT, Meier FE, Van Belle P, Hsu JY et al. Adenoviral gene transfer of beta3 integrin subunit induces conversion from radial to vertical growth phase in primary human melanoma. Am J Pathol 1998; 153:1435-1442.
37. Danen EH, Jansen KF, Van Kraats AA, Cornelissen IM, Ruiter DJ et al. Alpha v-integrins in human melanoma: Gain of alpha v beta 3 and loss of alpha v beta 5 are related to tumor progression in situ but not to metastatic capacity of cell lines in nude mice. Int J Cancer 1995; 61:491-496.
38. Marshall JF, Rutherford DC, Happerfield L, Hanby A, McCartney AC et al. Comparative analysis of integrins in vitro and in vivo in uveal and cutaneous melanomas. Br J Cancer 1998; 77:522-529.
39. Pasco S, Monboisse JC, Kieffer N. The alpha 3(IV)185-206 peptide from noncollagenous domain 1 of type IV collagen interacts with a novel binding site on the beta 3 subunit of integrin alpha Vbeta 3 and stimulates focal adhesion kinase and phosphatidylinositol 3-kinase phosphorylation. J Biol Chem 2000; 275:32999-33007
40. Hsu MY, Meier FE, Nesbit M, Hsu JY, Van Belle P et al. E-cadherin expression in melanoma cells restores keratinocyte-mediated growth control and down-regulates expression of invasion-related adhesion receptors. Am J Pathol 2000; 156:1515-1525.

41. Klein CE, Dressel D, Steinmayer T, Mauch C, Eckes B et al. integrin alpha 2 beta 1 is upregulated in fibroblasts and highly aggressive melanoma cells in three-dimensional collagen lattices and mediates the reorganization of collagen I fibrils. J Cell Biol 1991; 115:1427-1436.
42. Natali PG, Nicotra MR, Bartolazzi A, Cavaliere R, Bigotti A. integrin expression in cutaneous malignant melanoma: association of the alpha 3/beta 1 heterodimer with tumor progression. Int J Cancer 1993; 54:68-72.
43. Yoshinaga IG, Vink J, Dekker SK, Mihm MC Jr., Byers HR. Role of alpha 3 beta 1 and alpha 2 beta 1 integrins in melanoma cell migration. Melanoma Res 1993; 3:435-441.
44. Etoh T, Thomas L, Pastel-Levy C, Colvin RB, Mihm MC Jr. et al. Role of integrin alpha 2 beta 1 (VLA-2) in the migration of human melanoma cells on laminin and type IV collagen. J Invest Dermatol 1993; 100:640-647.
45. Danen EH, van Muijen GN, van de Wiel-van Kemenade E, Jansen KF, Ruiter DJ et al. Regulation of integrin-mediated adhesion to laminin and collagen in human melanocytes and in nonmetastatic and highly metastatic human melanoma cells. Int J Cancer. 1993; 54:315-321.
46. Melchiori A, Mortarini R, Carlone S, Marchisio PC, Anichini A et al. The alpha 3 beta 1 integrin is involved in melanoma cell migration and invasion. Exp Cell Res 1995; 219:233-242.
47. Hieken TJ, Ronan SG, Farolan M, Shilkaitis AL, Das Gupta TK. Beta 1 integrin expression: a marker of lymphatic metastases in cutaneous malignant melanoma. Anticancer Res 1996; 16:2321-2324.
48. Vihinen P, Nikkola J, Vlaykova T, Hahka-Kemppinen M, Talve L et al. Prognostic value of beta1 integrin expression in metastatic melanoma. Melanoma Res 2000; 10:243-251.
49. Hieken TJ, Ronan SG, Farolan M, Shilkaitis AL, Das Gupta TK. Molecular prognostic markers in intermediate-thickness cutaneous malignant melanoma. Cancer 1999; 85:375-382.
50. Hangan D, Morris VL, Boeters L, von Ballestrem C, Uniyal S et al. An epitope on VLA-6 (alpha6beta1) integrin involved in migration but not adhesion is required for extravasation of murine melanoma B16F1 cells in liver. Cancer Res 1997; 57:3812-3817.
51. Ruiz P, Dunon D, Sonnenberg A, Imhof BA. Suppression of mouse melanoma metastasis by EA-1, a monoclonal antibody specific for alpha 6 integrins. Cell Adhes Commun 1993; 1:67-81.
52. Thorne RF, Marshall JF, Shafren DR, Gibson PG, Hart IR et al. The integrins alpha3beta1 and alpha6beta1 physically and functionally associate with CD36 in human melanoma cells. Requirement for the extracellular domain OF CD36. J Biol Chem 2000; 275:35264-35275.
53. Ziober BL, Chen YQ, Ramos DM, Waleh N, Kramer RH. Expression of the alpha7beta1 laminin receptor suppresses melanoma growth and metastatic potential. Cell Growth Differ 1999; 10:479-490.
54. Mortarini R, Gismondi A, Santoni A, Parmiani G, Anichini A. Role of the alpha 5 beta 1 integrin receptor in the proliferative response of quiescent human melanoma cells to fibronectin. Cancer Res 1992; 52:4499-4506
55. Mould AP, Askari JA, Craig SE, Garratt AN, Clements J et al. integrin alpha 4 beta 1-mediated melanoma cell adhesion and migration on vascular cell adhesion molecule-1 (VCAM-1) and the alternatively spliced IIICS region of fibronectin. J Biol Chem 1994; 269:27224-27230.
56. Iida J, Meijne AM, Oegema TR Jr., Yednock TA, Kovach NL et al. A role of chondroitin sulfate glycosaminoglycan binding site in alpha4beta1 integrin-mediated melanoma cell adhesion. J Biol Chem 1998; 273:5955-5962.
57. Eisenmann KM, McCarthy JB, Simpson MA, Keely PJ, Guan JL et al. Melanoma chondroitin sulphate proteoglycan regulates cell spreading through Cdc42, Ack-1 and p130cas. Nat Cell Biol 1999; 1:507-513.
58. Burg MA, Grako KA, Stallcup WB. Expression of the NG2 proteoglycan enhances the growth and metastatic properties of melanoma cells. J Cell Physiol 1998; 177:299-312.
59. Clark EA, Golub TR, Lander ES, Hynes RO. Genomic analysis of metastasis reveals an essential role for RhoC. Nature 2000; 406:532-535.
60. Timar J, Bazaz R, Kimler V, Haddad M, Tang DG et al. Immunomorphological characterization and effects of 12-(S)-HETE on a dynamic intracellular pool of the alpha IIb beta 3-integrin in melanoma cells. J Cell Sci 1995; 108:2175-2186.
61. Trikha M, Timar J, Lundy SK, Szekeres K, Cai Y et al. The high affinity alphaIIb beta3 integrin is involved in invasion of human melanoma cells. Cancer Res 1997; 57:2522-2528.
62. Zutter MM, Mazoujian G, Santoro SA. Decreased expression of integrin adhesive protein receptors in adenocarcinoma of the breast. Am J Pathol 1990; 137:863-870.
63. Zutter MM, Krigman HR, Santoro SA. Altered integrin expression in adenocarcinoma of the breast. Analysis by in situ hybridization. Am J Pathol 1993; 142:1439-1448.

64. Zutter MM, Santoro SA, Staatz WD, Tsung YL. Reexpression of the alpha 2 beta 1 integrin abrogates the malignant phenotype of breast carcinoma cells. Proc Natl Acad Sci USA 1995; 92:7411-7415.
65. Sun H, Santoro SA, Zutter MM. Downstream events in mammary gland morphogenesis mediated by reexpression of the alpha2beta1 integrin: the role of the alpha6 and beta4 integrin subunits. Cancer Res 1998; 58:2224-2233.
66. Trusolino L, Cavassa S, Angelini P, Ando M, Bertotti A et al. HGF/scatter factor selectively promotes cell invasion by increasing integrin avidity. FASEB J 2000; 14:1629-1640.
67. Sung V, Stubbs JT III, Fisher L, Aaron AD, Thompson EW. Bone sialoprotein supports breast cancer cell adhesion proliferation and migration through differential usage of the alpha(v)beta3 and alpha(v)beta5 integrins. J Cell Physiol 1998; 176:482-494.
68. Wong NC, Mueller BM, Barbas CF, Ruminski P, Quaranta V et al. Alphav integrins mediate adhesion and migration of breast carcinoma cell lines. Clin Exp Metastasis 1998; 16:50-61.
69. Tuck AB, Elliott BE, Hota C, Tremblay E, Chambers AF. Osteopontin-induced, integrin-dependent migration of human mammary epithelial cells involves activation of the hepatocyte growth factor receptor (Met). J Cell Biochem 2000; 78:465-475.
70. Doerr ME, Jones JI. The roles of integrins and extracellular matrix proteins in the insulin-like growth factor I-stimulated chemotaxis of human breast cancer cells. J Biol Chem 1996; 271:2443-2447.
71. Mira E, Manes S, Lacalle RA, Marquez G, Martinez-A C. Insulin-like growth factor I-triggered cell migration and invasion are mediated by matrix metalloproteinase-9. Endocrinology 1999; 140:1657-1664.
72. Bakin AV, Tomlinson AK, Bhowmick NA, Moses HL, Arteaga CL. Phosphatidylinositol 3-kinase function is required for transforming growth factor beta-mediated epithelial to mesenchymal transition and cell migration. J Biol Chem 2000; 275:36803-36810.
73. von Schlippe M, Marshall JF, Perry P, Stone M, Zhu AJ et al. Functional interaction between E-cadherin and alphav-containing integrins in carcinoma cells. J Cell Sci 2000; 113:425-437
74. Adelsman MA, McCarthy JB, Shimizu Y. Stimulation of beta1-integrin function by epidermal growth factor and heregulin-beta has distinct requirements for erbB2 but a similar dependence on phosphoinositide 3-OH kinase. Mol Biol Cell 1999; 10:2861-2878
75. Ignatoski KM, Maehama T, Markwart SM, Dixon JE, Livant DL et al. ERBB-2 overexpression confers PI 3' kinase-dependent invasion capacity on human mammary epithelial cells. Br J Cancer. 2000; 82:666-674
76. Sugiura T, Berditchevski F. Function of alpha3beta1-tetraspanin protein complexes in tumor cell invasion. Evidence for the role of the complexes in production of matrix metalloproteinase 2 (MMP-2). J Cell Biol 1999; 146:1375-1389.
77. Keely PJ, Westwick JK, Whitehead IP, Der CJ, Parise LV. Cdc42 and Rac1 induce integrin-mediated cell motility and invasiveness through PI(3)K. Nature 1997; 390:632-636.
78. Morini M, Mottolese M, Ferrari N, Ghiorzo F, Buglioni S et al. The alpha 3 beta 1 integrin is associated with mammary carcinoma cell metastasis, invasion, and gelatinase B (MMP-9) activity. Int J Cancer 2000; 87:336-342.
79. Saad S, Bendall LJ, James A, Gottlieb DJ, Bradstock KF. Induction of matrix metalloproteinases MMP-1 and MMP-2 by coculture of breast cancer cells and bone marrow fibroblasts. Breast Cancer Res Treat 2000; 63:105-115.
80. Glukhova M, Koteliansky V, Sastre X, Thiery JP. Adhesion systems in normal breast and in invasive breast carcinoma. Am J Pathol 1995; 146:706-716.
81. Hendrix MJ, Seftor EA, Seftor RE, Trevor KT. Experimental coexpression of vimentin and keratin intermediate filaments in human breast cancer cells results in phenotypic interconversion and increased invasive behavior. Am J Pathol 1997; 150:483-495.
82. Gui GP, Wells CA, Browne PD, Yeomans P, Jordan S et al. integrin expression in primary breast cancer and its relation to axillary nodal status. Surgery 1995; 117:102-108.
83. Shaw LM, Chao C, Wewer UM, Mercurio AM. Function of the integrin alpha 6 beta 1 in metastatic breast carcinoma cells assessed by expression of a dominant-negative receptor. Cancer Res 1996; 56:959-963.
84. Seftor RE, Seftor EA, Sheng S, Pemberton PA, Sager R et al. Maspin suppresses the invasive phenotype of human breast carcinoma. Cancer Res 1998; 58:5681-5685.
85. Jones JL, Royall JE, Critchley DR, Walker RA. Modulation of myoepithelial-associated alpha6beta4 integrin in a breast cancer cell line alters invasive potential. Exp Cell Res 1997; 235:325-333.
86. Bachelder RE, Marchetti A, Falcioni R, Soddu S, Mercurio AM. Activation of p53 function in carcinoma cells by the alpha6beta4 integrin. J Biol Chem 1999; 274:20733-20737.

87. O'Connor KL, Shaw LM, Mercurio AM. Release of cAMP gating by the alpha6beta4 integrin stimulates lamellae formation and the chemotactic migration of invasive carcinoma cells. J Cell Biol 1998; 143:1749-1760.
88. Felding-Habermann B, O'Toole TE, Smith JW, Fransvea E, Ruggeri ZM et al. integrin activation controls metastasis in human breast cancer. Proc Natl Acad Sci USA 2001; 98:1853-1858.
89. Zhou Q, Sherwin RP, Parrish C, Richters V, Groshen SG et al. Contortrostatin, a dimeric disintegrin from Agkistrodon contortrix contortrix, inhibits breast cancer progression. Breast Cancer Res Treat 2000; 61:249-260.
90. Carriero MV, Del Vecchio S, Capozzoli M, Franco P, Fontana L et al. Urokinase receptor interacts with alpha(v)beta5 vitronectin receptor, promoting urokinase-dependent cell migration in breast cancer. Cancer Res 1999; 59:5307-5314.
91. Nguyen DH, Catling AD, Webb DJ, Sankovic M, Walker LA et al. Myosin light chain kinase functions downstream of Ras/ERK to promote migration of urokinase-type plasminogen activator-stimulated cells in an integrin-selective manner. J Cell Biol 1999; 146:149-164.
92. Cos S, Fernandez R, Guezmes A, Sanchez-Barcelo EJ. Influence of melatonin on invasive and metastatic properties of MCF-7 human breast cancer cells. Cancer Res 1998; 58:4383-4390.
93. Bonkhoff H, Stein U, Remberger K. Differential expression of alpha 6 and alpha 2 very late antigen integrins in the normal, hyperplastic, and neoplastic prostate: simultaneous demonstration of cell surface receptors and their extracellular ligands. Hum Pathol 1993; 24:243-248.
94. Romanov VI, Goligorsky MS. RGD-recognizing integrins mediate interactions of human prostate carcinoma cells with endothelial cells in vitro. Prostate 1999; 39:108-118.
95. Edlund M, Miyamoto T, Sikes RA, Ogle R, Laurie GW et al. integrin expression and usage by prostate cancer cell lines on laminin substrata. Cell Growth Differ 2001; 12:99-107.
96. Zheng DQ, Woodard AS, Fornaro M, Tallini G, Languino LR. Prostatic carcinoma cell migration via alpha(v)beta3 integrin is modulated by a focal adhesion kinase pathway. Cancer Res 1999; 59:1655-1664.
97. Zheng DQ, Woodard AS, Tallini G, Languino LR. Substrate specificity of alpha(v)beta(3) integrin-mediated cell migration and phosphatidylinositol 3-kinase/AKT pathway activation. J Biol Chem 2000; 275:24565-24574.
98. Delcommenne M, Tan C, Gray V, Rue L, Woodgett J et al. Phosphoinositide-3-OH kinase-dependent regulation of glycogen synthase kinase 3 and protein kinase B/AKT by the integrin-linked kinase. Proc Natl Acad Sci USA 1998; 95:11211-11216.
99. Persad S, Attwell S, Gray V, Delcommenne M, Troussard A et al. Inhibition of integrin-linked kinase (ILK) suppresses activation of protein kinase B/Akt and induces cell cycle arrest and apoptosis of PTEN-mutant prostate cancer cells. Proc Natl Acad Sci USA 2000; 97:3207-3212.
100. Graff JR, Konicek BW, McNulty AM, Wang Z, Houck K et al. Increased AKT activity contributes to prostate cancer progression by dramatically accelerating prostate tumor growth and diminishing p27Kip1 expression. J Biol Chem 2000; 275:24500-24505
101. Rabinovitz I, Nagle RB, Cress AE. integrin alpha 6 expression in human prostate carcinoma cells is associated with a migratory and invasive phenotype in vitro and in vivo. Clin Exp Metastasis 1995; 13:481-491
102. Davis TL, Cress AE, Dalkin BL, Nagle RB. Unique expression pattern of the alpha6beta4 integrin and laminin-5 in human prostate carcinoma. Prostate 2001; 46:240-248.
103. Knox JD, Cress AE, Clark V, Manriquez L, Affinito KS et al. Differential expression of extracellular matrix molecules and the alpha 6-integrins in the normal and neoplastic prostate. Am J Pathol 1994; 145:167-174.
104. Bello-DeOcampo D, Kleinman HK, Deocampo ND, Webber MM. Laminin-1 and alpha6beta1 integrin regulate acinar morphogenesis of normal and malignant human prostate epithelial cells. Prostate 2001; 46:142-153.
105. Nagle RB, Hao J, Knox JD, Dalkin BL, Clark V et al. Expression of hemidesmosomal and extracellular matrix proteins by normal and malignant human prostate tissue. Am J Pathol 1995; 146:1498-1507.
106. Allen MV, Smith GJ, Juliano R, Maygarden SJ, Mohler JL. Downregulation of the beta4 integrin subunit in prostatic carcinoma and prostatic intraepithelial neoplasia. Hum Pathol 1998; 29:311-318.
107. Dedhar S, Saulnier R, Nagle R, Overall CM. Specific alterations in the expression of alpha 3 beta 1 and alpha 6 beta 4 integrins in highly invasive and metastatic variants of human prostate carcinoma cells selected by in vitro invasion through reconstituted basement membrane. Clin Exp Metastasis 1993; 11:391-400
108. Bonaccorsi L, Carloni V, Muratori M, Salvadori A, Giannini A et al. Androgen receptor expression in prostate carcinoma cells suppresses alpha6beta4 integrin-mediated invasive phenotype. Endocrinology 2000; 141:3172-3182

109. Jewell K, Kapron-Bras C, Jeevaratnam P, Dedhar S. Stimulation of tyrosine phosphorylation of distinct proteins in response to antibody-mediated ligation and clustering of alpha 3 and alpha 6 integrins. J Cell Sci 1995; 108:1165-1174.
110. Rokhlin OW, Cohen MB. Expression of cellular adhesion molecules on human prostate tumor cell lines. Prostate 1995; 26:205-212.
111. Haywood-Reid PL, Zipf DR, Springer WR. Quantification of integrin subunits on human prostatic cell lines—Comparison of nontumorigenic and tumorigenic lines. Prostate 1997; 31:1-8.
112. MacCalman CD, Brodt P, Doublet JD, Jednak R, Elhilali MM et al. The loss of E-cadherin mRNA transcripts in rat prostatic tumors is accompanied by increased expression of mRNA transcripts encoding fibronectin and its receptor. Clin Exp Metastasis 1994; 12:101-107.
113. Livant DL, Brabec RK, Pienta KJ, Allen DL, Kurachi K et al. Anti-invasive, antitumorigenic, and antimetastatic activities of the PHSCN sequence in prostate carcinoma. Cancer Res 2000; 60:309-320.
114. Fornaro M, Tallini G, Bofetiado CJ, Bosari S, Languino LR. Down-regulation of beta 1C integrin, an inhibitor of cell proliferation, in prostate carcinoma. Am J Pathol 1996; 149:765-773.
115. Fornaro M, Manzotti M, Tallini G, Slear AE, Bosari S et al. Beta1C integrin in epithelial cells correlates with a nonproliferative phenotype: Forced expression of beta1C inhibits prostate epithelial cell proliferation. Am J Pathol 1998; 153:1079-1087.
116. Fornaro M, Tallini G, Zheng DQ, Flanagan WM, Manzotti Met al. p27(kip1) acts as a downstream effector of and is co-expressed with the beta1C integrin in prostatic adenocarcinoma. J Clin Invest 1999; 103:321-329.
117. Perlino E, Lovecchio M, Vacca RA, Fornaro M, Moro L et al. Regulation of mRNA and protein levels of beta1 integrin variants in human prostate carcinoma. Am J Pathol 2000; 157:1727-1734.
118. Hobeika AC, Etienne W, Cruz PE, Subramaniam PS, Johnson HM. IFNgamma induction of p21WAF1 in prostate cancer cells: role in cell cycle, alteration of phenotype and invasive potential. Int J Cancer 1998; 77:138-145.
119. Kostenuik PJ, Singh G, Orr FW. Transforming growth factor beta upregulates the integrin-mediated adhesion of human prostatic carcinoma cells to type I collagen. Clin Exp Metastasis 1997; 15:41-52.
120. Festuccia C, Bologna M, Gravina GL, Guerra F, Angelucci A et al. Osteoblast conditioned media contain TGF-beta1 and modulate the migration of prostate tumor cells and their interactions with extracellular matrix components. Int J Cancer 1999; 81:395-403.
121. Kostenuik PJ, Sanchez-Sweatman O, Orr FW, Singh G. Bone cell matrix promotes the adhesion of human prostatic carcinoma cells via the alpha 2 beta 1 integrin. Clin Exp Metastasis 1996; 14:19-26.
122. Lang SH, Clarke NW, George NJ, Testa NG. Primary prostatic epithelial cell binding to human bone marrow stroma and the role of alpha2beta1 integrin. Clin Exp Metastasis 1997; 15:218-227.
123. Trikha M, Timar J, Lundy SK, Szekeres K, Tang K et al. Human prostate carcinoma cells express functional alphaIIb(beta)3 integrin. Cancer Res 1996; 56:5071-5078.
124. Trikha M, Raso E, Cai Y, Fazakas Z, Paku S et al. Role of alphaII(b)beta3 integrinin prostate cancer metastasis. Prostate 1998; 35:185-192
125. Matsuoka T, Yashiro M, Nishimura S, Inoue T, Fujihara T et al. Increased expression of alpha2beta1-integrin in he peritoneal dissemination of human gastric carcinoma. Int J Mol Med 2000; 5:21-25.
126. Chan BM, Matsuura N, Takada Y Zetter BR, Hemler ME. In vitro and in vivo consequences of VLA-2 expression on rhabdomyosarcoma cells. Science 1991; 251:1600-1602.
127. Hangan D, Uniyal S, Morris VL, McDonald IC, von Ballestrem C et al. integrin VLA-2 (alpha2beta1) function in postextravasation movement of human rhabdomyosarcoma RD cells in the liver. Cancer Res 1996; 56:3142-3149.
128. Vihinen P, Riikonen T, Laine A, Heino J. integrin alpha 2 beta 1 in tumorigenic human osteosarcoma cell lines regulates cell adhesion, migration, and ivasion by interaction with type I collagen. Cell Growth Differ 1996; 7:439-447.
129. Palecek SP, Loftus JC, Ginsberg MH, Lauffenburger DA, Horwitz AF. integrin-ligand binding properties govern ell migration speed through cell-substratum adhesiveness. Nature 1997; 385:537-540.
130. Seftor RE, Seftor EA, Stetler-Stevenson WG, Hendrix MJ. The 72 kDa type IV collagenase is modulated via differential expression of alpha v beta 3 and alpha 5 beta 1 integrin during human melanoma cell invasion. Cancer Res 1993; 53:3411-3415
131. Bafetti LM, Young TN, Itoh Y, Stack MS. Intact vitronectin induces matrix metalloproteinase-2 and tissue inhibitor of metalloproteinases-2 expression and enhanced cellular invasion by melanoma cells. J Biol Chem 1998; 273:143-149.

132. Esparza J, Vilardell C, Calvo J, Juan M, Vives J, Urbano-Marquez A et al. Fibronectin upregulates gelatinase B (MMP-9) and induces coordinated expression of gelatinase A (MMP-2) and its activator MT1-MMP (MMP-14) by human T lymphocyte cell lines. A process repressed through RAS/MAP kinase signaling pathways. Blood 1999; 94:2754-2766.
133. Kurschat P, Zigrino P, Nischt R, Breitkopf K, Steurer P et al. Tissue inhibitor of matrix metalloproteinase-2 regulates matrix metalloproteinase-2 activation by modulation of membrane-type 1 matrix metalloproteinase activity in high and low invasive melanoma cell lines. J Biol Chem 1999; 274:21056-21062.
134. Koistinen P, Pulli T, Uitto VJ, Nissinen L, Hyypia T et al. Depletion of alphaV integrins from osteosarcoma cells by intracellular antibody expression induces bone differentiation marker genes and suppresses gelatinase (MMP-2) synthesis. Matrix Biol 1999; 18:239-251.
135. Ivaska J, Heino J. Adhesion receptors and cell invasion: Mechanisms of integrin-guided degradation of extracellular matrix. Cell Mol Life Sci 2000; 57:16-24.
136. Riikonen T, Westermarck J, Koivisto L, Broberg A, Kahari VM et al. integrin alpha 2 beta 1 is a positive regulator of collagenase (MMP-1) and collagen alpha 1(I) gene expression. J Biol Chem 1995; 270:13548-13552
137. Langholz O, Rockel D, Mauch C, Kozlowska E, Bank I et al. Collagen and collagenase gene expression in three-dimensional collagen lattices are differentialy regulated by alpha 1 beta 1 and alpha 2 beta 1 integrins. J Cell Biol 1995; 131:1903-1915.
138. Werb Z, Tremble PM, Behrendtsen O, Crowley E, Damsky CH. Signal transduction through the fibronectin receptor induces collagenase and stromelysin gene expression. J Cell Biol 1989; 109:877-889
139. Huhtala P, Humphries MJ, McCarthy JB, Tremble PM, Werb Z et al. Cooperative signaling by alpha 5 beta 1 and alpha 4 beta 1 integrins regulates metalloproteinase gene expression in fibroblasts adhering to fibronectin. J Cell Biol 1995; 129:867-879.
140. Pender SL, Salmela MT, Monteleone G, Schnapp D, McKenzie C et al. Ligation of alpha4ss1 integrin on human intestinal mucosal mesenchymal cells selectively Up-regulates membrane type-1 matrix metalloproteinase and confers a migratory phenotype. Am J Pathol 2000; 157:1955-1962.
141. Clark EA, Brugge JS. Integrins and signal transduction pathways: the road taken. Science 1995; 268:233-239.
142. Yamada KM, Miyamoto S. integrin transmembrane signaling and cytoskeletal control. Curr Opin Cell Biol 1995; 7:681-689.
143. Miyamoto S, Akiyama SK, Yamada KM. Synergistic roles for receptor occupancy and aggregation in integrin transmembrane function. Science 1995; 267:883-885.
144. Schaller MD, Otey CA, Hildebrand JD, Parsos JT. Focal adhesion kinase and paxillin bind to peptides mimicking beta integrin cytoplasmic domains. J Cell Biol 1995; 130:1181-1187.
145. Hildebrand JD, Schaller MD, Parsons JT. Paxillin, a tyosine phosphorylated focal adhesion-associated protein binds to the carboxyl terminal domain of focal adhesion kinase. Mol Biol Cell 1995; 6:637-647.
146. Chen HC, Appeddu PA, Parsos JT, Hildebrand JD, Schaller MD et al. Interaction of focal adhesion kinase with cytoskeletal protein talin. J Biol Chem 1995; 270:16995-16999.
147. Wood CK, Turner CE, Jackson P, Critchley DR. Characterisation of the paxillin-binding site and the C-terminal focal adhesion targeting sequence in vinculin. J Cell Sci 1994; 107:709-717.
148. Liu S, Thomas SM, Woodside DG, Rose DM, Kiosses WB et al. Binding of paxillin to alpha4 integrins modifies integrin-dependent biological responses. Nature 1999; 402:676-681.
149. Parsons JT, Parsons SJ. Src family protein tyrosine kinases: cooperating with growth factor and adhesion signaling pathways. Curr Opin Cell Biol 1997; 9:187-192.
150. Thompson EW, Yu M, Bueno J, Jin L, Maiti SN et al. Collagen inducd MMP-2 activation in human breast cancer. Breast Cancer Res Treat 1994; 31:357-370.
151. Ellerbroek SM, Fishman DA, Kearns AS, Bafetti LM, Stack MS. Ovarian carcinoma regulation of matrix metalloproteinase-2 and membrane type 1 matrix metalloproteinase through beta1 integrin. Cancer Res 1999; 59:1635-1641.
152. Chintala SK, Sawaya R, Gokaslan ZL, Rao JS. Modulation of matrix metalloprotease-2 and invasion in human glioma cells by alpha 3 beta 1 integrin. Cancer Lett 1996; 103:201-208.
153. Kubota S, Ito H, Ishibashi Y, Seyama Y. Anti-alpha3 integrin antibody indues the activated form of matrix metalloprotease-2 (MMP-2) with concomitant stimulation of invasion through matrigel by human rhabdomyosarcoma cells. Int J Cancer 1997; 70:106-111.
154. Sugiura T, Berditchevski F. Function of alpha3beta1-tetraspanin protein complexes in tumor cell ivasion. Evidence for the role of the complexes in production of matrix metalloproteinase 2 (MMP-2). J Cell Biol 1999; 146:1375-1389.

Chapter 3

Keratinocyte Interactions with Fibronectin During Wound Healing

H. Larjava, L. Koivisto and L. Häkkinen

Abstract

Reepithelialization of wounds is critical for survival. After injury, fibronectin-fibrin clot is formed. Keratinocytes become activated and start migrating into the clot. Migration involves coordinate expression of several new extracellular matrix molecules, fibronectin receptors and proteinases. Migrating keratinocytes express alternatively spliced EDA fibronectin while both EDA and EDB fibronectin isoforms are found in the granulation tissue. Fibronectins are assembled into polymerized matrix that is further cross-linked to fibrin and these multimeric complexes of fibronectins have their own specific effects on cell signaling. Upon wounding, keratinocytes express three new fibronectin receptors, namely $\alpha5\beta1$, $\alpha v\beta6$ and $\alpha v\beta1$ integrins. integrin $\alpha5\beta1$ recognizes the RGD sequence and the synergy site of fibronectin and plays a critical role in the early migration. Interestingly, an important role for the synergy site in keratinocyte migration has been proposed. In addition to cell migration, $\alpha5\beta1$ integrin also regulates expression of dozens of genes important for a variety of cell functions including cell growth. Integrin $\alpha v\beta6$ also mediates cell adhesion on fibronectin and it can functionally replace $\alpha5\beta1$ integrin. The most important function of $\alpha v\beta6$ may be, however, to activate TGFβ during late wound healing. Laminin-5 and tenascin-C are also expressed by the migrating keratinocytes. These cells have three laminin-5 and two putative tenascin-C receptors and their function needs to be spatially and temporally coordinated with the three fibronectin receptors. Elucidation of the mechanism of this coordination is crucial for better understanding of reepithelialization.

Reepithelialization: A Controlled Migration/Invasion of Keratinocytes

Tissue loss or damage in adult animals initiates the wound healing process that involves a series of controlled events resulting in tissue repair or scarring. Complete regeneration is only seldom observed and the site of the original wound is distinguished from the non-damaged tissue even weeks, months or sometimes years after the initial event. In cases where inflammation continues or healing does not follow the ordinary sequence, the wound becomes chronic and healing can lead to localized fibrosis. The roles of epithelial cells in the wound healing process have been expanded from mechanical coverage of the wound to active participation in the control of the inflammatory process. There is some evidence that epithelial cells may contribute to the formation of granulation tissue during the later phase of wound healing and also play a role in the formation of nonhealing wounds. This is evidenced from clinical observations demonstrating that skin wounds that have delayed reepithelialization are more likely to become chronic nonhealing wounds than those in which epithelial migration proceeds at the normal rate. The purpose of this review is to cover the mechanisms of epithelial cell migration in

Cell Invasion, edited by Jyrki Heino and Veli-Matti Kähäri.

wound provisional matrix with a special emphasis on epithelial cell interactions with provisional matrix fibronectin. Since epithelial (keratinocyte) migration in wound provisional matrix involves several matrix molecules and their receptors, their role in wound healing will also be briefly discussed as the context requires.

After the epithelium has been disrupted by tissue injury, reepithelialization starts rapidly in order to reestablish tissue integrity.[1,2] Reepithelialization involves migration of epithelial cells from the edges of the wound or, in the case of skin, from appendages such as hair follicles and sweat glands.[1,2] The migratory cells are in close contact with the wound provisional extracellular matrix that is mainly composed of fibrin, fibronectin and vitronectin.[1-3] Initially keratinocytes begin moving into the defect about 24 hours after the injury.[2] During the early migration, cell proliferation does not contribute to the migration. Rather, epithelial cells dissolve their hemidesmosomes, detach from basement membrane and move quickly into the wound defect. Proliferation contributes to the reepithelialization process at the later stages when it seeds more cells into the wound. The morphology of keratinocytes changes when they assume the migratory phenotype. Normally polarized basal keratinocytes become flattened and elongated. Long, α-actinin and actin-rich, cytoplasmic processes called lamellipodia, along with ruffling cytoplasmic processes, are observed in migrating keratinocytes in the provisional matrix.[1,4,5] Cell-cell connections through desmosomes are also reduced and there are no hemidesmosomes underneath the migrating keratinocytes. The numbers of cap-junctions, however, increase.[6] During the last few years, extensive progress has been made in the understanding of the signaling mechanisms which regulate lamellipodia formation in various cell types.[7] Intracellular signaling events are, however, cell and matrix specific and only scattered information is available about the signals that lead to lamellipodia formation in migrating keratinocytes.[8] It is conceivable that the small GTPases are important in the lamellipodia extension, migration and invasion.[7] Recent results from our laboratory suggest that the function of the small GTPase Rac is crucial for lamellipodia formation in HaCaT keratinocytes that have been stimulated by staurosporine (Koivisto et al, unpublished). Rac also seems to be essential in cell crawling from edges of wounded MDCK epithelial sheets[8] and it can participate in recruitment of high affinity integrin receptors into the lamellipodia.[9] Many cytokines that promote cell migration are able to cause alterations in the organization of cytoskeleton and cell shape. The effects of one such growth factor/cytokine, namely keratinocyte growth factor (KGF), on keratinocyte spreading on fibronectin was recently investigated.[10] KGF is a fibroblast-derived polypeptide that was originally found to promote keratinocyte growth. It was shown to promote keratinocyte migration[10] and participate in regulation of keratinocyte differentiation, together with β1 integrins.[11] KGF was able to promote the organization of actin cytoskeleton and lamella formation on the fibronectin matrix and cause alterations in integrin avidity,[10] demonstrating that KGF may have more functions in wound healing than originally anticipated. It can be expected that several wound fluid cytokines may be involved in regulation of lamellipodia formation in migrating keratinocytes in vivo.

Keratinocyte Invasion into the Clot

It has commonly been described that epithelial cells migrate on the exposed connective tissue matrix underneath the fibrin-fibronectin clot. In small oral mucosal wounds, however, the keratinocytes cut their way directly through the clot and may not interact with the connective tissue matrix at all (Fig. 1).[12] Migrating keratinocytes are highly phagocytic, allowing them to penetrate through tissue debris or the clot.[2] Degradation of the fibrin-fibronectin clot appears to be critical for wound healing since wounds in animals that lack the plasminogen gene do not reepithelialize.[13] Interestingly, it was recently demonstrated that keratinocytes migrate though tunnels of digested fibrin-fibronectin in vitro.[14] This process requires plasminogen activation at the leading edge. Although it was not addressed in this paper, it seems likely that integrins play a role in fibrin clot removal. Migrating cells must be able to focalize the proteolysis into the leading edge.[15] This could be done by activation of proteolytic enzymes at specific

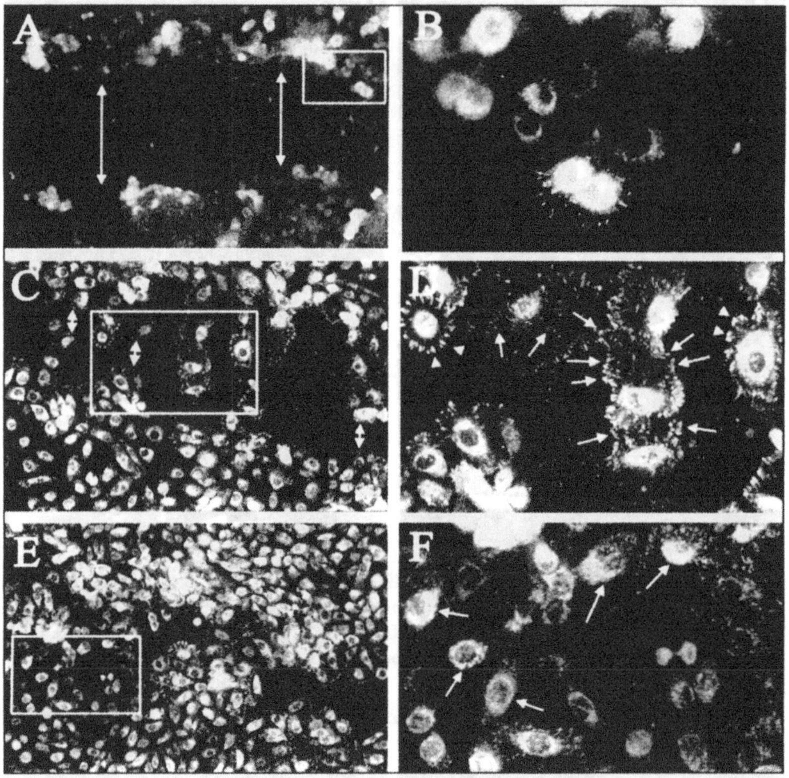

Fig. 4. Deposition of EDA fibronectin is induced during keratinocyte migration by TGFβ and becomes removed after migration stops. Confluent human skin keratinocyte cultures (HaCaT cells) were wounded and allowed to migrate in the absence (A and B) or in the presence of 10 ng/ml TGFβ1 (C-F) in their normal growth medium (KGM, Clonetics) before fixation, permeabilization and immunostaining with a monoclonal antibody against EDA fibronectin (clone IST-9; Accurate Chemical & Scientific Corp, Westbury, N.Y.). A) In control cultures, wound space remained open and very little cell migration occurred. Fibronectin EDA staining was weak and mostly perinuclear. Only some cells at the wound margin organized fibronectin into pericellular streaks. B) Higher magnification of the rectangular area shown in A. C) TGFβ1 induced cell migration and expression of EDA fibronectin. D) Higher magnification of the area shown in C. Migrating keratinocytes organized EDA fibronectin into focal adhesion-like structures underneath the cells and they left tracks of EDA fibronectin behind (white arrows). In some cells, EDA fibronectin localized also at the focal adhesion-like structures at the edges of the lamellar cytoplasm (arrowheads). E and F) After the wound was closed, EDA fibronectin trails were removed from underneath the cells and fibronectin showed predominantly perinuclear localization (small arrows in F). F shows a higher magnification of the rectangular area indicated in E.

binding domain of fibronectin and the PHSRN peptide were able to induce keratinocyte motility and invasion through the basement membrane and extracellular matrix of sea urchin embryos. Keratinocytes do not normally invade through the basement membrane matrix, but rather use that as a template to establish new, stratified epithelium. On a molar basis, the PHSRN peptide was surprisingly over tenfold more active than the cell binding domain. Even more surprisingly, this small peptide was able to accelerate wound closure and reepithelialization dramatically in diabetic animals and slightly in normal mice.[49] It was speculated that PHSRN is proteolytically released from fibronectin in normal healing wounds, providing chemotactic signals for cells to induce cell migration. This process could be impaired in diabetic animals. Alternatively, PHSRN peptide could become immobilized in the wound and acts as a matrix molecule.[50] A similar peptide (PHSCN) was able to block metastasis in rats.[49] Small peptides like this do not usually possess a strong biological effect because of rapid clearance in circulation.[50] These papers remain controversial with such dramatic results and need to be confirmed by other investigators.[50] If true, the use of this synergy region of the cell binding domain of fibronectin would be very valuable for millions of people suffering from impaired wound healing. Interpretation of the results of the above study demonstrating improved healing with the synergy peptide in ob/ob mice is even more difficult in light of a recent paper about leptin effects on wound healing in these mice.[52] Impaired wound healing in the leptin-deficient ob/ob mice has been explained for years by the diabetic phenotype of the animals. Administration of leptin, either systemically or locally, however, appears to normalize the wound healing defect in these mice.[52] The impaired healing response can be, therefore, explained entirely by the genetic background of the animals and not by, e.g., keratinocyte chemotaxis towards fibronectin fragments with the synergy region. Leptin effects as such on wound healing are interesting. Leptin seems to enhance the reepithelialization by stimulating keratinocyte proliferation through activation of STAT3 transcription factor by ObRb leptin receptor that is temporarily regulated during wound repair.[52] This is in agreement with findings with STAT3-deficient animals that possess an impaired wound repair phenotype.[53]

Biological Functions of Fibronectin Matricryptins

Fibronectin fragments have been demonstrated in wound fluids from chronic wounds,[54,55] and they have several potential functions that are different from the intact fibronectin molecule. This extracellular matrix fragmentation is considered crucially important for wound repair because it produces biologically active polypeptides in which novel cryptic sites are exposed.[56] These sites were recently termed as matricryptic sites and the polypeptides matricryptins, respectively.[56] Fibronectin has potentially several of these matricryptic sites that function in cell adhesion, cell proliferation, matrix assembly and matrix degradation (see below).[56] Chronic skin and corneal wounds show advanced degradation of fibronectin, probably via the plasmin pathway. Fibronectin supplementation into these nonhealing chronic skin and corneal ulcers seems to improve reepithelialization.[2] Fibronectin fragments containing the cell-binding site are involved with regulation of MMP-1 expression in vitro.[57] This mechanism has not been proven for keratinocytes but is potentially important when the keratinocytes face the connective tissue collagen matrix because the migration of keratinocytes on type I collagen requires MMP-1 activity.[27] The EDA domain, but not the native fibronectin, can also induce IL-1 expression, at least in connective tissue cells, which can lead to increased MMP-1 synthesis.[58] The role of RGD- and PHSRN-containing peptides and other fibronectin-derived matricryptins in keratinocyte migration and signaling clearly warrants further investigation.

Fibronectin Matrix Assembly and Cross-Linking to Fibrin

The fibronectin matrix assembly is likely to play an important role in cell adhesion and signaling during wound repair. Interaction of cells with the extracellular matrix conveys information about the nature of matrix proteins and how the matrix has been assembled into three-dimensional structures.[59] Cells polymerize fibronectin by assembling soluble fibronectin from cells or plasma into insoluble fibrils in a multistep pericellular process. Fibronectin fibrillogenesis requires the integrin $\alpha5\beta1$ receptors that bind to the cell binding domain to initiate the process.[60] Contacts of $\alpha5\beta1$ integrins with both of the RGD and synergy sites as well as interaction of the so-called matrix assembly site at the cell surface with the amino terminal region (first five type I repeats) of fibronectin are needed for fibril polymerization.[61-66] The process is completed by fibronectin self-assembly.[67] Activation of integrins can overcome the requirement of the synergy site.[66] A novel RGD-independent fibronectin assembly site was recently discovered in the alternatively spliced V region of the molecule.[68] This assembly site requires $\alpha 4\beta 1$ integrin expression. The heparin binding domain of fibronectin is also important for polymerized fibril formation.[69] Polymerization of fibronectin has functional significance because polymerized fibronectin matrix promotes cell growth. This promotion appears to involve cell surface proteoglycans.[70]

Interestingly, an additional site at the amino terminus of the fibronectin molecule also contributes to polymerization. This first type III module contains a 76 amino acid sequence (termed III1-C) that can, when added to soluble fibronectin, polymerize it into high molecular mass multimers, so called superfibronectin fibrils, in vitro.[67] The isolated III1-C fragments are believed to interfere with the intramolecular interactions that keep fibronectin in solution.[71] Alternatively, they could change the conformation of fibronectin, opening cryptic sites essential for fibril self-self assembly.[65] Polymerized superfibronectin is much more potent in promoting cell adhesion than the nonpolymerized form.[67] The structure of superfibronectin and whether superfibronectin fibrils resemble or compose natural fibrils is not known. Interestingly, systemic admininstration of polymerized superfibronectin has a strong antimetastatic effect.[72] The III1-C fragment appears also to be a potent inhibitor of tumor growth, angiogenesis and metastasis.[73] The effects of polymerized superfibronectin in keratinocyte adhesion, migration and growth have not been investigated.

The active role of cells in fibronectin fibrillogenesis directs the deposition of the extracellular matrix in the desired location and prevents unwanted fibrillogenesis in circulation. Although keratinocytes express EDA fibronectin in vitro they don't deposit it into the traditional fibronectin matrix seen in fibroblast cultures. Migrating keratinocytes leave tracks of cellular fibronectin behind, but these tracks are no longer present when the cells reach confluency (Fig. 4). It is likely that fibronectin that is deposited to support migration is proteolytically removed rather than organized into extensive fibrils when keratinocytes stop migrating. Lack of $\alpha 5\beta 1$ integrin does not explain the inability to make polymerized fibronectin fibrils because this integrin is expressed abundantly by cultured keratinocytes.[12,74] The inability of keratinocytes to organize a polymerized fibronectin matrix may also relate to alterations in integrin activity.[75] Finally, the small GTPase Rho-mediated cell contractility is needed for fibronectin polymerization, possibly because stretching of the fibronectin molecule by the cells exposes cryptic self-assembly sites necessary for polymerization.[76] Structurally, the RGD site resides in a flexible loop and therefore its exposure may potentially also be modulated by the tension of the fibronectin molecule.[77] Visualization of green fluorescent protein-tagged fibronectin in real time in living cell cultures has elegantly demonstrated that fibronectin fibrils are highly elastic and are continuously under tension produced by cells fibroblasts (www.pnas.org[78]). It is possible that keratinocytes are unable to polymerize fibronectin because they are less contractile cells than fibroblasts.

The turnover rate of integrins may also be implicated in fibronectin fibrillogenesis since acceleration of $\beta 1$ integrin turnover rates by antisense cDNA construct appears to block fibronectin fibrillogenesis in MG-63 cells.[79] In human keratinocytes, the turnover rate of $\beta 1$ integrin is much faster than in fibroblasts.[74] Also many tumor cells have rapid integrin turnover rates and fail to assemble the fibronectin matrix.[80] It can be hypothesized that with high turnover of integrin receptors, cells fail to initiate the conformational change in fibronectin required for fibrillogenesis. The inability to assemble the fibronectin matrix may be beneficial for keratinocytes that don't normally deposit this type of matrix in vivo. Fibrillar fibronectin matrix could impair the deposition of the basement membrane after migration is complete.

Fibrin is cross-linked to fibronectin by factor XIII in the clot. The ratio of fibronectin to fibrin in the clot and the degree of clot cross-linking may also regulate keratinocyte behavior.[81] Keratinocytes do not express the $\alpha v\beta 3$ integrin that is usually required for fibrin binding.[12,82] While adding fibronectin into fibrin increases keratinocyte spreading, the elevated amounts of fibrin reduces keratinocyte adhesion.[81] Cross-linking of fibrin to fibronectin by factor XIII is known to increase cell adhesion in many cell types including keratinocytes.[83,84] Organization of the cytoskeleton is also regulated by fibronectin-fibrin cross-linking.[85] Fibronectin in the clot can be recognized by the $\alpha5\beta1$ integrin and this will promote clot retraction by various cell types.[86] Interestingly, cross-linking fibrin alone by factor XIII appears to promote some adhesion of keratinocytes.[81] The mechanism of this induction is not known but it provides evidence that keratinocytes in vivo are able to adhere not only to fibronectin in the clot but also to cross-linked fibrin. Cellular fibronectin produced by migrating keratinocytes also interacts with fibrin but with different kinetics and interactions from that produced from plasma.[87]

Expression of alternatively spliced variants of fibronectin is regulated by cytokines. Fibronectin expression in keratinocytes can be greatly stimulated by certain growth factors and cytokines. TGFβ is obviously the key cytokine involved in keratinocyte migration. TGFβ1 is the major isoform of the members of the TGFβ-family in wound keratinocytes. TGFβ1 inhibits the growth of keratinocytes in vitro and in vivo and it seems to stimulate keratinocyte motility by switching the cells from the differentiating to regenerative phenotype.[88] TGFβ is able to stimulate the production of fibronectin[89] and the expression of its fibronectin receptors (see below)[82,90] as well as specifically stimulate the expression of EDA type fibronectin.[91] In the keratinocyte scratch-wound model, TGFβ1 stimulates the expression of EDA fibronectin in cells adjacent to the wound (Fig. 4.4). Wounding alone appears also to slightly stimulate EDA fibronectin expression. In MDCK cells, TGFβ appears to also regulate the ratio of EDA-containing fibronectin mRNA to EDA negative fibronectin mRNA.[46] In this system, migrating cells seem to express a lesser amount of total fibronectin but relatively more of the EDA containing variant. Induction of TGFβ1 in migrating keratinocytes is thought to be crucial for the successful reepithelialization of skin wounds because reduced amounts of TGFβ1 are associated with impaired wound healing and administration of TGFβ1 has been shown to promote wound healing.[92] Surprisingly, there is no defect in reepithelialization in TGFβ1 deficient animals.[93] It is possible that other cytokines can compensate for TGFβ1 when it is absent. Epidermal growth factor and hepatocyte growth factor/scatter factor, that are also abundantly present in wounds, stimulate keratinocyte migration and secretion of cellular fibronectin.[89,91] Alternatively, TGFβ1 could have its major function in regulation of inflammation and granulation tissue formation (see below).

Fibronectin Receptor Integrin Expression During Reepithelialization: $\alpha5\beta1$ Integrin

As discussed above, integrins α/β heterodimers are responsible for most cell-extracellular matrix interactions in various cell types including epithelial keratinocytes.[94-96] During wound healing, keratinocytes have to adjust their integrin receptors to be able to adhere and migrate on the extracellular matrix proteins of the provisional matrix. Wounding causes a change in the expression levels and/or distribution of the existing integrins and induces expression of three new fibronectin receptor integrins, namely $\alpha5\beta1$, $\alpha v\beta1$ and $\alpha v\beta6$ (Fig. 5).[12,30] These integrins are usually absent from normal healthy adult epithelium.[12,96,97] Additionally, wound keratinocytes in vivo express at least syndecan-1 and CD44 heparan sulfate proteoglycans that are able to participate in fibronectin binding.[28] Integrin $\alpha3\beta1$, that localizes to basal keratinocytes in the stationary epithelium and to migrating keratinocytes in vivo,[12,98] has been reported to function as a fibronectin receptor in some cell types, but in keratinocytes it preferably acts as a receptor for laminin-5.[99,100] In culture, skin keratinocytes become activated and start to express $\alpha5\beta1$ integrin.[101] The mechanism of induction of $\alpha5\beta1$ expression both in vivo and in vitro in keratinocytes is still largely unknown. It can be speculated that local cytokines such as TGFβ or exposure to serum components during early wound repair could initiate the expression that is further upregulated when keratinocytes contact fibronectin. For example, expression of $\alpha5\beta1$ integrin can be either induced or stimulated by TGFβ1 in keratinocytes.[90,82] Alternatively, fibronectin itself could participate in the induction of its own receptor. In corneal epithelial cells, fibronectin is known to upregulate $\alpha5\beta1$ integrin expression through specific transcriptional activation.[102]

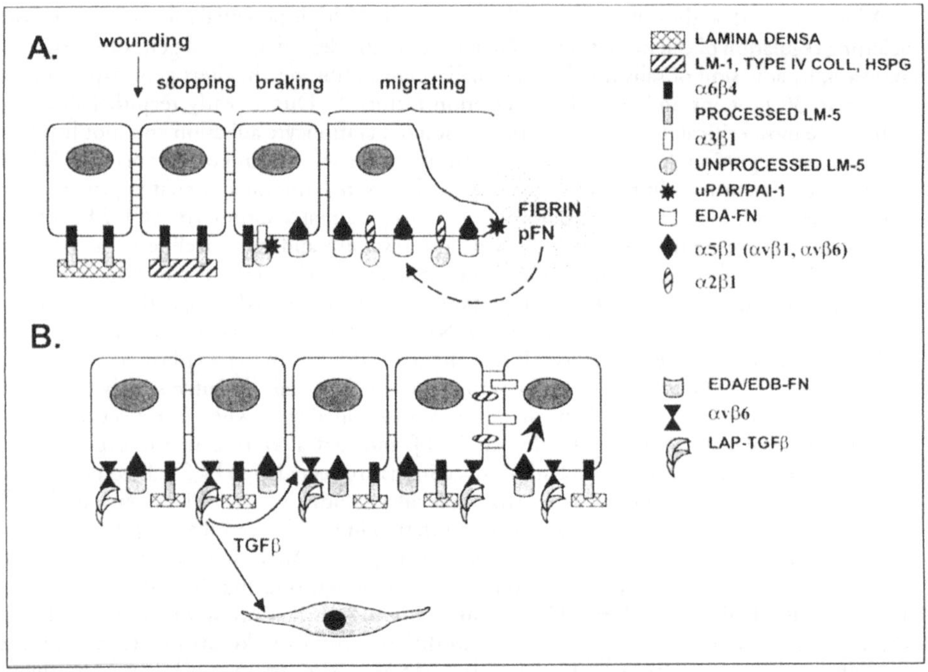

Fig. 6. A hypothetical model illustrating some of the integrin-mediated cell-matrix interactions regulating keratinocyte function during wound reepithelialization. A) The epithelial cells at the tip of the migrating epithelial cell sheet send out cytoplasmic extensions (lamellipodia) into the fibrin-fibronectin clot. Receptor for urokinase type plasminogen activator (uPAR) in complex with plasminogen activator inhibitor (PAI-1) is expressed at the tip of the lamellipodia promoting focalized plasminogen activation, fibrinolysis and migration. The migrating cells deposit a matrix that is rich in EDA fibronectin (EDA-FN) and unprocessed form of laminin-5 (LM-5). The cells use integrins α5β1 and α2β1 to attach and migrate on EDA-fibronectin and laminin-5, respectively. In addition to the major fibronectin receptor α5β1, cells can also use αvβ1 and αvβ6 integrins for adhesion and migration on fibronectin. Degradation of fibronectin present in the blood clot releases fibronectin fragments (matricryptins; arrow with a dashed line) that may modulate the function of migrating keratinocytes. Keratinocyte migration slows down when the cells use α2β1 integrin to bind to the unprocessed laminin-5. Cells also use the focalized plasminogen activation complex to proteolytically process laminin-5. The processed form of laminin-5 supports integrin α6β4 mediated hemidesmosome formation and retards migration. Once the cells have become stationary, they start to deposit basement membrane proteins. B) When the migrating epithelial sheets have joined to completely cover the wound space, the expression of αvβ6 integrin is strongly upregulated. This integrin can be used for cell adhesion on EDA and EDB fibronectins, but its major role is in binding to the latency associated peptide in the latent TGFβ (LAP-TGFβ). This binding can activate TGFβ and present it to the neighboring epithelial cells and/or to the fibroblasts (long arrows) to promote matrix deposition. Cells also express α5β1 integrin that can be used to bind fibronectin. This interaction can induce signaling cascades regulating cell proliferation involved in epithelial stratification and differentiation (short arrow). At this point, the basement membrane, including lamina densa, is regenerated.

factors for keratinocytes.[119] EGFR ligands are membrane bound and released (shed) by enzymatic cleavage from the cells. Expression of HB-EGF is rapidly induced at the wound edge of scrape-wounded epithelial monolayers in vitro.[120] HB-EGF shedding appears to be important for reepithelialization in vivo.[121] It is likely, therefore, that α5β1 integrin collaborates with EGFR signaling, not only in regulation of cell growth but also in cell migration during wound repair. This was evidenced by the finding that EGF differentially regulated expression of several gene clusters depending on whether cells were in contact with fibronectin or laminin.[122]

Function of αvβ6 Integrin

As discussed above, wound keratinocytes express another fibronectin-binding receptor, αvβ6 integrin, which is not normally present in keratinocytes of resting epithelium (Fig. 5). The expression of αvβ6 integrin can be induced by wounding or placing epithelial cells in culture.[97,123,124] The kinetics of expression of α5β1 and αvβ6 integrins is different in wound keratinocytes. Induction of α5β1 fibronectin receptor expression occurs at early stages of wound healing when the epithelium starts to migrate across the wound provisional matrix.[12] Expression of αvβ6 integrin is low or undetectable during early migration and maximal expression coincides with basement membrane reorganization and granulation tissue formation when the migrating edges of wound epithelium have already joined.[30,97,125] However, both α5β1 and αvβ6 integrins are sometimes coexpressed in migrating keratinocytes (Fig. 5).[97]

Function of fibronectin binding integrins was recently investigated in HaCaT keratinocytes, which are spontaneously transformed keratinocytes that have acquired an immortal but non-malignant growth type.[126,127] Integrin expression pattern in HaCaT cells is similar to that in primary keratinocytes although they express less α5β1 integrin.[106,128,129] Even after TGFβ1 treatment, the main receptor for fibronectin in NHEK cells is α5β1 integrin (unpublished). HaCaT keratinocytes constitutively express αvβ6 integrin which can be further upregulated by TGFβ1.[82] HaCaT cells also used αvβ6 integrin as their main fibronectin receptor for cell spreading. In untreated cells, both α5β1 and αvβ6 integrins were needed for maximal cell spreading, whereas in TGFβ1-treated cells the increased expression of αvβ6 integrin alone was adequate for supporting maximal cell spreading on fibronectin. This agrees with findings of αvβ6 integrin transfected cells in which αvβ6 integrin could replace α5β1 in fibronectin binding.[124] Treatment of HaCaT cells with TGFβ1 also promoted keratinocyte migration. Interestingly, keratinocyte integrins had different functions depending on whether or not deformation of cell body was required for locomotion.[82] When HaCaT cells were allowed to migrate through fibronectin-coated membranes in a short-term Boyden chamber assay that required cell body deformation for translocation, αvβ6 integrin was found to be the main individual mediator of cell movement. Comparable results were recently obtained using a similar assay by Huang et al.[130] When keratinocytes migrated laterally from a cell cluster, migration appeared to be mediated mainly by β1 integrins. Therefore, depending on the mechanics of cell migration, different integrins may be involved.

Although the exogenously provided fibronectin dose-dependently supports keratinocyte migration, it is unlikely that fibronectin is the only ligand for the migrating cells. As detailed above, during wound healing, migrating keratinocytes continuously express laminin-5,[12] and inhibitory antibodies against both α3 integrin subunit and laminin-5 are able to reduce keratinocyte migration on fibronectin.[131] It is, therefore, possible that, along with the induction of fibronectin receptors and fibronectin isoforms, TGFβ could induce other receptors and matrix molecules conducive for cell migration. Based on our data, keratinocyte migration on fibronectin involved at least α3β1, α5β1 and αvβ6 integrins. In addition, cultured keratinocytes can switch between αvβ6 and α5β1 integrins in fibronectin binding if the other receptor is rendered unfunctional since they are both able to mediate cell migration on fibronectin.[82] A reservoir of fibronectin receptors is likely to be evolutionarily beneficial to protect the survival of the organism, and different fibronectin receptors may at least partially replace functions of each other when necessary. Support to the idea that αvβ6 integrin can mediate keratinocyte migration on fibronectin comes from the studies on β6 deficient keratinocytes in culture. Migration on fibronectin was reduced in keratinocytes from β6 -/- mice.[130] Interestingly, keratinocytes from β6 -/- mice also demonstrated reduced migration on vitronectin, suggesting that vitronectin can also serve as a ligand for αvβ6 integrin. In the same study, Huang et al[130] suggested a critical role for PKC in enhancement of αvβ6-mediated cell migration. It is not known how keratinocytes in vivo could coordinately use both fibronectin and vitronectin for their migration into the wound clot matrix.

In embryonic cells expressing $\alpha v\beta 1$ integrin as their major αv integrin, this receptor mediated cell attachment and spreading on fibronectin when these cells were made deficient of $\alpha v\beta 1$ integrin but not when $\alpha 5\beta 1$ integrin was present.[132] Integrins $\alpha 5\beta 1$, $\alpha v\beta 1$ and $\alpha v\beta 6$ all bind to the same RGD sequence in fibronectin molecules.[123,133,134] Interestingly, binding of $\alpha v\beta 6$ integrin to fibronectin has recently been shown to be inhibited, in addition to the RGD-peptides, by non-RGD peptide (DLXXL) present in several molecules but not in fibronectin itself,[135] suggesting that there are distinct structural requirements for interaction of the integrins $\alpha 5\beta 1$ and $\alpha v\beta 6$ with fibronectin. This was already proposed after demonstration that $\alpha v\beta 6$ integrin does not require the synergy site for its full activity in fibronectin binding.[136] Because the cell migration rate is dependent on substratum ligand levels, cell integrin expression, as well as integrin-ligand affinity,[105] utilization of a combination of receptors of different affinities can be beneficial to migration versatility. Integrin receptors may also regulate different aspects of cell motility e.g., migration speed and motile cell phenotype as has been shown for $\alpha 4\beta 1$ and $\alpha 5\beta 1$ integrins.[137] In addition, fibronectin binding to different integrins is likely to induce distinct cellular signaling pathways in cells.

Although $\alpha v\beta 6$ integrin was the major fibronectin-binding integrin in HaCaT cells, it showed a minor role in lateral cell migration, and its expression mainly at late stages of adult wound healing indeed suggests that $\alpha v\beta 6$ integrin does not function exclusively in mediating cell migration. Inactivation of the $\beta 6$ integrin gene has had no influence on short-term wound healing but resulted in inflammatory changes in the skin and lungs, suggesting a role for $\alpha v\beta 6$ integrin in modulation of inflammation.[138] Neutrophil recruitment appears to be associated with $\alpha v\beta 6$ integrin expression.[139] Furthermore, it has recently been shown that in cultured lung epithelial cells $\alpha v\beta 6$ integrin can bind latency-associated protein (LAP) of the TGFβ protein complex.[140] This binding appears to open the structure and lead to activation of TGFβ that can induce cells in the immediate neighborhood. integrin $\alpha v\beta 6$ binds to LAP with a much higher affinity than to fibronectin.[140] $\beta 6$-deficient mice seem to be protected from experimental lung fibrosis,[140] suggesting that $\alpha v\beta 6$ integrin could participate in the regulation of connective tissue formation. Because TGFβ also regulates the differentiation of myofibroblasts[141] and deposition of extracellular matrix in the granulation tissue,[142] it is conceivable that keratinocyte $\alpha v\beta 6$ integrin could have much wider functions than just mediating fibronectin binding during wound reepithelialization. Our findings that the expression of $\alpha v\beta 6$ integrin in wounds matches with the production of the extracellular matrix bridge by fibroblasts under the fused epithelium,[12] differentiation of myofibroblasts (Häkkinen et al, unpublished) and peak expression of biologically active TGFβ in the wound[143] support the spatio-temporal requirement for this putative mechanism.

Role of $\alpha v\beta 6$ Integrin in Malignant Transformation of Keratinocytes

As reviewed above, normal wound healing is a complex process that needs to be strictly regulated. An impairment in these regulatory processes can potentially lead to formation of hypertrophic scarring or chronic, unhealing wounds and even contribute to epithelial tumorigenesis. Cellular functions in malignant transformation resemble those associated with keratinocyte migration/invasion into the clot except that the cells have lost the coordination of the process. This failure may lead to unregulated growth, to proteolytic degradation of the basement membrane and to cell invasion through the fibronectin- and collagen-containing connective tissue.

We and others have previously shown that $\alpha v\beta 6$ integrin is induced in oral squamous cell carcinoma (SCC) in vivo.[144-146] Premalignant oral lesions that express $\alpha v\beta 6$ integrin are more likely to progress,[146] but the role of $\alpha v\beta 6$ integrin in this process has not been established. Adhesion to fibronectin, regulation of cell growth inside collagen matrix,[147] activation of TGFβ and modulation of MMP expression are all possible functions for $\alpha v\beta 6$ integrins in tumorigenesis. Recently, we examined the role of fibronectin binding integrin receptors in SCC cell lines expressing different amounts of αvβ6 integrin on cell surfaces.[148] The cell lines expressing αvβ6 integrin demonstrated β1 integrin-independent cell spreading on fibronectin, unlike the cell line with minimal αvβ6 integrin expression. When β6 integrin was transfected into these cells, their spreading on fibronectin also became partially β1 integrin-independent. The data from these experiments suggest that SCC cells in vitro can cooperatively use αvβ6 as well as α5β1 and α5β1 integrin receptors for interaction with fibronectin matrix and also adapt by switching to another receptor of the same ligand. This may also have some in vivo relevance since both αvβ6 and α5β1 integrins appear to be expressed in the primary SCC tumors.[144,149,150] The presence of αvβ1 integrin in SCC tumors in vivo is difficult to show because there are no specific antibodies available recognizing the αvβ1 integrin complex. In contrast to the results with SCC cells, in HaCaT keratinocytes and in primary epidermal keratinocytes also expressing αvβ1 integrin as their major αv integrin, this integrin failed to support cell spreading on fibronectin (unpublished results from our laboratory).[82] In embryonic cells, αvβ1 integrin was functional in cell attachment and spreading on fibronectin only when these cells were made deficient of α5β1 integrin,[132] suggesting that the role of αvβ1 integrin may be different in other cell types and even linked to the malignant cell phenotype. The ability to migrate on fibronectin may be crucial for tumor invasion and metastasis. SCC cells seemed to prefer β1 integrins as their main migration receptors on fibronectin. However, αvβ6, αvβ1 and α5β1 integrins all appeared to act cooperatively and interchangeably in SCC migration on fibronectin. In addition to mediating cell migration on fibronectin, αvβ6 integrin may have some other important roles in tumor cell invasion and cancer progression e.g. in mediating migration on tenascin-C[151] or in inducing MMP expression.[152] Integrin αvβ6 may also preferably serve as a migration receptor for other fibronectin isoforms expressed by SCC tumors.[153,154]

One possible function for αvβ6 integrin in malignant transformation could be in TGFβ activation. TGFβ has been implicated to play a role in cell transformation and carcinogenesis.[155] In transgenic mice, continuous overexpression of TGFβ1 by epithelial cells enhanced the malignant progression rate and phenotype and induced high incidence of particularly malignant fibroblastoid spindle cell carcinomas.[156] TGFβ can also directly stimulate reversible epithelial-mesenchymal transformation in cultured keratinocytes.[157-160]

How do Keratinocytes Migrate on a Composite Matrix?

It is obvious that, in addition to fibronectin, keratinocytes can use alternative extracellular matrix molecules for their migration. Tenascin-C is also expressed under the migrating wound epithelial cells and it may modulate interactions of integrins with their ligands.[161] Keratinocytes may use α9β1 and αvβ6 integrins for binding to tenascin-C[162-164] and both of these integrins are expressed by migrating wound keratinocytes in vivo.[30] Laminin-5 is always deposited underneath the migrating cells both in vitro and in vivo regardless of the wound-type.[12,29] Some studies provide evidence that laminin-5 is the key extracellular matrix molecule mediating migration regardless of which matrix the cells have been originally seeded on.[165] Laminin-5 is recognized by three integrin receptors of keratinocytes, namely α2β1, α3β1 and α6β4. There is controversial data regarding whether laminin-5 promotes or retards keratinocyte migration and which integrins are involved.[165,166] A hypothetical view describing how keratinocytes utilize fibronectin and laminin-5 during migration is presented in Figure 6. Recent evidence points to the role of α2β1 integrin in transient binding of endogenously produced unprocessed laminin-5 which promotes keratinocyte migration.[165] In contrast, binding of newly synthesized laminin-5 by α3β1 integrin could retard cell migration.[165] Finally, α6β4 integrin could then bind to plasmin-cleaved laminin-5 that could induce hemidesmosome formation.[167] Although laminin-5 appears to be a crucial extracellular matrix molecule in regulation of keratinocyte migration and hemidesmosome formation, it is controversial whether it is absolutely essential for reepithelialization in vivo.[166] Cutaneous wounds in some individuals with defects

97. Haapasalmi K, Zhang K, Tonnesen M et al. Keratinocytes in human wounds express αvβ6 integrin. J Invest Dermatol 1996; 106:42-48.
98. Peltonen J, Larjava H, Jaakkola S et al. Localization of integrin receptors for fibronectin, collagen, and lamimin in human skin: Variable expression in basal and squamous cell carcinomas. J Clin Invest 1989; 84:1916-1923.
99. Carter WG, Ryan MC, Gahr PJ. Epiligrin, a new cell adhesion ligand for integrin α3β1 in epithelial basement membranes. Cell 1991; 65:599-610.
100. Elices MJ, Urry LA, Hemler ME. Receptor functions for the integrin VLA-3: Fibronectin, collagen, and laminin binding are differentially influenced by Arg-Gly-Asp peptide and by divalent cations. J Cell Biol 1991; 112:169181.
101. Grinnell F. The activated keratinocyte: up regulation of cell adhesion and migration during woun healing. J Trauma 1990; 30:S144-149.
102. Larouche K, Leclerc S, Salesse C et al. Expression of the α5 integrin subunit gene promoter is positively regulated by the extracellular matrix component fibronectin through the transcription facto Sp1 in corneal epithelial cells in vitro. J Biol Chem 2000; 275:39182-39192.
103. Miyamoto S, Katz BZ, Lafrenie RM et al. Fibronectin an integrins in cell adhesion, signaling, and morphogenesis. Ann N Y Acad Sci 1998; 857:119-129.
104. Lafrenie RM, Yamada KM. Integrins and matrix molecules in salivary gland cell adhesion, signaling, and gene expression. Ann N Y Acad Sci 1998; 842:42-48.
105. Palecek SP, Loftus JC, Ginsberg MH et al. integrin-ligand binding properties govern cell migration speed throug cell-substratum adhesiveness. Nature 1997; 385:537-540.
106. Sung CC, O'Toole EA, Lannutti BJ et al. integrin α5β1 expression is required for inhibition of keratinocyte migration by ganglioside GT1b. Exp Cell Res 1998; 239:311-319.
107. Paller AS, Arnsmeier SL, Chen JD et al. Ganglioside GT1b inhibits keratinocyte adhesion and migration on a fibronectin matrix. J Invest Dermatol 1995; 105:237-242.
108. Wang X, Sun P, Al-Qamari A et al. Carbohydrate-carbohydrate binding of ganglioside to integrin α5 modulates α5β1 function. J Biol Chem 2000.
109. Aplin AE, Howe AK, Jliano RL. Cell adhesion molecules, signal transduction and cell growth. Curr Opin Cell Biol 1999; 11:737-744.
110. Jones PH, Watt FM. Separation of human epidermal stem cellsfrom transit amplifying cells on the basis of differences in integrin function and expression. Cell 1993; 73:713-724.
111. Pellegrini G, De Luca M, Orecchia G et al. Expression, topography, and function of integrin receptors are severely altered in keratinocytes from involved and uninvolved psoriatic skin. J Clin Invest 1992; 89:1783-1795.
112. Ting KM, Rothaupt D, McCormick TS et al. Oerexpression of the oncofetal Fn variant containing the EDA splice-in segment in the dermal-epidermal junction of psoriatic uninvolved skin. J Invest Dermatol 2000; 114:706-711.
113. Carroll JM, Romero MR, Watt FM. Suprabasal integrin expression in the epidermis of transgenic mice results in developmental defects nd a phenotype resembling psoriasis.
114. Bata-Csorgo Z, Cooper KD, Ting KM et al. Fibronectin and α5 integrin regulate keratinocyte cell cyclin. A mechanism for increased fibronectin potentiation of T cell lymphokine-driven keratinocyte hyperproliferation in psoriasis. J Clin Invest 1998; 101:1509-1518.
115. Brakebusch C, Grose R, Quondamatteo F et al. Skin and hair follicle integrity is crucially dependent on beta 1 integrin expression on keratinocytes. EMBO J 2000; 19:3990-4003.
116. Assoian RK, Schwartz MA. Coordinate signaling by integrins and eceptor tyrosine kinases in the regulation of G1 phase cell-cycle progression. Curr Opin Genet Dev 2001; 11:48-53.
117. Kuwada SK, Li X. integrin α5β1 mediates fibronectin-dependent epithelal cell proliferation through epidermal growth factor receptor activation. Mol Biol Cell 2000; 11:2485-2496.
118. Nanney LB, King LE. Epidermal growth factor and transforming growth factor-b. In: Clark RAF, ed. The Molecular and Cellular Biology of Wound Repair. Plenum Press, 1996:171-194.
119. Hashimoto K, Higashiyam S, Asada H et al. Heparin-binding epidermal growth factor-like growth factor is an autocrine growth factor for human keratinocytes. J Biol Chem 1994; 269:20060-20066.
120. Ellis PD, Hadfield KM, Pascall JC et al. Heparin-binding epidermal-growth-factor-like growth facto gene expression is induced by scrape-wounding epithelial cell monolayers: involvement of mitogen-activated protein kinase cascades. Biochem J 2001; 354:99-106.
121. Tokumaru S, Higashiyama S, Endo T et al. Ectodomain shedding of epidermal growth factor receptor ligands is required for keratinocyte migration in cutaneous wound healing. J Cell Biol 2000; 151:209-220.
122. Yarwood SJ, Woodgett JR. Extracellular matrix composition determines the transcriptional response to epidermal growth factor receptor activation. Proc Natl Acad Sci U S A 2001; 98:4472-4477.
123. Busk M, Pytela R, Sheppard D. Characterization of the integrin αvβ6 as a fibronectin-binding protein. J Biol Chem 1992; 267:5790-5796.

124. Weinacker A, Chen A, Agrez M et al. Role of integrin avb6 in cell attachment to fibronectin: heterologous expression of intact and secreted forms of the receptor. J Biol Chem 1994; 269:6940-6948.
125. Clark RAF, Spencer J, Larjava H et al. Reepithelialization of normal human excisional wounds is associated with a switch from αvβ5 to αvβ6 integrins. Br J Dermatol 1996; 135:46-51.
126. Boukamp P, Petrusevska RT, Breitkreut D et al. Normal keratinization in a spontaneously immortalized aneuploid human keratinocyte line. J Cell Biol 1988; 106:761-771.
127.Breitkreutz D, Schoop VM, Mirancea N et al. Epidermal differentiation and basement membrane formation by HaCaT cells in surface tranplants. Eur J Cell Biol 1998; 75:273-286.
128. Adams JC, Watt FM. Expression of β1, β3, β4, and β5 integrins by human epidermal keratinocytes and non-differentiating keratinocytes. J Cell Biol 1991; 115:829-841.
129. Boukamp P, Fuseng NE. "Trans-differentiation" from epidermal to mesenchymal/myogenic phenotype is associated with a drastic change in cell-cell and cell-matrix adhesion molecules. J Cell Biol 1993; 120:981-993.
130. Huang X, Wu J, Spong S et al. integrin αvβ6 is critical for keratinocyte migration on both its known ligand, fibronectin, and on vitronectin. J Cell Sci 1998; 111:2189-2195.
131. Zhang K, Kramer RH. Laminin 5 deposition promotes keratinocyte motility. Exp Cell Res 1996; 227:309-322.
132. Yang JT, Hynes RO. Fibronectin receptor functions in embryonic cells deficient in α5β1 integrin can be replaced by av integrins. Mol Biol Cell 1996; 7:1737-1748.
133. Pytela R, Pierschbacher MD, Ruoslahti E. Identification and isolation of a 140 kd cell surface glycoprotein with properties expected of a fibronectin receptor. Cell 1985; 40:191-198.
134. Vogel BE, Tarone G, Giancotti FG e al. A novel fibronectin receptor with an unexpected subunit composition (αvβ1). J Biol Chem 1990; 265:5934-5937.
135. Kraft S, Diefenbach B, Mehta R et al. Definition of an unexpected ligand recognition motif for αvβ6 integrin. J Biol Chem 1999; 274:1979-1985.
136. Chen J, Maeda T, Sekiguchi K et al. Distinct structural requirements for interction of the integrins α5β1, αvβ5, and αvβ6 with the central cell binding domain of fibronectin. Cell Adhes Commun 1996; 4:237-250.
137. Chon JH, Netzel R, Rock BM et al. α5β1 and α5β1 control cell migration on fibronectin by differentially regulating cell speed and motile cell phenotype. Ann Biomed Eng 1998; 26:1091-1101.
138. Huang XZ, Wu JF, Cass D et al. Inactivation of the integrin β6 subunit gene reveals a role of epithelial integrins in regulating inflammation in the lung and skin. J Cell Biol 1996; 133:92-928.
139. Miller LA, Barnett NL, Sheppard D et al. Expression of the β6 integrin subunit is associated with sites of neutrophil influx in lung epithelium. J Histochem Cytochem 2001; 49:41-48.
140. Munger JS, Huang X, Kawakatsu H et al. The integrin αvβ6 binds and activates latent TGFβ1: a mechanism for regulating pulmonary inflammation and fibrosis. Cell 1999; 96:319-328.
141. Desmouliere A, Geinoz A, Gabbiani F et al. Transforming growth factor-β1 induces smooth muscle actin expression in granulation tssue myofibroblasts and in quiescent and growing cultured fibroblasts. J Cell Biol 1993; 122:103-111.
142. Massague J. The transforming growth factor-b family. Annu Rev Cell Biol 1990; 6:597-641.
143. Yang L,Qiu CX, Ludlow A et al. Active transforming growth factor-β in wound repair: determination using a new assay. Am J Pathol 1999; 154:105-111.
144. Breuss JM, Gallo J DeLisser HM et al. Expression of the α6 integrin subunit in development, neoplasia and tissue repair suggests a role in epithelial remodelling. J Cell Sci 1995; 108:2241-2251.
145. Jones J, Watt FM, Speight PM. Changes in the expression of αv integrins in oral squamous cell carcinomas. J Oral Pathol Med 1997; 26:63-68.
146. Hamidi S, Salo T, Kainulainen T et al. Expression of αvβ6 integrin in oral leukoplakia. Br J Cancer 2000; 82:1433-1440.
147. Dixit RB, Chen A, Chen J et al. Identification of a sequence within the integrin b6 subunit cytoplasmic domain that is required to support the specific effect of αvβ6 on proliferation in three-dimensional culture. J Biol Chem 1996; 27:25976-25980.
148. Koivisto L, Grenman R, Heino J et al. Integrins α5β1, αvβ1, and αvβ6 collaborate in squamous carcinoma cell spreading and migration on fibronectin. Exp Cell Res 2000; 255:10-17.
149. Kosmehl H, Berndt A, Ketenkamp D et al. integrin receptors and their relationship to cellular proliferation and differentiation of oral squamus cell carcinoma. A quantitative immunohistochemical study. J Oral Pathol Med 1995; 24:343-348.
150. Shinohara M, Nakamura S, SasakiM et al. Expression of integrins in squamous cell carcinoma of the oral cavity. Correlations with tumor invasion and metastasis. Am J Clin Pathol 1999; 111:75-88.
151. Ramos DM, Chen BL, Boylen K et al. Stromal fibroblasts influence oral squamous-cell carcinoma cell interactions with tenascin-C. Int J Cancer 1997; 72:369-376.

152. Niu J, Gu X, Turton J et al. integrin-mediated signalling of gelatinase B ecretion in colon cancer cells. Biochem Biophys Res Commun 1998; 249:287-291.
153. Liu H, Chen B, Zardi L et al. Soluble fibronectin promotes migration of oral squamous-cell carcinoma cells. Int J Cancer 1998; 78:261-267.
154. Mandel U, Gaggero B, Reibel J et al. Oncofetal fibronectins in oral carcinomas: correlation of two different types. APMIS 1994; 102:695-702.
155. Akhurst RJ, Balmain A. Genetic events and the role of TGFβ in epithelial tumour progression. J Pathol 1999; 187:82-90.
156. Cui W, Fowlis DJ, Bryson S et al. TGFβ1 inhibits the formation of benign skin tumors, but enhances prgression to invasive spindle carcinomas in transgenic mice. Cell 1996; 86:531-542.
157. Miettinen PJ, Ebner R, Lopez AR et al. TGF-β induced transdifferentiation of mammary epithelial cells to mesenchymal cells: involvement of type I receptors. J Cell Biol 1994; 127:2021-2036.
158. Caulin C, Scholl FG, Frontelo P et al. Chronic exposure of cultured transformed mouse epidermal cells to transforming growth factor-β1 induces an epithelial-mesenchymal transdifferentiation and a spindle tumoral phenotype. Cell Growth Differ 1995; 6:1027-1035.
159. Frontelo P, Gonzalez-Garrigues M, Vilaro S et al. Transforming growth factor β1 induces squamous carcinoma cell variants with increased metastatic abilities and a disorganized cytoskeleton. Exp Cell Res 1998; 244:420-432.
160. Portella G, Cumming SA, Liddell J et al. Transforming growth factor b is essential for spindle cell conversion of mouse skin carcinoma in vivo: implications for tumor invasion. Cell Growth & Differ 1998; 9:393-404.
161. Crossin KL. Tenascin: a multifunctional extracellular matrix protein with a restricted distribution in development and disease. J Cell Biochem 1996; 61:592-598.
162. Prieto AL, Andersson-Fisone C, Crossin KL. Characterization of multiple adhesive and counteradhesive domains in the extracellular matrix protein cytotactin. J Cell Biol 1992; 119:663-678.
163. Yokosaki Y, Palmer EL, Prieto AL et al. The integrin α9β1 mediates cell attachment to a non-RGD site in the third fibronectin type III repeat of tenascin. J Biol Chem 1994; 269:26691-26696.
164. Yokosaki Y, Matsuura N, Higashiyama S et al. Identification of the ligand binding site for the integrin α9β1 in the third fibronectin type III repeat of tenascin-C. J Biol Chem 1998; 273:11423-11428.
165. Decline F, Rousselle P. Keratinocyte migration requires α9β1 integrin-mediated interaction with the laminin 5 gamma2 chain. J Cell Sci. 2001;114:811-823.
166. Nguyen BP, Ryan MC, Gil SG et al. Deposition of laminin 5 in epidermal wounds regulates integrin signaling and adhesion. Curr OpinCell Biol 2000; 12:554-562.
Cell 1995; 83:957-968.
167. Goldfinger LE, Stack MS, Jones JC. Processing of laminin-5 and its functional consequences: role of plasmin and tissue-type plasminogen activator. J Cell Biol 1998; 141:255-2.

CHAPTER 4

Matrix Metalloproteinases and Cell Migration in the Development of Cardiovascular Disease

Sarah J. George, Andrew C. Newby and Andrew H. Baker

The matrix metalloproteinase (MMP) family of enzymes has the ability to degrade all components of the extracellular matrix. For this fundamental reason, the MMPs have been implicated in the initiation, development and progression of diverse clinical conditions ranging from cancer and arthritis to renal and cardiovascular diseases. Due to the continuing isolation of novel MMPs and MMP-related enzymes, combined with the diversity within these protease families, the characterization of MMP involvement in disease states is an ongoing but rapidly evolving field with wide ranging implications in both basic science and related clinical applications. This review will focus on the involvement of MMPs in cardiovascular disease with particular emphasis on MMP regulation within the blood vessel wall, MMP involvement in the development of atherosclerosis and vessel injury associated with acute vascular intervention. We also discuss the implications of these findings on MMP-inhibition, either using synthetic MMP inhibitors or gene therapy by overexpression of tissue inhibitor of metalloproteinases (TIMPs), as a potential route to therapy for treatment of cardiovascular disease.

Extracellular Matrix Composition of the Blood Vessel Wall

The blood vessel wall is a highly ordered and complex structure that regulates blood flow and pressure, vascular tone and rigidity. The extracellular matrix plays a central role in maintaining vessel wall homeostasis and hence its composition is critical in elucidating how matrix turnover is regulated in both normal and diseased conditions. The structure of the normal blood vessel wall, as depicted in Figure 1, emphasises the presence of three major types of vascular extracellular matrix, namely basement membranes, an interstitial collagenous matrix and elastic fibres.

Basement Membrane

A basement membrane lies under the luminal endothelial cell monolayer and surrounds all quiescent, contractile SMCs, in contrast to adventitial fibroblasts, which lack a basement membrane. Basement membranes consist of a meshwork of the interrupted fibrillar, type IV collagen, the multi-adhesive glycoprotein, laminin and a variety of highly sulphated proteoglycans (PGs), amongst which those with heparan sulphate PGs (HSPGs) predominate.[1] These components of basement membranes appear appropriate to control the passive permeability of aqueous solutes. However, their function around SMC may be principally to maintain the contractile MMP-13 phenotype, since plating freshly isolated smooth muscle cells onto type IV collagen or laminin retards transition to the activated, synthetic phenotype of SMC that is characteristic of vascular repair.[2] The signalling mechanisms underlying these effects of type IV collagen and laminin are unknown, although binding to constitutively expressed cell surface integrins is likely to be involved.[3] This diverse family of heterodimeric membrane spanning receptors contains members that can bind to collagens and several multi-adhesive

Cell Invasion, edited by Jyrki Heino and Veli-Matti Kähäri. ©2002 Eurekah.com.

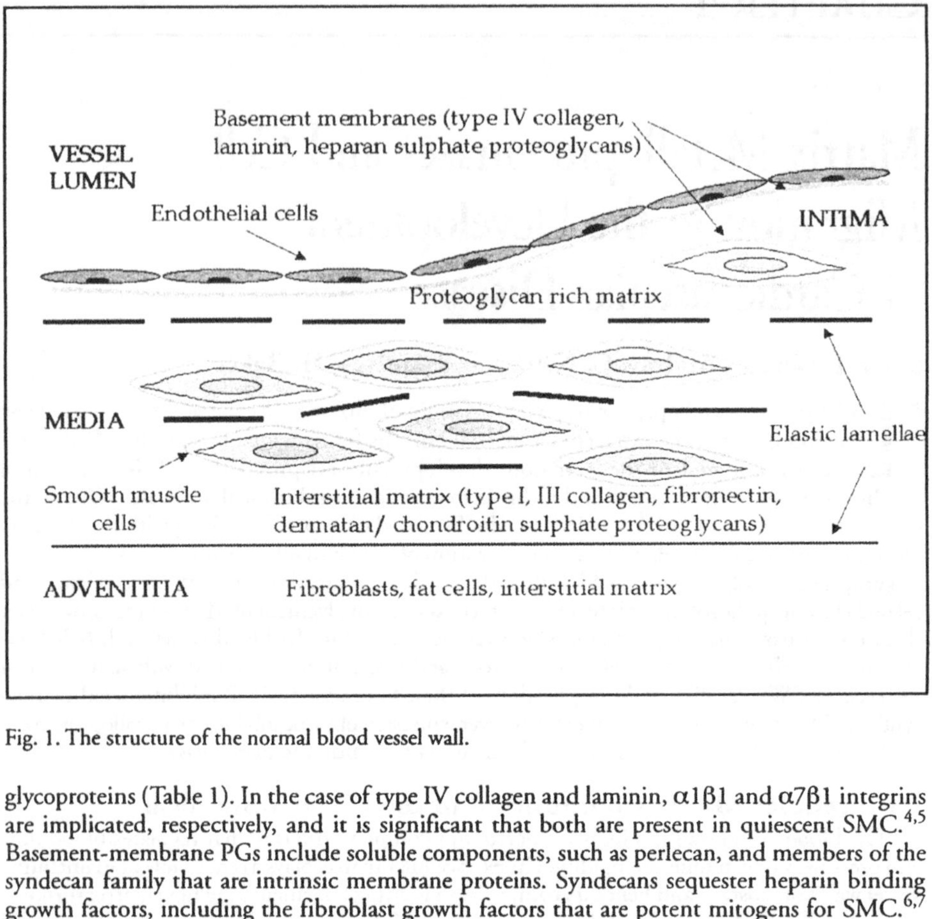

Fig. 1. The structure of the normal blood vessel wall.

glycoproteins (Table 1). In the case of type IV collagen and laminin, $\alpha1\beta1$ and $\alpha7\beta1$ integrins are implicated, respectively, and it is significant that both are present in quiescent SMC.[4,5] Basement-membrane PGs include soluble components, such as perlecan, and members of the syndecan family that are intrinsic membrane proteins. Syndecans sequester heparin binding growth factors, including the fibroblast growth factors that are potent mitogens for SMC.[6,7] HSPGs also directly inhibit SMC proliferation[8] by interacting with protein kinase C dependent pathways,[9] inhibitory actions that can be reversed by heparanases.[10] HSPGs are also implicated in activation of latent TGF-β, a cytokine that inhibits SMC proliferation.[11]

Interstitial Matrix

The interstitial collagenous matrix of the normal blood vessel wall contains mainly fibrillar types-I and -III collagen, smaller quantities of other fibril-associated collagens and the multi-adhesive glycoprotein fibronectin, and PGs of the chondroitin sulphate and dermatan sulphate types (CS/DSPGs).[1] The triple helical coil of fibrillar collagens provides most of the tensile strength, while abundant CS/DSPGs, such as versican and biglycan, are highly hydrated and hence contribute to the turgor of the blood vessel wall. Turnover rates of matrix components are low in contractile SMCs,[12] and hence most cell matrix contacts with quiescent SMCs are made by basement membrane components. Moreover, receptors for binding of fibronectin ($\alpha v\beta3$ and $\alpha5\beta1$ integrins) are downregulated (see Table 1).

Elastic Components

In the larger arteries, the SMC normally reside within lacunae formed by a honey comb of the hydrophobic protein, elastin.[1] The spherical elastin monomers are covalently cross-linked into close packed arrays assembled along fibrillar proteins, such as fibrillin. The elastic lammela is therefore not so much an elastic band but a sheet of rubber balls welded together. Smaller

Table 1. Integrins expressed in vascular smooth muscle cells

Integrin	Ligands	Expression	References
α1β1	Collagen (I, II, III, IV, VIII), laminin	High in blood vessels, downregulated in culture	(5)
α2β1	Collagen (I, IV) laminin	Increased in cultured cells	(5)
α4β1	Fibronectin, osteopontin	In fetal aorta, re-expressed in atherosclerosis	(178)
α5β1	Fibronectin	Increased by injury	(25, 26)
α7β1	Laminin	Present in blood vessels, decreased by culture	(4)
αvβ3	Fibronectin, thrombospondin, osteopontin	Expressed in arteries but upregulated by culture	(27-29)
αvβ5	Osteopontin	Increased by injury	(28, 179)

The table shows some of the integrins expressed in vascular SMC and their major ligands. For a more complete list see the review by Boudreau and Jones (3).

elastic fibres are present in all arteries, and recent research suggests they too may play a role in inhibiting SMC proliferation, by binding to a non-integrin elastin laminin receptor.[13]

Extracellular Matrix Disruption in Vessel Wall Remodelling

Vessel wall remodelling occurs as an adaptation to alterations in pressure and flow (e.g., in vein grafts) and in response to mechanical (e.g., angioplasty) or biochemical injury (e.g., atherosclerosis). Common features include phenotypic modulation, migration and proliferation of smooth muscle cell accompanied and arguably orchestrated by changes in the extracellular matrix.[14] The resulting changes are depicted in Figure 2. Phenotypic modulation involves a dramatic increase in the level of protein (including matrix protein) synthesis[12,15] and a change in the spectrum of expressed genes. This includes upregulation of connective tissue growth factors, including platelet derived growth factor[16] and TGF-β.[17] Partly as a consequence, matrix components, including hyaluronan, fibrillar collagens (which are now prevalent in their monomeric forms), versican and fibronectin are increased.[15,18,19] Type VIII collagen[20] and multi-adhesive glycoproteins, such as tenascin,[21] thrombospondin[22,23] and osteopontin,[24] which are not present during quiescence, are also upregulated. At the same time integrin receptors for fibronectin (α5β1)[25,26] and the other glycoproteins (αvβ3)[27-29] appear on the SMC surface (see Table 1).

As discussed above, matrix remodelling may be required for phenotypic modulation of SMC, and this is promoted by fibronectin.[2,30] Monomeric collagen also profoundly alters the pattern of gene expression in SMC.[31] Interstitial matrix components, including collagen and fibronectin also upregulate the expression of some matrix degrading metalloproteinases,[32-34] which suggests the possibility of positive feed back. Furthermore, a recent study has demonstrated that type VIII collagen, a matrix protein only expressed during intimal thickening and in atherosclerotic plaques, stimulates SMC migration and MMP-2 and MMP-9 production.[35] In addition, MMP synthesis was not affected by collagen type I suggesting that this feedback regulation of matrix degradation is a property unique to matrix molecules that are upregulated

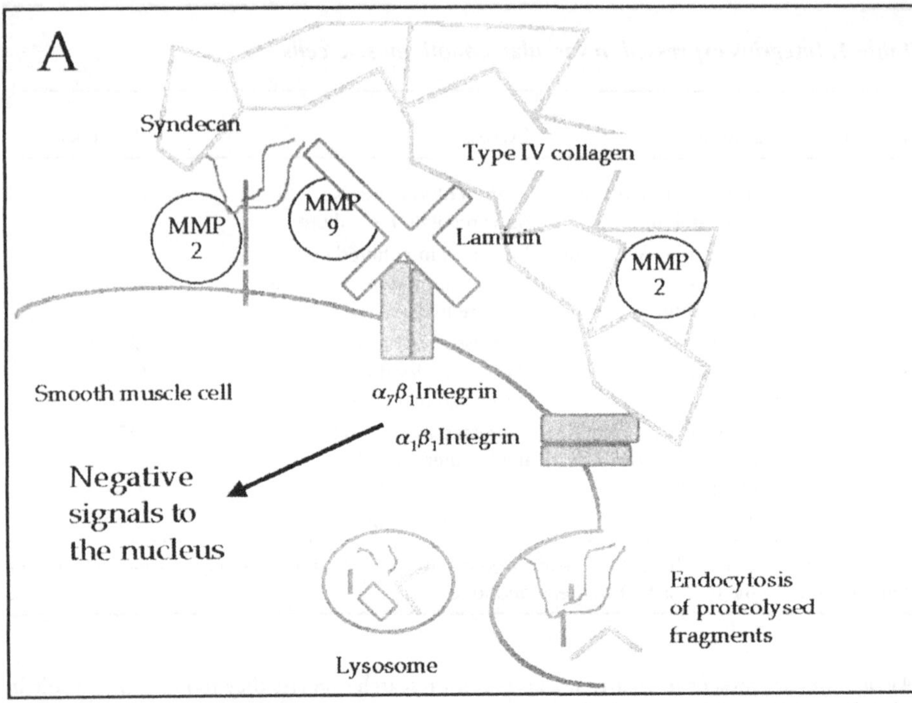

Fig. 2A. Cell matrix interactions in quiescent and activated vascular smooth muscle cells. Quiescent SMC are surrounded by a basement membrane consisting of type IV collagen, laminin and heparan sulphate proteoglycans, such as syndecan. Cell contacts through integrins maintain cells in a quiescent state, reversal of which requires matrix remodelling. Matrix metalloproteinases (MMPs) –2 and –9 can remodel all components of the basement membrane, preparing fragments for internalisation and further breakdown.

after injury. Matrix remodelling may be needed prior to SMC proliferation simply to detach cells from firm contacts with the matrix (the cage hypothesis). Remodelling matrix components, such as collagen I, may also reveal binding sites necessary for migration.[36] In addition, binding to α5β1 appears to be critical for fibronectin polymerisation, cytoskeletal rearrangement and the promotion of SMC migration.[25] Engagement of αvβ3 facilitates migration on other glycoproteins,[27] such as thrombospondin and osteopontin. This appears necessary to allow activation of calcium calmodulin activated kinase-II,[37] a key signalling event in SMC migration. Similar mechanisms may underlie regulation of smooth muscle cell proliferation.[14] Engagement of integrins activates focal adhesion kinase and this facilitates growth factor-induced proliferation.[38] Binding to αvβ3 also allows downregulation of cyclin-dependent kinase inhibitors,[39] thereby removing a brake on SMC proliferation. Downregulation of these inhibitors is a necessary step in the proliferative response of SMC cultured on monomeric collagen[40] and a key difference underlying the suppression of mitogenic responses to serum of SMC in rat aorta.[41]

Cell migration is a complex process stimulated by chemotaxis, requiring the combined action of chemoattractants and remodeling of the extracellular matrix.[42] Matrix remodeling may take two forms, namely removal of physical restraints and the formation of productive cell matrix interactions that are necessary for movement.[43] Cells must attach to a matrix to gain traction for migration, but they also have to detach from the matrix to translocate. This process also requires activation of cell surface receptors for matrix molecules, including integrins, and the reorganization of the matrix. Attachment and release are precisely coordinated by focal

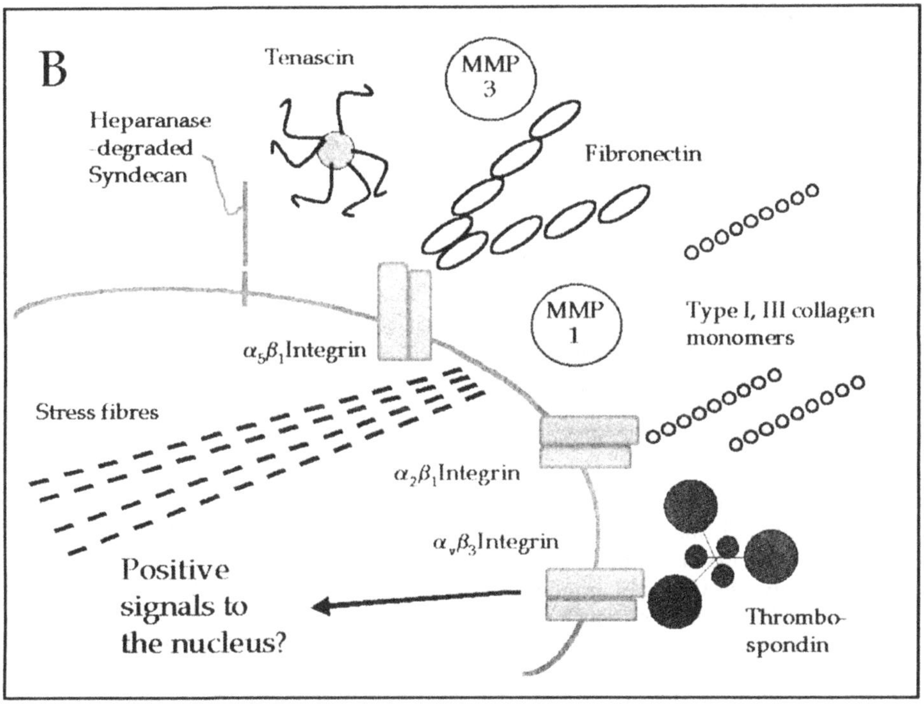

Fig. 2B. Cell matrix interactions in quiescent and activated vascular smooth muscle cells. Activated SMC show degraded basement membrane and increased interstitial matrix protein expression, with collagen I and III (in monomeric forms), tenascin and polymerised fibronectin as prominent components. SMC express different integrins that promote activation, cytoskeletal rearrangement and secretion of MMPs-1 and -3 that have activity on the interstitial matrix components.

localization of matrix-degrading proteinases on the cell surface, and MMPs also facilitate cell invasion by clearing a path through complex, three-dimensional extracellular matrices.[44]

This regulation of cell phenotype and fate evoked by complex interplay between vascular cells and the vessel wall extracellular matrix underlies the need to determine the array of protease enzymes responsible for these actions. MMPs are central to this understanding.

Matrix Metalloproteinases

Classification of MMPs

The classification and basic characteristics of MMPs are documented by Reunanen and Kähäri (in this book) and TIMPs in Table 2. Developmental and homeostatic remodeling of the extracellular matrix is a highly regulated process orchestrated by MMPs, a large family of zinc containing, calcium-dependent neutral proteases. MMPs can collectively degrade all structural proteins of the extracellular matrix including interstitial collagens (I, II, III and V), basement membrane collagens (IV), fibronectin, laminin, proteoglycan and elastin. The members of the MMP family are divided into four main classes based on their structure (Reunanen and Kähäri, in this book). Although each MMP is a product of a different gene, there is a high degree of sequence and structural domain homology between the MMPs. All MMPs have a short signal sequence and a pro-peptide region at the N-terminus, containing a cysteine residue that ligates with zinc at the catalytic domain and maintains the enzyme in the inactive or pro-

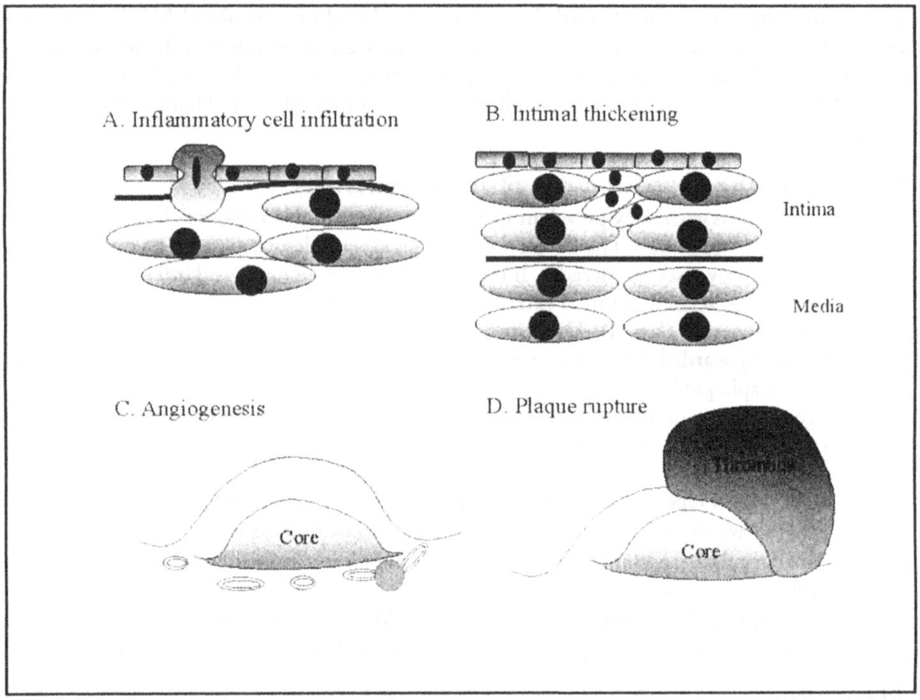

Fig. 4. The involvement of metalloproteinases during atherosclerotic plaque development. (A) The degradation of extracellular matrix may be required for transendothelial migration of inflammatory cells. (B) The release of cytokines and MMPs by macrophages in the fatty streak may facilitate SMC migration to the intima (C) MMP activity is essential for angiogenesis that can nourish the plaque as well as cause intraplaque haemorrhage. (D) The weakening of the shoulder regions by MMP-stimulated extracellular matrix degradation may lead to plaque rupture.

vitro findings. The expression of MMP-2 was thought to be similar in normal and diseased arteries,[63] but another study detected a 4-fold elevation of MMP-2 in diseased arteries.[84] MMP expression is not exclusive to macrophages, MMPs have also been detected in SMC, lymphocytes and endothelial cells.[63,78,79]

The above studies have the disadvantage that they are purely descriptive and provide no direct evidence for an involvement in plaque instability. Direct evidence has been provided in recent studies. Firstly, over-expression of MMP-1 has been shown to be associated with increased circumferential tensile stress in the fibrous cap.[85] Secondly, there is a strong correlation between the percentage of the lipid core occupied with haemorrhage and the percentage of the lipid core perimeter positive for MMP-1.[76] Thirdly, a recent demonstration that enhanced cleavage of collagen by collagenases is associated with increased presence of macrophage-derived active forms of MMP-1 and MMP-13 in human lipid-rich atherosclerotic plaques compared to fibrous plaques[82] provides additional supporting evidence. Finally, adenoviral delivery of TIMP-1 to apolipoprotein E knockout (apo$E^{-/-}$) mice fed a lipid-rich diet reduced lesion area by approximately 32%.[86] In this study, histological and immunohistochemical analysis revealed that over-expression of TIMP-1 resulted in enhanced collagen, elastin and SMC α-actin content and reduced macrophage number. This group proposed that, in addition to inhibiting MMP activity and reducing matrix degradation, this over-expression of TIMP-1 reduces lesion size by reducing SMC migration and protecting HDL particles and thereby promoting efficient reverse cholesterol transport from macrophages. In contrast, synthetic MMP inhibitors failed to reduce plaque development in LDL-receptor deficient mice.[87] This discrepancy may be due to fundamental differences in the use of synthetic MMP inhibitors or gene-based TIMPs involving bioavailability, tissue distribution, half-lives of drugs and genes in vivo, metabolism and degradation. Further studies are required to address these issues.

The hypothesis that augmented matrix degradation leads to atherosclerotic plaque rupture therefore remains largely unproven. This goal has been hampered in part by the lack of an animal model of plaque rupture. The recent establishment of a mouse model of MMP-13 plaque rupture[88] may now permit this hypothesis to be tested using TIMP and MMP knock-out mice. It is thought that their release of MMPs from the numerous macrophages in the shoulder regions may lead to activation of other MMPs and increased matrix degradation. These sites are subjected to maximal circumferential stress[89] and are prone to clinical rupture,[90] which may cause the matrix degradation promoting plaque weakening, fissure, or rupture.[91,92] In addition to direct degradation of the extracellular matrix, some MMPs may indirectly affect matrix degradation. For example, although MMP-11 only weakly degrades extracellular matrix it has been suggested that it may play a role in atherosclerotic plaque development by the reducing the levels of serpins, serine proteinase inhibitors such as α2-macroglobulin, α2-antiplasmin and α1-antitrypsin.[81] Reduction in serpins will cause increased activity of cathepsins, plasmin and leukocyte elastase, which can degrade extracellular matrix. A recent study using MMP-9-overexpressing cells in vitro and in vivo indicates that MMP-9 may also play a beneficial role in adaptive vessel remodeling.[93] MMP-9 activity increased vessel circumference, thinning the vessel wall and decreasing intimal matrix content, which may provide the space for the plaque to grow by inducing vessel dilation and increasing the luminal area. Stenosis or plaque development may therefore only occur if this adaptive process becomes maximised or overcome. It should therefore be considered that the activity of MMPs, including MMP-9 in atherosclerotic plaques will not always be detrimental. Interventional studies will elucidate precise mechanisms.

Inflammatory Cell Infiltration

Plaque development is thought to be initiated by the entry and oxidation of low-density lipoprotein (LDL) in the arterial intima.[94] This leads to the activation of endothelial cells, which express adhesion molecules including vascular cell adhesion molecule-1 (VCAM-1), promoting infiltration of circulating monocytes and T lymphocytes.[95] It has been proposed that adhesion of T lymphocytes to endothelial cells via VCAM-1 induces MMP-2 production and release, which in turn, may facilitate the extracellular matrix degradation required for migration through the endothelial cell layer and basement membrane.[96,97] However, this proposal has been challenged by a recent study demonstrating that neither TIMP-1 nor synthetic MMP inhibitors reduced migration of neutrophils across pulmonary endothelial cells.[98] Furthermore, normal neutrophil transendothelial migration has been reported in both gelatinase B[99] and elastase[100] knockout mice. The contact of inflammatory cells with matrix components during migration may also initiate extracellular matrix degradation. Contact with type I collagen and laminin increase the expression of MMP-9 in monocytes.[101,102] Interaction of monocytes with type I collagen may also induce proteolysis by stimulation of superoxide production[103] and secretion of interleukin-1 (IL-1)[104] which in turn can activate latent MMP-2 and MMP-9[105] and upregulate MMP-9, MMP-1 and MMP-3 production in SMC.[75,106] Although it is still unclear whether induction of MMPs is mediated directly by extracellular matrix composition[107] or via effects of substrate on cell shape and spreading,[108] it is clear that MMPs assist monocyte and T lymphocyte infiltration through the extracellular matrix during atherosclerotic plaque development.

Angiogenesis

Newly formed blood vessels may play an important role in atherosclerotic plaque development either by providing blood components to "nourish" the developing plaques or by causing repeated intimal haemorrhages that can eventually cause artery thrombosis.[109,110] Intimal neovascularisation largely originates from the adventitia and is closely associated with the extent of coronary stenosis and the inflammatory reaction.[110,111] Angiogenesis requires degradation of vascular basement membrane prior to migration and proliferation of endothelial cells and hence MMPs are essential in this process. The pivotal role of MMPs in angiogenesis was highlighted by a study showing that endothelial cells secrete high concentrations of MMPs.[112] Furthermore exogenous TIMP-1 and TIMP-2 and anti-MMP-2 and MMP-9 antibodies inhibit endothelial cell tube formation in vitro,[113,114] MMP inhibitors reduce corneal angiogenesis in vivo[115] and gene transfer of TIMP-1 inhibits endothelial cell migration.[116] Angiogenesis may depend on direct co-operation between adhesive and proteolytic mechanisms because MMP-2 binds to integrin $\alpha v\beta 3$ on angiogenic blood vessels.[117] Localisation of macrophages expressing MMPs, including MMP-9, to areas of adventitial vasa vasorum suggests that in addition to the involvement of macrophage-derived MMP-9 in matrix alterations associated with neovascularisation in atherosclerosis,[80] macrophages can release cytokines which stimulate MMP secretion by endothelial cells.[118] Other agents such as thrombin may play an important role in angiogenesis since it activates proMMP-2 in endothelial cells[119] and increases MMP-1 and MMP-3 expression through G-protein-coupled thrombin receptor.[120]

Regulation of MMP Activity in Atherosclerotic Plaques

As alluded to in previous sections, the proteolytic activity of MMPs is regulated at the level of MMP/TIMP gene transcription, latent enzyme activation and binding of TIMPs to pro- and activated MMPs. Potential modulators of MMP expression in SMC and macrophages have been examined in numerous in vitro studies. Several soluble inflammatory cytokines, including tumor necrosis factor-α (TNF-α) and IL-1, secreted by activated T lymphocytes increase MMP-1, MMP-9, MMP-3 and MT1-MMP expression in SMC and macrophages from atherosclerotic plaques.[63,83,106,121] The synthesis of MMP-1, MMP-3 and MMP-9 however is inhibited by other cytokines IL-4, interferon-γ and IL-10.[122-125] Although TIMP expression appears to be less regulated by inflammatory cytokines,[75,122] TIMP-1 is induced by IL-10[124] and TIMP-3 is induced by platelet-derived growth factor and transforming growth factor-β.[106] Factors other than cytokines may also regulate MMP activity since histamine released from mast cells induces MMP-1 production,[126] PDGF released from SMC or macrophages can increase MMP-1 expression[127] and basic fibroblast growth factor released from injured SMC can increase MMP-1 production whilst reducing TIMP-2 production.[128] Furthermore, the combined presence of cytokines and growth factors synergistically increase MMP production in SMC.[48,106] Oxidized low-density lipoprotein (LDL) may also play an important role in the regulation of MMPs in atherosclerosis because it upregulates MMP activity by inducing MMP-9 expression while reducing TIMP-1 expression in macrophages.[129] This is supported by the observation that lipid-lowering can reduce MMP activity in atherosclerotic plaques in rabbits.[130]

It has also been proposed that cell-cell contact may be an additional factor in the regulation of MMP expression in atherosclerotic plaques. Activated T cells may play a pivotal role in the induction of MMP expression since CD40 ligation induces MMP-9, MMP-1, MMP-3 and MMP-11 expression in human macrophages[81,131,132] and MMP-1, MMP-3, MMP-9 proMMP-2 and MMP-11 in SMC.[81,133] This was supported by the detection of CD40 and CD40 ligand (CD40L) in atherosclerotic plaques[132] and the observation that administration of an antibody against mouse CD40L to mice lacking the LDL receptor reduced aortic atherosclerotic lesions by nearly 60%.[134]

The mode of MMP activation in atherosclerotic plaques in vivo is unknown, however several mechanisms have been suggested. Firstly, it has proposed that chymase and tryptase released from activated mast cells that accumulate in the shoulder regions of atherosclerotic plaques initiate the cascade of MMP activation.[135] Secondly, it has been suggested that release of reactive oxygen radicals by lipid-laden macrophages may activate MMP-2 and MMP-9 activation by disrupting thiol reactions in the pro-enzyme[105] and by inactivating TIMP-1.[136] Thirdly, urokinase generated plasmin may be a pathophysiological activator of pro-MMPs.[137] Finally, pro-MMPs may be activated by other MMPs.[138]

TIMPs like MMPs are produced by SMCs in vitro[75,106] and are present in atherosclerotic plaques at higher levels than in normal arteries.[63,79,139] TIMPs are co-located with MMPs in the macrophage-rich shoulder regions and between the fibrous cap and necrotic core.[63,79,139] This clearly demonstrates that the TIMPs play an important role in controlling MMP activity and maintaining plaque stability in vivo. Despite the co-location of TIMPs and MMPs, in situ zymography studies have revealed that an excess of MMPs to TIMPs is present in the shoulder and core regions of human[135,140] and rabbit atherosclerotic plaques[141] (see Fig. 5 for an example).

MMPs and Acute Vascular Injury

Vessel Wall Damage

The requirement for intervention to alleviate obstructed coronary arteries in the heart leads to the use of balloon angioplasty to re-open the coronary artery or coronary bypass grafting (CABG) usually using either internal mammary artery or saphenous vein (dependent on the requirement for single or multiple bypasses). Both these interventions result in acute injury to the vessel wall. In the case of balloon angioplasty injury results from physical trauma due to balloon dilatation including rupture of the internal elastic lamina (IEL). Whilst this leads to restoration of blood flow to the occluded coronary artery, often the beneficial effect is short-lived due to post-angioplasty restensosis, which occurs in approximately 30-50% of patients, for whom further interventions are required.[142] In the case of CABG, injury results from surgical preparation of the graft and, in the case of autologous saphenous vein, exposure to higher arterial blood pressure in the coronary circulation. This ultimately leads to CABG failure due to neointimal hyperplasia and superimposed atherosclerosis.[143] The injury causes endothelial cell removal leading to reduced vascular responsiveness and tone, exposure of the vessel wall to pro-thrombotic stimuli, as well as damage to the deeper vessel architecture resulting in increased migration and proliferation of SMC as well as adventitial remodelling. Whilst the interventions and mechanisms of angioplasty restenosis and CABG failure are clearly different, there are a number of common factors that ultimately define the high failure rates observed:

1. Acute injury to the vascular endothelium resulting in loss of vessel tone and exposure of the vessel wall to a pro-thrombotic environment.
2. Infiltration of leukocytes resulting in release of cytokine mediators that can regulate MMP activity at the transcriptional level.
3. Physical damage to the medial SMC layer of the vessel wall leading to SMC migration and proliferation.

MMPs and In Vitro Studies

The involvement of MMPs in the migration of SMC through extracellular matrices has been studied extensively in cell culture, in appropriate animal models and in human specimens. Under basal conditions, SMC can secrete MMP-2 and TIMPs-1 and -2.[75,144] However, stimulation of SMC with cytokines such as IL-1 or TNF-α leads to expression and secretion of a repertoire of MMP enzymes reflecting the inducible nature of many of the MMPs.[75,106] The primary MMPs induced in this manner are MMP-1, -3 and -9.[75] This not only leads to a net elevation in MMP levels in the extracellular space but also an increase in the activated form of MMPs resulting from both elevated synthesis and lack of a compensatory elevation of TIMP secretion under identical conditions. Interestingly, Fabunmi et al[106]demonstrated that a fundamental requirement for cell stimulation by both growth factors and cytokines was required

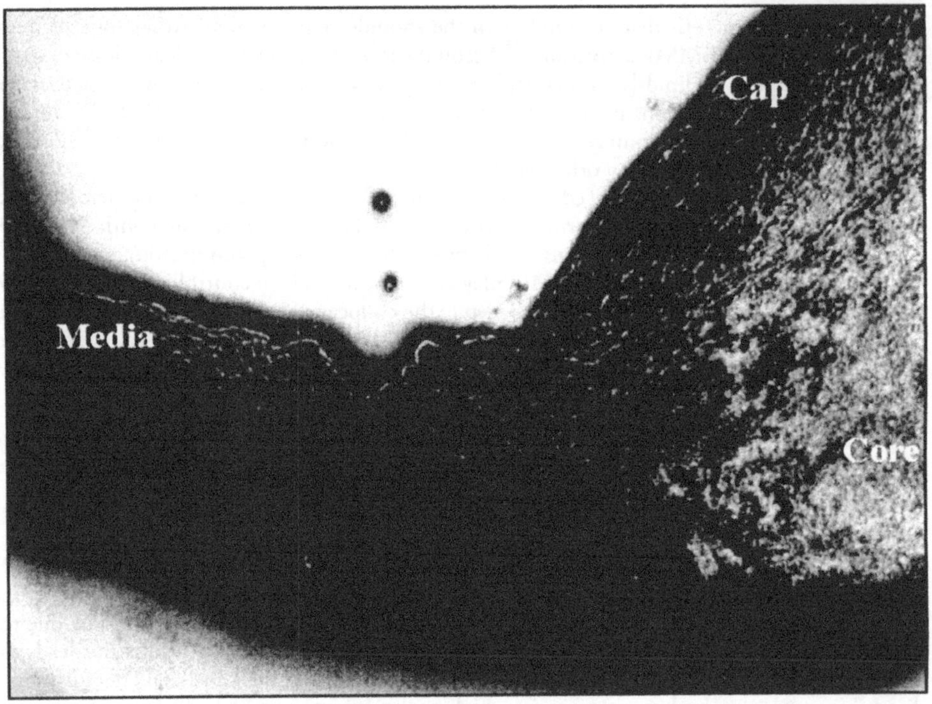

Fig. 5. MMP activity detected in an atherosclerotic plaque by in situ zymography. White areas indicative of MMP-mediated gelatin degradation on frozen sections represent areas with active gelatinolytic MMPs. Conversely black regions represent areas without net gelatinolytic activity. The activity of MMPs in the fibrous cap and normal vessel media are negative or have low MMP activity whilst high levels are detected in the atherosclerotic core.

for synergistic induction of MMP-9, conditions which had no effect on TIMP secretion. This is of particular importance at sites of vascular injury as SMC are exposed to a plethora of stimuli including growth factors released upon injury and cytokines released by infiltrating leukocytes. Follow-up studies demonstrated that MMPs-1, -3 and -9, which all contain similar transcription factor binding sites in their promoters are upregulated synergistically in response to growth factors and cytokines, a phenotype involving the NF-κB pathway.[48] Similarly, in the rabbit SMC line, Rb-1, mechanical injury of confluent cultures results in the rapid induction of mRNA expression of MMPs-1 and -3.[145] Furthermore as discussed previously, other classes of molecule important in the vasculature such as thrombin, which is generated at sites of vascular injury, can lead to activation of MMP-2 leading to potentiation of MMP activity.[146] Together, these studies demonstrate that steady-state expression of MMPs can be rapidly upregluated upon exposure to acute vascular injury involving growth factors and cytokines present in the vessel wall under injurious conditions.

Studies in Models of Disease

The use of animal models of acute vascular injury has enhanced our understanding of these complex human disease mechanisms and provided excellent experimental tools to evaluate and understand MMP involvement in these pathologies. These studies have illustrated that injury induces SMC migration and proliferation and increased MMP-9, MMP-2, MMP-3, MMP-8, MMP-12 and MT1-MMP expression.[147-151] The increased expression of MT1-MMP mRNA coincides with MMP-2 activation in the neointima after balloon injury of rat and rabbit arteries,[151,152] suggesting MT1-MMP plays a role in MMP-2 activation during intimal thickening. However, it is possible that proMMP-2 may also be activated by the high local concentrations of thrombin,[146] caused by platelet adhesion and activation of the coagulation cascade after vascular wall injury.[153] The effect of injury on TIMP expression is less clear. One study could not detect any change in TIMP-1, TIMP-2 nor TIMP-3 expression after injury[149] although increased TIMP-1, TIMP-2 and TIMP-4 has been detected in other studies.[154-157]

The carotid artery is stripped of endothelium leading to a consistent and pronounced neointima formed over 14-21 days in the rat balloon angioplasty model. This model is particularly suited for studies involving SMC migration and proliferation, phenotypes that implicitly involve MMPs. SMC proliferation begins 24 h post-denudation of the endothelium and SMCs begin to migrate from the media to the intima at 4 days. Following this, a second wave of proliferation leads to a consistent and reproducible neointima formation at 14-21 days.[158] In similarity to in vitro data discussed above, control rat carotid arteries expressed pro-MMP-2 constitutively.[147,148] However, MMP-9 is induced rapidly following injury and is detectable during the migration phase of the response to injury. Elevation of MMP-2 levels do not appear until 4-5 days post-injury but appear to persist during neointima formation.[147,148] These studies therefore strongly associate MMP-2 and -9 with the time course of SMC migration and neointima formation post-injury. Intervention studies were clearly required to corroborate MMP involvement. Zempo et al[159] demonstrated that the broad-spectrum MMP inhibitor BB-94 (marimistat, British Biotech PLC) blocked platelet-derived growth factor (PDGF)-induced SMC migration through complex extracellular matrix (Matrigel™) in vitro, without an effect on proliferation. Further, daily i.p. injections of BB-94 decreased intimal thickening by reduction in SMC migration and proliferation. Using a different, but albeit equally broad spectrum MMP inhibitor GM6001 (Glycomed Inc.), Bendeck et al[160] described an elegant set of experiments showing, in slight contrast to Zempo et al,[148] conclusive evidence for MMP-mediated regulation of SMC migration in the absence of an effect on proliferation but only at early time points. Inhibition of MMPs using this inhibitor resulted in a substantial reduction in early migration but also a "catch-up" phenomenon where the resulting lesion sizes were no different from controls due to enhanced proliferation in the GM6001-treated group between days 4 and 14. Although these two studies undermined any potential for MMP-inhibition in the treatment of post-angioplasty in humans (that occurs over a prolonged time period), their usefulness cannot be ruled out until suitable studies in humans have been completed. Clearly the expression profiles of MMP-2 and -9 following balloon injury suggests that they may be involved in diverse mechanisms or that the regulator(s) of MMP-2 and -9 expression post-injury are induced rapidly (in the case of MMP-9) or more delayed (for MMP-2). The rapid, transient induction of MMP-9 post-injury and the rigorous control of MMP-9 expression under basal conditions in SMC suggest an important role in disease progression. To evaluate the effect of direct MMP-9 overexpression in SMC vitro and in vivo Mason et al[93] used a tetracycline-inducible system to regulate MMP-9 levels. MMP-9 overexpression induced SMC migration and invasion in vitro and in vivo. In vivo this resulted in an increase in vessel circumference, vessel wall thinning and reduced matrix composition in the intima, hence providing direct evidence for MMP-9 involvement in vessel wall disruption in the rat carotid injury model.

In recent years, the use of vascular injury to arteries of knockout mice has provided information on the role of MMPs in vascular remodelling. For example, electrical injury of femoral arteries in mice which stimulates intimal thickening, caused enhanced expression of MMP-2 and MMP-9[161] and in TIMP-1 deficient mice (TIMP-1-/-), intimal thickening was significantly increased compared to wild type mice.[162]

Inhibitor Studies

To examine directly whether the shift in the proteolytic balance after injury contributes to intimal thickening several studies using gene transfer have been carried out. Using the antisense approach, inhibition of stromelysin inhibited intimal thickening in the carotid artery 7 days after injury.[163] In this study it was also noted that the stromelysin antisense oligonucleotides inhibited injury-induced phenotypic modulation and therefore they suggest that this caused decreased migration and proliferation and intimal thickening. Over-expression of TIMP-1 in balloon injured rat carotid arteries using transduced SMC[164] and adenoviruses[157] reduced intimal thickening by 40% and 30% respectively, 14 days after injury. In both studies TIMP-1 over-expression did not block SMC proliferation in the media and intima at 2 and 14 days.[157,164] However, Dollery et al[157] demonstrated that inhibition of SMC migration by 60% was at least partly responsible for the reduced intimal thickening. In contrast, although over-expression of TIMP-2 using adenoviruses significantly reduced intimal thickening by 53% eight days after injury, no effect was observed by 21 days after injury.[165] In this study, over-expression of TIMP-2 reduced SMC migration by 36% but proliferation was not affected. Taken together these studies clearly demonstrate the involvement of MMPs in injury induced intimal thickening in the rat.[165]

In other systems, further association of MMPs and disease progression has been documented. For example, in a double injury iliac artery model of post-angioplasty restenosis, MMP-2 is strongly upregulated in the injured vessel wall, in fact persisting for at least 12 weeks post-injury.[166] However, although MMP inhibition, again using GM6001, reduced collagen content, the overall effect on neointima formation was not significantly affected. In the pig carotid artery (a model more indicative of human post-angioplasty restenosis than the rat due to the deeper nature of the injury including rupture of the internal elastic lamina) both MMP-2 and -9 levels are elevated post-balloon injury.[167] MMP-2 and -9 were elevated 3 and 7 days post-injury, and were localised to medial cells, particularly around areas of necrosis and within the neointima itself. Expression profiles of both gelatinases paralleled the time course of SMC proliferation and migration in this model. However, as MMP inhibitor studies using the same model have not been forthcoming, a definite role for MMPs cannot yet be assigned.

Fewer studies have addressed the expression and role of MMPs in coronary artery bypass graft failure (CABG). Two basic models have thus far been used to define MMP involvement in vein graft failure, the human saphenous vein organ culture model[168] and the porcine-to-carotid-interposition graft model in vivo.[169] George and co-workers[170] observed evaluated MMP-2 and -9 levels in freshly isolated (removed from patient but not prepared for grafting) and surgically prepared (by distension with heparinised blood prior to grafting) human saphenous vein using the organ culture model. In the damaged vein increases in pro-MMP-2, active MMP-2 and pro-MMP were observed, which was further enhanced upon culturing for 14 days, a period where neointima formation progresses. Interestingly, and consistent with models of balloon injury discussed above, MMP-9 induction was rapid (3 h) and expression was both within the medial SMC at early time points and later in highly proliferative neointimal SMC consistent with MMP association with SMC migration and proliferation. These studies have been complemented by evaluation of MMPs in experimental vein grafts in pigs.[171] Using the saphenous vein-to-carotid interposition graft model, MMP expression levels were examined in relation to neointima formation over an extended time period (6 months). In contrast to control vein, which expressed pro- and active MMP-2, vein grafts had upregulated MMP-2 levels and induced MMP-9. In both cases, and in similarity to post-angioplasty models, MMP levels were elevated during active SMC proliferation and migration and hence neointima formation (up to 1 month) but returned to basal levels at later time points (6 months).[171]

The use of TIMP gene transfer has directly demonstrated the involvement of MMPs in SMC proliferation and migration associated with vein graft failure. The use of human models such as the ex vivo organ culture model is clearly important for the development and assessment of effective gene-based therapies for future human use. This model was used for TIMP overexpression studies using adenoviral vectors. Prior studies using adenoviral vectors overexpressing either TIMP-1, -2 or -3 defined fundamental differences in the phenotypes of SMC infected with each adenovirus.[172] As expected overexpression of TIMP-1, -2 or -3 inhibited SMC migration and invasion through reconstituted basement membrane, an effect reproduced by synthetic MMP inhibitors.[173] However, unlike TIMP-1, TIMP-2 overexpression also inhibited SMC proliferation, a phenotype that would be additionally advantageous for applications in vivo. TIMP-3 overexpression however promoted SMC death by activation of apoptosis,[172] an effect mediated by the N-terminus of the TIMP-3 protein.[173] In human saphenous vein organ cultures, overexpression of TIMPs-1, -2 and -3 reduced neointima formation by 545, 84% and 90% respectively.[174-176] As this model is ex vivo the effect of over-expression is determined in the absence of an immune response to the adenovirus thus allowing continual high-level overexpression of each transgene. Although, all three TIMPs ablated MMP activity in the vein wall, TIMP-3, in agreement with in vitro data, also promoted SMC apoptosis.[176] These studies imply that MMP inhibition alone is entirely sufficient to effectively inhibit neointima formation through modulation of SMC migration, and the induction of apoptosis may show additional efficacy. When these studies were repeated in vivo in the porcine interposition graft model,[176] only TIMP-3 had a significant effect on neointima formation at 28 days.[176] It is not clear whether the effect of TIMP-3 is due to promotion of apoptosis and inhibition of MMP activity, apoptosis alone, or due to the fact that the recombinant TIMP-3 is maintained for longer than other TIMPs in the vessel wall following gene transfer. The persistence of recombinant TIMP-3 is due to its ability to bind to the extracellular matrix.[177]

From the wealth of data documenting MMP expression and activity profiles following acute vascular injury, it is clear that MMPs play a fundamental role in disease progression leading to failure of either angioplasty or CABG procedures. Interventional studies using either synthetic inhibitors of gene-based overexpression of TIMPs have, albeit variably, corroborated these data and further highlighted a potential route to therapy. Further studies are however required to define whether long term overexpression of TIMPs or persistent localized MMP inhibition using synthetic inhibitors provides a strategy for use in clinical scenarios.

Concluding Remarks

In summary, we have highlighted the structure of the normal blood vessel wall and documented current understanding of the role of MMPs in the disruption of fundamental cell-matrix events observed in both chronic and acute vascular diseases. This understanding has led to the use of MMP inhibitors for interventional studies that have proven a central role for these important enzymes in disease progression. Further research in appropriate animal models will document the intimate interactions that MMPs mediate in the normal vessel wall, the trail of destruction that they can mediate when deregulated and the potential for MMP inhibition as a long-term therapy for diverse cardiovascular diseases.

References

1. Raines EW. The extracellular matrix can regulate vascular cell migration, proliferation, and survival: Relationships to vascular disease. Int J Exp Pathol 2000; 81:173-182.
2. Hedin U, Bottger BA, Forsberg E et al. Diverse effects of fibronectin and laminin on phenotypic properties of cultured smooth muscle cells. J Cell Biol 1988; 107:307-319.
3. Boudreau NJ, Jones PL. Extracellular matrix and integrin signalling: the shape of things to come. Biochem J 1999; 339:481-488.
4. Yao CC, Breuss J, Pytela R et al. Functional expression of the alpha 7 integrin receptor in differentiated smooth muscle cells. J Cell Sci 1997; 110:1477-1487.
5. Skinner MP, Raines EW, Ross R. Dynamic expression of alpha-1-beta-1 and alpha-2-beta-1 integrin receptors by human vascular smooth-muscle cells—alpha-2-beta-1 integrin is required for chemotaxis across type-I collagen-coated membranes. Am J Pathol 1994; 145:1070-1081.
6. Yayon A, Klagsbrun M, Esko JD et al. Cell surface, heparin-like molecules are required for binding of basic fibroblast growth factor to its high affinity receptor. Cell 1991; 64:841-848.
7. Woods A, Oh ES, Couchman JR. Syndecan proteoglycans and cell adhesion. Matrix Biol 1998; 17:477-483.

8. Bingley J, Hayward I, Campbell J et al. Arterial heparan sulfate proteoglycans inhibit vascular smooth muscle cell proliferation and phenotype change in vitro and neointimal formation in vivo. J Vasc Surg 1998; 28:308-318.
9. Au YPT, Dobrowolska G, Morris DR et al. Heparin decreases activator protein-1 binding to DNA in part by posttranslational modification of Jun B. Circ Res 1994; 75:15-22.
10. Campbell JH, Rennick RE, Kalevitch SG et al. Heparan sulphate degrading enzymes induce modulation of smooth muscle phenotype. Exp Cell Res 1992; 200:156-167.
11. Grainger DJ, Witchell CM, Watson JV et al. Heparin decreases the rate of proliferation of rat vascular smooth muscle cells by releasing transforming growth factor β-like activity from serum. Cardiovasc Res 1993; 27:2238-2247.
12. Ang AH, Tachas G, Campbell JH et al. Collagen synthesis by cultured rabbit aortic smooth-muscle cells: Alteration with phenotype. Biochem J 1990; 265:461-469.
13. Spofford CM, Chilian WM. The elastin-laminin receptor functions as a mechanotransducer in vascular smooth muscle. Am J Physiol Heart Circ Physiol 2001; 280(3):H1354-1360.
14. Newby A, Zaltsman AB. Fibrous cap formation or destruction—The critical importance of vascular smooth muscle cell proliferation, migration and matrix formation. Cardiovasc Res 1999; 41:345-360.
15. Rekhter MD. Collagen synthesis in atherosclerosis: Too much and not enough. Cardiovasc Res 1999; 41:376-384.
16. Nilsson J, Sjölund M, Palmberg L et al. Arterial smooth muscle cells in primary culture produce a platelet-derived growth factor-like protein. Proc Natl Acad Sci USA 1985; 82:4418-4422.
17. Nikol S, Isner JM, Pickering JG et al. Expression of transforming growth factor-b1 is increased in human vascular restenosis lesions. J Clin Invest 1992; 90:1582-1592.
18. Thyberg J, Blomgren K, Roy J et al. Phenotypic modulation of smooth muscle cells after arterial injury is associated with changes in the distribution of laminin and fibronectin. J Histochem Cytochem 1997; 45(6):837-846.
19. Evanko SP, Angello JC, Wight TN. Formation of hyaluronan- and versican-rich pericellular matrix is required for proliferation and migration of vascular smooth muscle cells. Arterioscler Thromb Vasc Biol 1999; 19:1004-1013.
20. Plenz G, Dorszewski A, Volker W et al. Cholesterol-induced changes of type VIII collagen expression and distribution in carotid arteries of rabbit. Arterioscler Thromb Vasc Biol 1999; 19:2395-2404.
21. Jones PL, Jones FS. Tenascin-C in development and disease: Gene regulation and cell function. Matrix Biol 2000; 19:581-596.
22. Yabkowitz R, Mansfield PJ, Ryan US et al. Thrombospondin mediates migration and potentiates platelet-derived growth factor-dependent migration of calf pulmonary artery smooth muscle cells. J Cell Physiol 1993; 157(1):24-32.
23. Chen D, Guo K, Yang J et al. Vascular smooth muscle cell growth arrest on blockage of thrombospondin-1 requires p21 (Cip1/WAF1). Am J Physiol Heart Circ Physiol 1999; 277:H1100-H1106.
24. Shanahan CM, Cary NRB, Metcalfe JC et al. High expression of genes for calcification-regulating proteins in human atherosclerotic plaques. J Clin Invest 1994; 93:2393-2402.
25. Pickering JG, Chow LH, Li S et al. Alpha5beta1 integrin expression and luminal edge fibronectin matrix assembly by smooth muscle cells after arterial injury. Am J Pathol 2000; 156(2):453-465.
26. Barillari G, Albonici L, Incerpi S et al. Inflammatory cytokines stimulate vascular smooth muscle cells locomotion and growth by enhancing alpha5beta1 integrin expression and function. Atherosclerosis 2001; 154(2):377-385.
27. Srivatsa SS, Fitzpatrick LA, Tsao PW et al. Selective alpha v beta 3 integrin blockade potently limits neointimal hyperplasia and lumen stenosis following deep coronary arterial stent injury: Evidence for the functional importance of integrin alpha v beta 3 and osteopontin expression during neointima formation. Cardiovasc Res 1997; 36(3):408-428.
28. Corjay MH, Diamond SM, Schlingmann K et al. Alphavbeta 3, alphav beta5, and osteopontin are coordinately upregulated at early time points in a rabbit model of neointima formation. J Cell Biochem 1999; 75:492-504.
29. Stouffer GA, Hu ZY, Sajid M et al. Beta(3) integrins are upregulated after vascular injury and modulate thrombospondin- and thrombin-induced proliferation of cultured smooth muscle cells. Circulation 1998; 97:907-915.
30. Hedin U, Bottinger BA, Lutham J et al. A substrate of the cell attachment sequence of fibronectin (Arg-Gly-Asp-Ser) is sufficient to promote transition of arterial smooth muscle cells from a contractile to a synthetic phenotype. Dev Biol 1989; 133:489-501.
31. Ichii T, Koyama H, Tanaka S et al. Fibrillar collagen specifically regulates human vascular smooth muscle cell genes involved in cellular responses and pericellular matrix environment. Circ Res 2001; 88:460-467.

32. Werb Z, Tremble PM, Behrendtsen O et al. Signal transduction through the fibronectin receptor induces collagenase and stromelysin gene expression. J Cell Biol 1989; 109:877-889.
33. Huhtala P, Humphries MJ, McCarthy JB et al. Cooperative signaling by alpha-5-beta-1 and alpha-5-beta-1 integrins regulates metalloproteinase gene-expression in fibroblasts adhering to fibronectin. J Cell Biol 1995; 129:867-879.
34. Sudbeck BD, Pilcher BK, Welgus HG et al. Induction and repression of collagenase-1 by keratinocytes is controlled by distinct components of different extracellular matrix compartments. J Biol Chem 1997; 272:22103-22110.
35. Hou G, Mulholland D, Gronska MA et al. Type VIII collagen stimulates smooth muscle cell migration and matrix metalloproteinase synthesis after arterial injury. Am J Pathol 2000; 156:467-476.
36. Stringa E, Knauper V, Murphy G et al. Collagen degradation and platelet-derived growth factor stimulate the migration of vascular smooth muscle cells. J Cell Sci 2000; 113:2055-2064.
37. Bilato C, Curto KA, Monticone RE et al. Inhibition of vascular smooth muscle cell migration by peptide and antibody antagonists of the alpha(v)beta(3) integrin complex is reversed by activated calcium/calmodulin-dependent protein kinase II. J Clin Invest 1997; 100(3):693-704.
38. Hauck CR, Hsia DA, Schlaepfer DD. Focal adhesion kinase facilitates platelet-derived growth factor-BB-stimulated ERK2 activation required for chemotaxis migration of vascular smooth muscle cells. J Biol Chem 2000; 275:41092-9.
39. Assoian RK, Marcantonio EE. The extracellular matrix as a cell cycle control element in atherosclerosis and restenosis. J Clin Invest 1996; 98:2436-2439.
40. Koyama H, Raines EW, Bornfeldt KE et al. Fibrillar collagen inhibits arterial smooth muscle cell proliferation through regulation of Cdk2 inhibitors. Cell 1996; 87:1069-1078.
41. Izzard T, Taylor CT, Birkett S et al. Mechanism of inhibition of smooth muscle cell proliferation by the native extracellular matrix. J Submicroscopic Cytol Pathol 2000; 32:446(abstract).
42. Van Leeuwen RTJ. Extracellular proteolysis and the migrating vascular smooth muscle cell. Fibrinolysis 1996; 10:59-74.
43. Abedi H, Zachary I. Signalling mechanisms in the regulation of vascular cell migration. Cardiovasc Res 1995; 30:544-556.
44. Murphy G, Gavrilovic J. Proteolysis and cell migration: Creating a path? Curr Opin Cell Biol 1999; 11(5):614-621.
45. Shofuda K, Moriyama K, Nishihashi A et al. Role of tissue inhibitor of metalloproteinases-2 (TIMP-2) in regulation of pro-gelatinase A activation catalyzed by membrane-type matrix metalloproteinase-1 (MT1-MMP) in human cancer cells. J Biochem 1998; 124(2):462-470.
46. Sato H, Seiki M. Regulatory mechanism of 92-kDa Type-IV collagenase gene expression which is associated with invasiveness of tumor cells. Oncogene 1993; 8(2):395-405.
47. Fini ME, Strissel KJ, Girard MT et al. Interleukin-1 alpha mediates collagenase synthesis stimulated by phorbol 12-myristate 13-acetate. J Biol Chem 1994; 269(15):11291-11298.
48. Bond M, Fabunmi RP, Baker AH et al. Synergistic upregulation of metalloproteinase-9 by growth factors and inflammatory cytokines: an absolute requirement for transcription factor NF-κB. FEBS Lett 1998; 435:29-34.
49. Pei D, Weiss SJ. Furin-dependent intracellular activation of the human stromelysin-3 zymogen. Nature 1995; 375:244-247.
50. Sato H, Takino T, Kinoshita T et al. Cell surface binding and activation of gelatinase A induced by expression of membrane-type-1-matrix metalloproteinase (MT1-MMP). FEBS Lett 1996; 385(3):238-240.
51. Cao JA, Rehemtulla A, Bahou W et al. Membrane type matrix metalloproteinase 1 activates pro-gelatinase A without furin cleavage of the N-terminal domain. J Biol Chem 1996; 271(47):30174-30180.
52. Nagase H. Activation mechanisms of matrix metalloproteinases. Biol Chem 1997; 378:151-160.
53. Van Wart HE, Birkedal-Hansen H. The cysteine switch: A principle of regulation of metalloproteinase activity with potential applicability to the entire matrix metalloproteinase gene family. Proc Natl Acad Sci USA 1990; 87:5578-5582.
54. Sato H, Takin T, Okada Y et al. A matrix metalloproteinase expressed on the surface of invasive tumour cells. Nature 1994; 370:61-65.
55. Takino T, Sato H, Shinagawa A et al. Identification of the second membrane-type matrix metalloproteinase (MT-MMP-2) gene from a human placenta cDNA library—MT-MMPs form a unique membrane-type subclass in the MMP family. J Biol Chem 1995; 270(39):23013-23020.
56. Strongin AY, Collier I, Bannikov G et al. Mechanism of cell-surface activation of 72 kDa type IV collagenase—Isolation of the activated form of membrane metalloprotease. J Biol Chem 1995; 270(10):5331-5339.
57. Willenbrock F, Murphy G. Structure-function relationships in the tissue inhibitors of metalloproteinases. Am J Respir Crit Care Med 1994; 150:S165-S170.

58. Strongin AY, Collier I, Bannikov BL et al. Mechanism of cell surface activation of 72 kDa type IV collagenase. J Biol Chem 1995; 270(5331-5338):5331-5338.
59. Murphy G, Willenbrock F, Ward RV et al. The C-terminal domain of 72kDa gelatinase A is not required for catalysis, but is essential for membrane activation and modulates interactions with tissue inhibitors of metalloproteinases. Biochem J 1992; 283:637-641.
60. Howard EW, Bullen E C, J BM. Regulation of the autoactivation of human 72-kDa progelatinase by tissue inhibitor of metalloproteinases-2. J Biol Chem 1991; 266:13064-13069.
61. Ross R. The pathogenesis of atherosclerosis: a perspective for the 1990s. Nature 1993; 362:801-809.
62. Davies MJ, Thomas AC. Plaque fissuring—The cause of acute myocardial infarction, sudden ischaemic death, and crescendo angina. Br Heart J 1985; 53:363-373.
63. Galis ZS, Sukhova GK, Lark MW et al. Increased expression of matrix metalloproteinases and matrix degrading activity in vulnerable regions of human atherosclerotic plaques. J Clin Invest 1994; 94:2493-2503.
64. Nikkari ST, Obrien KD, Ferguson M et al. Interstitial collagenase (MMP-1) expression in human carotid atherosclerosis. Circulation 1995; 92(6):1393-1398.
65. Stary HC, Chandler AB, Dinsmore RE et al. A definition of advanced types of atherosclerotic lesions and a histological classification of atherosclerosis: A report from the Committee on Vascular Lesions of the Council on Arteriosclerosis, American Heart Association. Circulation 1995; 92(5):1355-1374.
66. Falk E. Why do plaques rupture? Circulation 1992; 86:III-30-III-42.
67. Shah PK. Pathophysiology of plaque rupture and the concept of plaque stabilization. Cardiol Clin 1996; 14:17-29.
68. Arroyo LH, Lee RT. Mechanisms of plaque rupture: mechanical and biologic interactions. Cardiovasc Res 1999; 41:369-375.
69. van der Wal AC, Becker AE. Atherosclerotic plaque rupture—Pathologic basis of plaque stability and instability. Cardiovasc Res 1999; 41:334-344.
70. George SJ. Tissue inhibitors of metalloproteinases and metalloproteinases in atherosclerosis. Curr Opin Lipidol 1998; 9:413-423.
71. Newby AC, Zaltsman AB. Fibrous cap formation or destruction—The critical importance of vascular smooth muscle cell proliferation, migration and matrix formation. Cardiovasc Res 1999; 41:345-360.
72. Kockx MM, Knaapen MWM. The role of apoptosis in vascular disease. J Pathol 2000; 190:267-280.
73. Shah PK, Falk E, Badimon JJ et al. Human monocyte-derived macrophages induce collagen breakdown in fibrous caps of atherosclerotic plaques: Potential role of matrix-degrading metalloproteinases and implications for plaque rupture. Circulation 1995; 92(6):1565-1569.
74. Shapiro SD, Campbell EJ, Senior RM et al. Proteinases secreted by human mononuclear phagocytes. J Rheumatol 1991; 18(Suppl 27):95-98.
75. Galis ZS, Muszynski M, Sukhova GK et al. Cytokine-stimulated human vascular smooth muscle cells synthesize a complement of enzymes required for extracellular matrix digestion. Circ Res 1994; 75:181-189.
76. Nikkari ST, Geary RL, Hatsukami T et al. Expression of collagen, interstitial collagenase, and tissue inhibitor of metalloproteinases-1 in restenosis after carotid endarterectomy. Am J Pathol 1996; 148(3):777-783.
77. Henney AM, Wakely PR, Davies MJ et al. Localization of stromelysin gene expression in atherosclerotic plaques by in-situ hybridization. Proc Natl Acad Sci USA 1991; 88:8154-8158.
78. Halpert I, Sires UI, Roby JD et al. Matrilysin is expressed by lipid-laden macrophages at sites of potential rupture in atherosclerotic lesions and localizes to areas of versican deposition, a proteoglycan substrate for the enzyme. Proc Natl Acad Sci USA 1996; 93:9748-9753.
79. Brown DL, Hibbs MS, Kearney M. Identification of 92-kDa gelatinase in human coronary atherosclerotic lesions: Association of active enzyme synthesis with unstable angina. Circulation 1995; 91:2125-2131.
80. McMillan WD, Patterson BK, Keen RR et al. In situ localisation and quantification of mRNA for 92-kD type IV collagenase and its inhibitor in aneurysmal, occlusive, and normal aorta. Arterioscler Thromb Vasc Biol 1995; 15:1139-1144.
81. Schönbeck U, Mach F, Sukhova GK et al. Expression of stromelysin-3 in atherosclerotic lesions: Regulation via CD40-CD40 ligand signaling in vitro and in vivo. J Exp Med 1999; 189(5):843-853.
82. Sukhova GK, Schonbeck U, Rabkin E et al. Evidence for increased collagenolysis by interstitial collagenases-1 and -3 in vulnerable human atheromatous plaques. Circulation 1999; 99(19):2503-2509.
83. Rajavashisth TB, Xu XP, Jovinge S et al. Membrane type 1 matrix metalloproteinase expression in human atherosclerotic plaques—Evidence for activation by proinflammatory mediators. Circulation 1999; 99(24):3103-3109.
84. Li Z, Li L, Zielke R et al. Increased expression of 72-kd type IV collagenase (MMP-2) in human aortic atherosclerotic lesions. Am J Path 1996; 148:121-128.

85. Lee RT, Schoen FJ, Loree HM et al. Circumferential stress and matrix metalloproteinase 1 in human coronary atherosclerosis—Implications for plaque rupture. Arterioscler Thromb Vasc Biol 1996; 16(8):1070-1073.
86. Rouis M, Adamy C, Duverger N et al. Adenovirus-mediated overexpression of tissue inhibitor of metalloproteinase-1 reduces atherosclerotic lesions in apolipoprotein E-deficient mice. Circulation 1999; 100(5):533-540.
87. Prescott MF, Sawyer WK, Von Linden-Reed J et al. Effect of matrix metalloproteinase inhibition on progression of atherosclerosis and aneurysm in LDL receptor-deficient mice overexpressing MMP-3, MMP-12, and MMP-13 and on restenosis in rats after balloon injury. Ann NY Acad Sci 1999; 878:179-190.
88. Johnson J, Jackson C. Atherosclerotic plaque rupture in the apolipoprotein E knockout mouse. Atherosclerosis 2001; 154:399-406.
89. Cheng GC, Loree HM, Kamm RD et al. Distribution of circumferential stress in ruptured and stable atherosclerotic lesions. A structural analysis with histopathological correlation. Ciruclation 1993; 87:1179-1187.
90. Richardson PD, Davies MJ, Born GVR. Influence of plaque configuration and stress distribution on fissuring of coronary atherosclerotic plaques. Lancet 1989; 2:941-944.
91. Lendon CL, Davies MJ, Born GVR et al. Atherosclerotic plaque caps are locally weakened when macrophage density is increased. Atherosclerosis 1991; 87:87-90.
92. Libby P. Molecular bases of the acute coronary syndromes. Circulation 1995; 91:2844-2850.
93. Mason DP, Kenagy RD, Hasenstab D et al. Matrix metalloproteinase-9 over-expression enhances vascular smooth muscle cell migration and alters remodelling in the injured rat carotid artery. Circ Res 1999; 85:1179-1185.
94. Gerrity RG. The role of the monocyte in atherogenesis. Migration of foam cells from atherosclerotic lesions. Am J Pathol 1981; 103:191-200.
95. O'Brien KD, Allen MD, McDonald TO et al. Vascular cell adhesion molecule-1 is expressed in human coronary atherosclerotic plaques. Implications for the mode of progresssion of advanced coronary atherosclerosis. J Clin Invest 1993; 92:945-951.
96. Romanic AM, Madri JA. The induction of 72-kD gelatinase in T cells upon adhesion to endothelial cells is VCAM-1 dependent. J Cell Biol 1994; 125(5):1165-1178.
97. Madri JA, Graesser D, Haas T. The roles of adhesion molecules and proteinases in lymphocyte transendothelial migration. Biochem Cell Biol 1996; 74:749-757.
98. Mackarel AJ, Cottell DC, Russell KJ et al. Migration of neutrophils across human pulmonary endothelial cells is not blocked by matrix metalloproteinase or serine protease inhibitors. Am J Resp Cell Mol Biol 1999; 20:1209-1219.
99. Betsuyaku T, Shipley J, Liu Z et al. Neutrophil emigration in the lungs, peritoneum, and skin does not require gelatinase B. Am J Resp Cell Mol Biol 1999; 20:1303-1309.
100. Belaaouaj A, McCarthy R, Baumann M et al. Mice lacking neutrophil elastase reveal impaired host defence against gram negative bacterial sepsis. Nat Med 1998; 4:615-618.
101. Wesley RB, Meng X, Godin D et al. Extracellular matrix modulates macrophage functions characteristic to atheroma. Collagen type I enhances acquistion of resident macrophage traits by human peripheral blood monocytes in vitro. Arterioscler Thromb Vasc Biol 1998; 18:432-440.
102. Khan KMF, Falcone DJ. Role of laminin in matrix induction of macrophage urokinase-type plasminogen activitator and 92-kDa metalloproteinase expression. J Biol Chem 1997; 272:8270-8275.
103. Gudewicz PW, Frewin MB, Heinel et al. Priming of human monocyte superoxide production and arachidonic acid metabolism by adherence to collagen- and basement membrane-coated surfaces. J Leukoc Biol 1994; 55:423-429.
104. Pacifici R, Basilico C, Roman J et al. Collagen-induced release of interleukin 1 from human blood mononuclear cells: Potentiation by fibronectin binding to the alpha5beta1 integrin. J Clin Invest 1992; 89:61-67.
105. Rajagopalan S, Meng XP, Ramasamy S et al. Reactive oxygen species produced by macrophage-derived foam cells regulate the activity of vascular matrix metalloproteinases in vitro—Implications for atherosclerotic plaque stability. J Clin Invest 1996; 98(11):2572-2579.
106. Fabunmi RP, Baker AH, Murray EJ et al. Divergent regulation by growth factors and cytokines of 95 kDa and 72 kDa gelatinases and tissue inhibitors of metalloproteinases-1, -2 and -3 in rabbit aortic smooth muscle cells. Biochem J 1996; 315(Part 1):335-342.
107. Jones PA, Werb Z. Degradation of connective tissue matrices by macrophages. Influence of composition on proteolysis of glycoproteins, elastin, and collagen by macrophages in culture. J Exp Med 1980; 152:1527-1536.
108. Aggeler J, Frisch SM, Werb Z. Changes in cell shape correlated with collagenase gene expression in rabbit synovial fibroblasts. J Cell Biol 1984; 98:1662-1671.

109. Barger AC, Beeuwkes R, Lainey LL et al. Hypothesis: vasa vasorum and neovascularisation of human coronary arteries. Possible role in the pathophysiology of atheroscelerosis. N Engl J Med 1984; 310:175.
110. Zhang Y, Cliff WJ, Schoefl GI et al. Immunohistochemical study of intimal microvessels in coronary atherosclerosis. Am J Pathol 1993; 143(1):164-172.
111. Kumamoto M, Nakashima Y, Sueishi K. Intimal neovascularization in human coronary atherosclerosis: Its origin and pathophysiological significance. Hum Pathol 1995; 26(4):450-456.
112. Herron GS, Banda MJ, Clark EJ et al. Secretion of metalloproteinases by stimulated capillary endothelial cells. Expression of collagenase and stromelysin activities is regulated by endogenous inhibitors. J Biol Chem 1986; 261:2814-2818.
113. Schnapper HW, Grant DS, Stetler-Stevenson WG et al. Type IV collagenase(s) and TIMPs modulate endothelial cell morphogenesis in vitro. J Cell Physiol 1993; 156:235-246.
114. Johnson MD, Kim H-RC, Chesler L et al. Inhibition of angiogenesis by tissue inhibitor of metalloproteinase. J Cell Physiol 1994; 160:194-202.
115. Galargy RE, Grobelny D, Foellmer HG et al. Inhibition of angiogenesis by the matrix metalloprotease inhibitor N-[2R-2-(hydroxamidocarbonymethyl)-4-methylpentanoyl)]-c-tryptophan methylmaide. Cancer Res 1994; 54:4715-4718.
116. Fernandez HA, Kallenbach K, Seghezzi G et al. Inhibition of endothelial cell migration by gene transfer of tissue inhibitor of metalloproteinases-1. J Surg Res 1999; 82:156-162.
117. Brooks PC, Stromblad S, Sanders LC et al. Localization of matrix metalloproteinase MMP-2 to the surface of invasive cells by interaction with integrin alpha v beta 3. Cell 1996; 85(5):683-693.
118. Sunderkötter C, Steinbrink K, Goebeler M et al. Macrophages and angiogenesis. J Leuk Biol 1994; 55:410-422.
119. Zucker S, Conner C, DiMassmo BL et al. Thrombin induces the activation of progelatinases A in vascular endothelial cells. Physiologic regulation of angiogenesis. J Biol Chem 1995; 270:23730-23738.
120. Duhamel-Clérin E, Orvain C, Lanza F et al. Thrombin receptor-mediated increase of two matrix metalloproteinases, MMP-1 and MMP-3 in human endothelial cells. Arterioscler Thromb Vasc Biol 1997; 17:1931-1938.
121. Lee E, Grodzinsky AJ, Libby P et al. Human vascular smooth muscle cell-monocyte interactions and metalloproteinase secretion in culture. Arterioscler Thromb Vasc Biol 1995; 15:2284-2289.
122. Shapiro SD, Campbell EJ, Kobayashi DK et al. Immune modulation of metalloproteinase production in human macrophages. Selective pretranslational suppression of interstitial collagenase and stromelysin biosynthesis by interferon-γ. J Clin Invest 1990; 86:1204-1210.
123. Xie B, Dong Z, Fidler IJ. Regulatory mechanisms for the expression of type IV collagenases/gelatinases in murine macrophages. J Immunol 1994; 152(7):3637-3644.
124. Lacraz S, Nicod LP, Chicheportiche R et al. IL-10 inhibits metalloproteinase and stimulates TIMP-1 production in human mononuclear phagocytes. J Clin Invest 1995; 96(5):2304-2310.
125. Sasaguri T, Arima N, Tanimoto A et al. A role for interleukin 4 in production of matrix metalloproteinase 1 by human aortic smooth muscle cells. Atherosclerosis 1998; 138:247-253.
126. Satoh T, Sugama K, Matsuo A et al. Histamine as an activator of cell growth and extracellular matrix reconstruction for human vascular smooth muscle cells. Atherosclerosis 1994; 110:53-61.
127. Yanagi H, Sasaguri Y, Sugama K et al. Production of tissue collagenase (matrix metalloproteinase 1) by human aortic smooth muscle cells in response to platelet-derived growth factor. Atherosclerosis 1992; 91:207-216.
128. Pickering JG, Ford CM, Tang B et al. Coordinated effects of fibroblast growth factor-2 on expression of fibrillar collagens, matrix metalloproteinases, and tissue inhibitors of matrix metalloproteinases by human vascular smooth muscle cells—Evidence for repressed collagen production and activated degradative capacity. Arterioscler Thromb Vasc Biol 1997; 17(3):475-482.
129. Xu XP, Meisel SR, Ong JM et al. Oxidized low-density lipoprotein regulates matrix metalloproteinase-9 and its tissue inhibitor in human monocyte-derived macrophages. Circulation 1999; 99(8):993-998.
130. Aikawa M, Rabkin E, Okada Y et al. Lipid lowering by diet reduces matrix metalloproteinase activity and increases collagen content of rabbit atheroma: A potential mechanism of lesion stabilization. Circulation 1998; 97(24):2433-2444.
131. Malik N, Greenfield BW, Wahl AF et al. Activation of human monocytes through CD40 induces matrix metalloproteinases. J Immunol 1996; 156:3952-3960.
132. Mach F, Schönbeck U, Bonnefoy J-Y et al. Activation of monocyte/macrophage functions related to acute atheroma complication by ligation of CD40. Induction of collagenase, stromelysin, and tissue factor. Circulation 1997; 96:396-399.
133. Schönbeck U, Mach F, Sukhova GK et al. Regulation of matrix metalloproteinase expression in human vascular smooth muscle cells by lymphocytes. A role for CD40 signaling in plaque rupture? Circ Res 1997; 81:448-454.

134. Mach F, Schonbeck U, Sukhova GK et al. Reduction of atherosclerosis in mice by inhibition of CD40 signalling. Nature 1998; 394(6689):200-203.
135. Johnson JL, Jackson CL, Angelini GD et al. Activation of matrix-degrading metalloproteinases by mast cell proteases in atherosclerotic plaques. Arterioscler Thromb Vasc Biol 1998; 18:1707-1715.
136. Frears ER, Zhang Z, Blake DR et al. Inactivation of tissue inhibitor of metalloproteinase-1 b peroxynitrite. FEBS Lett 1996; 381:21-34.
137. Carmeliet P, Moons L, Lijnen R et al. Urokinase-generated plasmin activates matrix metalloproteinases during aneurysm formation. Nature Genetics 1997; 17:439-444.
138. Nagase H. Activation mechanisms of matrix metalloproteinases. Biol Chem 1997; 378:151-160.
139. Fabunmi RP, Sukhova GK, Sugiyama S et al. Expression of tissue inhibitor of metalloproteinases-3 in human atheroma and regulation in lesion-associated cells. A potential protective mechanism in plaque stability. Circ Res 1998; 83:270-278.
140. Galis ZS, Sukhova GK, Kranzhofer R et al. Macrophage foam cells from experimental atheroma constitutively produce matrix degrading proteinases. Proc Natl Acad Sci USA 1995; 92:402-406.
141. Zaltsman AB, George SJ, Newby AC. Increased secretion of tissue inhibitors of metalloproteinases-1 and -2 from the aortas of cholesterol fed rabbits partially counterbalances increased metalloproteinase activity. Arterioscler Thromb Vasc Biol 1999; 19:1700-1707.
142. Glagov S. Intimal hyperplasia, vascular modeling, and the restenosis problem. Circulation 1994; 89(6):2888-2891.
143. Campeau L, Enjalbert M, Lesperance J et al. Atherosclerosis and late closure of aortocoronary saphenous vein grafts: Sequential angiographic studies at 2 weeks, 1 year, 5 to 7 years and 10 to 12 years after surgery. Circulation 1983; 68:II-1s-II-7s.
144. Pauly RR, Passaniti A, Bilato C et al. Migration of cultured vascular smooth muscle cells through a basement membrane barrier requires type IV collagenase activity and is inhibited by cellular differentiation. Circ Res 1994; 75:41-54.
145. James TW, Wagner R, White LA et al. Induction of Collagenase and Stromelysin Gene Expression by Mechanical Injury in a Vascular Smooth Muscle-Derived Cell Line. J Cell Physiol 1993; 157(2):426-437.
146. Galis ZS, Kranzhofer R, Fenton JW et al. Thrombin promotes activation of matrix metalloproteinase-2 produced by cultured vascular smooth muscle cells. Arterioscler Thromb Vasc Biol 1997; 17(3):483-489.
147. Bendeck MP, Zempo N, Clowes AW et al. Smooth muscle cell migration and matrix metalloproteinase expression after arterial injury in the rat. Circ Res 1994; 75:539-545.
148. Zempo N, Kenagy RD, Au T et al. Matrix metalloproteinases of vascular wall cells are increased in balloon-injured rat carotid. J Vasc Surg 1994; 20:209-217.
149. Webb KE, Henney AM, Anglin S et al. Expression of matrix metalloproteinases and their inhibitor TIMP-1 in the rat carotid artery after balloon injury. Arterioscler Thromb Vasc Biol 1997; 17(9):1837-1844.
150. Tai JTN, Brooks EE, Liang S et al. Determination of temporal expression patterns for multiple genes in the rat carotid artery injury model. Arterioscler Thromb Vasc Biol 2000; 20:2184-2191.
151. Jenkins GM, Crow MT, Bilato C et al. Increased expression of membrane-type matrix metalloproteinase and preferential localization of matrix metalloproteinase-2 to the neointima of balloon-injured rat carotid arteries. Circulation 1998; 97:82-90.
152. Wang H, Keiser JA. Expression of membrane-type matrix metalloproteinase in rabbit neointimal tissue and its correlation with matrix-metalloproteinase-2 activation. J Vasc Res 1998; 35:45-54.
153. Bar-Shavit R, Benezra M, Sabbah V et al. Thrombin as a multifunctional protein: Induction of cell adhesion and proliferation. Am J Respir Cell Mol Biol 1992; 6:123-130.
154. Wang M, Hu YJ, Shima I et al. Identification of positive and negative regulator elements for the tissue inhibitor of metalloproteinase 1 gene. Oncol Res 1998; 10(4):219-233.
155. Forough R, Hasenstab D, Zempo N et al. Induction of tissue inhibitor of metalloproteinase-2 in injured rat artery. In: Greenwald RA, Golub LM, editors. Inhibition of matrix metalloproteinases: Therapeutic potential. NY Acad Sci 1994:384-388.
156. Hasenstab D, Forough R, Clowes AW. Plasminogen activator inhibitor type 1 and tissue inhibitor of metalloproteinases-2 increase after arterial injury in rats. Circ Res 1997; 80(4):490-496.
157. Dollery CM, Humphries SE, McClelland A et al. Expression of tissue inhibitor of matrix metalloproteinases 1 by use of an adenoviral vector inhibits smooth muscle cell migration and reduces neointimal hyperplasia in the rat model of vascular balloon injury. Circulation 1999; 99(24):3199-3205.
158. Clowes AW, Reidy MA, Clowes MM. Kinetics of cellular proliferation after arterial injury. Smooth muscle growth in the absence of endothelium. Lab Invest 1983; 49:327-333.
159. Zempo N, Koyama N, Kenagy RD et al. Regulation of vascular smooth muscle cell migration and proliferation in vitro and injured rat arteries by a synthetic matrix metalloproteinase inhibitor. Arterioscler Thromb Vasc Biol 1996; 16:28-33.

160. Bendeck MP, Irvin C, Reidy MA. Inhibition of matrix metalloproteinase activity inhibits smooth muscle cell migration but not neointimal thickening after arterial injury. Circ Res 1996; 78:38-43.
161. Lijnen HR, Van Hoef B, Lupu F et al. Function of the plasminogen/plasmin and matrix metalloproteinase systems after vascular injury in mice with targetted inactivation of fibrinolytic system genes. Arterioscler Thromb Vasc Biol 1998; 18:1035-1045.
162. Lijnen HR, Soloway P, Collen D. Tissue inhibitor of matrix metalloproteinase-1 impairs arterial neointima formation after vascular injury in mice. Circ Res 1999; 85:1186-1191.
163. Lovdahl C, Thyberg J, Cereck B et al. Antisense oligonucleotides to stromelysin mRNA inhibit injury-induced proliferation of arterial smooth muscle cells. Histol Histopathol 1999; 14(4):1101-1112.
164. Forough R, Koyama N, Hasenstab D et al. Overexpression of tissue inhibitor of matrix metalloproteinase-1 inhibits vascular smooth muscle cell functions in vitro and in vivo. Circ Res 1996; 79:812-820.
165. Cheng L, Mantile G, Pauly R et al. Adenovirus-mediated gene transfer of the human tissue inhibitor of metalloproteinase-2 blocks vascular smooth muscle cell invasiveness in vitro and modulates neointimal development in vivo. Circulation 1998; 98(20):2195-2201.
166. Strauss BH, Robinson R, Batchelor WB et al. In vivo collagen turnover following experimental balloon angioplastyinjury and the role of matrix metalloproteinases. Circ Res 1996; 79:541-550.
167. Southgate K, Fisher M, Banning A et al. Upregulation of basement membrane degrading metalloproteinase secretion after balloon injury of pig carotid arteries. Circ Res 1996; 79:1177-1187.
168. Soyombo AA, Angelini G, Bryan A et al. Intimal proliferation in an organ culture of human saphenous vien. Am J Path 1990; 137:1401-1414.
169. Angelini G, Bryan A, Williams H et al. Time-course of medial and intimal thickening in pig venous arterial grafts: Relationship to endothelial injury and cholesterol accumulation. J Thorac Cardiovasc Surg 1992; 103:1093-1103.
170. George SJ, Zaltsman AB, Newby AC. Surgical preparative injury and neointima formation increase MMP-9 expression and MMP-2 activation in human saphenous vein. Cardiovasc Res 1997; 33:447-459.
171. Southgate K, Mehta D, Izzat M et al. Increased secretion of basement membrane degrading metalloproteinases in pig saphenous vein into carotid artery interposition grafts. Arterioscler Thromb Vasc Biol 1999; 19:1640-1649.
172. Baker AH, Zaltsman AB, George SJ et al. Divergent effects of tissue inhibitor of metalloproteinase-1, -2, or -3 overexpression on rat vascular smooth muscle cell invasion, proliferation, and death in vitro—TIMP-3 promotes apoptosis. J Clin Invest 1998; 101(6):1478-1487.
173. Bond M, Murphy G, Bennett M et al. Localisation of the death domain of TIMP-3 to the N-terminus: Metalloproteinase inhibition is associated with pro-apoptotic activity. J Biol Chem 2000; 275:41358-41363.
174. George SJ, Johnson JL, Angelini GD et al. Adenovirus-mediated gene transfer of the human TIMP-1 gene inhibits smooth cell migration and neointimal formation in human saphenous vein. Hum Gene Ther 1998; 9:867-877.
175. George SJ, Baker AH, Angelini GD et al. Gene transfer of tissue inhibitor of metalloproteinase-2 inhibits metalloproteinase activity and neointima formation in human saphenous veins. Gene Ther 1998; 5:1552-1560.
176. George SJ, Lloyd CT, Angelini GD et al. Inhibition of late vein graft neointima formation in human and porcine models by adenovirus-mediated overexpression of tissue inhibitor of metalloproteinase-3. Circulation 2000; 101:296-304.
177. Leco KJ, Khokha R, Pavloff N et al. Tissue inhibitor of metalloproteinases-3 (TIMP-3) is an extracellular matrix-associated protein with a distinctive pattern of expression in mouse cells and tissues. J Biol Chem 1994; 269:9352-9360.
178. Duplaa C, Couffinhal T, Dufourcq P et al. The integrin very late antigen-4 is expressed in human smooth muscle cell—Involvement of alpha(4) and vascular cell adhesion molecule-1 during smooth muscle cell differentiation. Circ Res 1997; 80:159-169.
179. Dufourcq P. H. L, Moreau C, Daret D et al. Vitronectin expression and interaction with receptors in smooth muscle cells from human atheromatous plaque. Arterioscler Thromb Vasc Biol 1998; 18:168-176

Chapter 5

Cancer Invasion-Related Genes

Anja Bosserhoff and Reinhard Buettner

Introduction

This review provides a brief overview on gene families involved in invasion. Mechanistically, these molecules are involved in deregulation of adhesive interaction of tumor cells with each other and with extracellular matrices, in synthesis and activation of proteases and other enzymes, and in locomotion of tumor cells and organization of the cytoskeleton. Examples for these gene families and their general role in invasion are discussed.

Conventional therapy of malignant tumors is frequently limited by acquisition of an invasive and metastatic phenotype and progression to a systemic disease. Thus, understanding and manipulating molecular events leading to systemic spread represent major challenges for current cancer research. Invasion involves a highly regulated and coordinated cascade of complex molecular processes including cell attachment, cell detachment, secretion of proteases, cell migration and exchanging signals with other cells in the local milieu. While invasion is regarded as a key signature of malignant tumors, it is also found as part of the normal behavior of inflammatory blood cells, in tissues engaged in morphogenetic movements of normal embryogenesis and in a number of instances of normal and pathological tissue remodeling in the adult. Remarkably different genes involved in invasion have been identified during the past years. Although a few, such as the family of matrix metalloproteinases (MMPs), have been analyzed in some detail, most of them have been recognized only recently and probably many are yet to be discovered.

In light of this largely preliminary knowledge our review summarizes data on cellular and molecular mechanisms underlying the phenotypic characteristics of tumor cells that determine their invasive capability. Further aspects of metastasis, i.e., angiogenesis, will not be considered here.

Some of the genes listed in Table 1 are reviewed intensively in other Chapters of this book and will therefore not be described here. The broad family of matrix metalloproteinases (MMPs), their inhibitors and activators are covered in Chapters 1, 4, 7 and 10, and the plasminogen activator system of proteinases in Chapter 8. Chapters 2, 3 and 6 detail various aspects of integrin-mediated cell-matrix interaction. The other genes listed in Table 1 and their roles in cancer cell invasion are discussed in further detail.

Proteases

Cathepsins

Cathepsin B, H , D and L, all belonging to the family of cysteine endopeptidases, have been shown to participate in processes of tumor growth, vascularization, invasion and metastasis. In several different types of cancer, such as malignant melanomas and colorectal carcinomas, cathepsins were found to be upregulated and useful as prognostic markers.[1-3] Cathepsin B was shown to degrade extracellular matrix proteins, like collagen IV and laminin, and to activate the precursor form of urokinase plasminogen activator (uPA), perhaps thereby initiating an extracellular proteolytic cascade.[4]

Cell Invasion, edited by Jyrki Heino and Veli-Matti Kähäri. ©2002 Eurekah.com.

Table 1. Overview on gene families involved in tumor cell invasion and examples

Protein function	Members involved in invasion
Proteases	MMPs, cathepsins, tPA, uPA, transmembranous proteases
Nonproteolytic enzymes	Heparanase, hyaluronidase
Cell-cell adhesion molecules	Cadherins, GM2, GM3, ICAM-1, Sialyl Lewis X, galectin
Cell-matrix adhesion molecules	Integrins, CD44, laminin receptor
Matrix molecules	Fibronectin, laminin, collagens, thrombospondin
Growth factors	EGF, HGF (scatter factor), TGF-β, Wnt
Cytokines	IL-1, TNF-α, IL-6
Signal transduction molecules	PI3-K, PLC-γ, Ras, FAK, β-catenin
Cytoskeleton	Rho, Rac, Cdc 42, keratins, actins
Recently identified molecules with complex functions	MIA, TIAM, amphoterin, SPARC

Transmembranous Proteases

Aminopeptidase N/CD13 (143 kD) and dipeptidyl peptidase IV/CD26 are Zn^{2+}-dependent ectopeptidases localized on the cell surface of a wide variety of cells. They are involved in tumor cell invasion and the formation of metastases. Results show that both aminopeptidase N and dipeptidyl peptidase play an active role in degradation and invasion of ECM and may be involved in the molecular mechanisms of blood-borne metastasis.[5,6] The up-regulated invasion of cancer cells was inhibited by bestatin, a specific inhibitor of aminopeptidase N. Cellular migration correlated highly with aminopeptidase N activity. These findings suggest that aminopeptidase N expression contributes to the invasive potential of human cancer cells.[7]

Non-Proteolytic Enzymes

Heparanase

Endo-β-D-glucuronidase is commonly referred to as heparanase. Enzymatic targets of the 50kD heparanases (543 aa) are heparan sulfate proteoglycans (HSPGs) in the extracellular matrix. Expression of heparanase correlates with the invasive potential of tumor cell lines.[8,9] Further, treatment with heparanase inhibitors markedly reduces the frequency of metastases in experimental animal models of tumor metastasis.[10,11]

Hyaluronidase

Hyaluronidases are broadly distributed enzymes with varying substrate specificities, a wide range of pH optima and different catalytic mechanisms.[12] One of their substrates, hyaluronic acid, represents a major component of the extracellular matrix in brain[13] and soft tissues and has been identified as the ligand of CD44. Disruption of basement membrane integrity by hyaluronidase during cell invasion has been implicated in the development of metastatic carcinoma. The amount of hyaluronidase expression correlates with prognosis in a variety of different cancer types.[14-16] Further, experimental overexpression of hyaluronidase in tumor cells leads to accelerated tumor growth and formation of highly vascularized and more invasive tumors.[17-19]

Cell-Cell Adhesion Molecules

Cadherins

E-cadherin and its associated catenin complexes have been identified as key molecules in cell adhesion. The transmembrane E-cadherin proteins form calcium-dependent homodimers via extracellular binding and concomitantly transduce signals from the micro-environment to other molecular complexes implicated in invasion. Both inherited and somatically acquired E-cadherin mutations have been identified to play important roles in human cancer. E-cadherin inactivating mutants were found in breast, colon and gastric carcinomas, and deregulation of E-cadherin expression was reported to occur in malignant melanomas[20,21] (Fig.1). Overexpression of E-cadherin significantly decreased the metastatic capacity of tumor cells measured in in vitro and in vivo experiments,[22] indicating that levels of E-cadherin expression may represent critical parameters for the invasive and metastatic tumor phenotype.

CAMs

Members of the immunoglobulin superfamily of adhesion molecules, including ICAM-1, ICAM-2, ICAM-3, VCAM-1 and MadCAM-1 (Muc-18), bind to integrins on leukocytes and mediate their flattening onto the blood vessel wall. This interaction is critical for subsequent extravasation and the further course of inflammatory reactions in the surrounding tissue. Interestingly, tumor cells have been shown to use similar mechanisms for extravasation during metastasis.[23-26]

GMs

Gangliosides are glycosphingolipids that are widely distributed in vertebrate tissues and body fluids and are abundant in neural tissues, where they have been associated with development and maturation of the brain, neuritogenesis, synaptic transmission, memory formation and synaptic aging. They are overexpressed on tissues of neuroectodermal origin and particularly in tumors such as melanomas, sarcomas, neuroblastomas, astrocytomas, and small cell lung cancers. Today, many data suggest that some of the effects exerted by gangliosides are due to interactions with proteins that participate in the transduction of signals through the membrane in caveolae and caveolae-like membrane microdomains. Monoclonal antibodies produced in several laboratories have defined at least five different gangliosides: GM3, GM2, GD2, GD3, 9-O-acetyl GD3. They were shown to be involved in promoting cell invasion through the basement membrane and to regulate cell attachment to matrix proteins.[27-29]

Sialyl Lewis A, X and Other Highly Glycosylated Structures

An increasing body of evidence suggests that changes in N-glycosylation of tumor cell proteins such as branching, increased sialylation, polysialylation, decreased fucosylation and enhanced formation of Lewis X, sialyl Lewis X (sLe(x)) and sialyl Lewis A antigens represent important parameters for measuring invasive potential of tumor cells.[30] Different epithelial tumors have been shown to express the entire enzymatic machinery for the synthesis of sLe(x). Thereby, adhesion to endothelial E- and P-selectin and extravasation from the blood via trafficking through the endothelium, important steps in hematogenous spread, are facilitated.[31-33] Intriguingly, the level of expression of carbohydrate surface ligands in cancer cells coincides with the frequency of metastasis and survival of cancer patients. Therefore current investigations of the molecular mechanisms leading to alteration of glycosyltransferase activities and enhanced expression of the surface carbohydrate ligands may provide novel insights into molecular carcinogenesis.[34-37]

Galectin

Galectins represent a family of mammalian lectins with specificity to β-galactosides, e.g., the poly-N-acetyllactosamine residue of laminin.

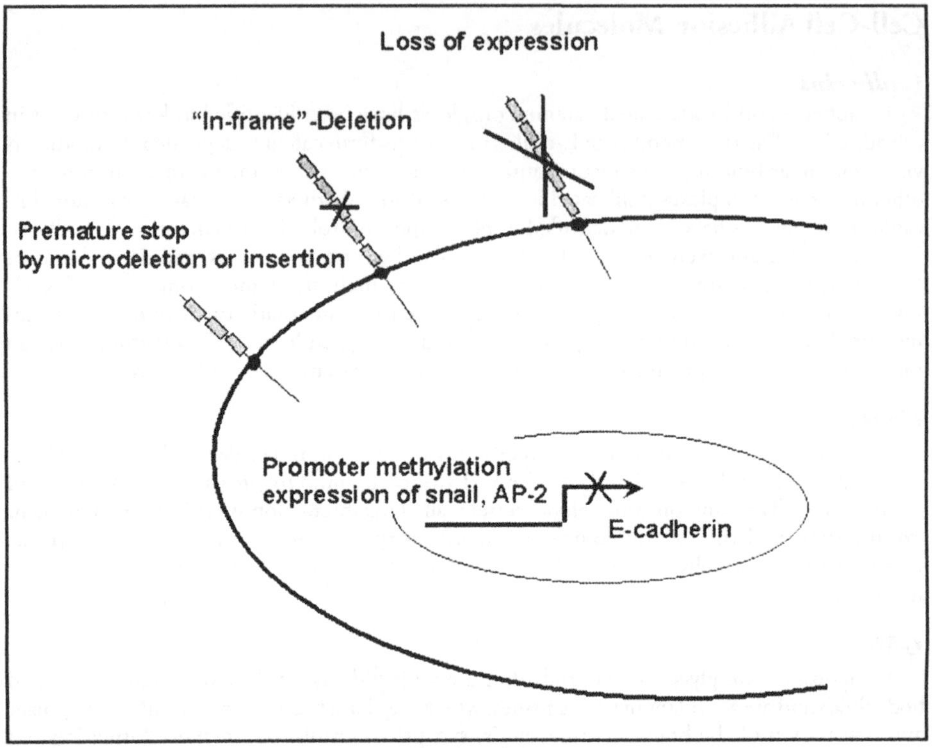

Fig. 1. Molecular changes leading to loss of functional E-cadherin expression in tumor cells. See section on Cadherins for details.

Galectins are involved in growth-regulatory mechanisms but also in cell-cell and cell-matrix adhesion. It has been suggested that the level of galectin expression, mainly galectin-3, parallels malignancy in different cancer types.[38-40] In vitro studies suggested that galectin-3 protects cancer cells from apoptosis by promoting cell adhesion properties[41,42] and also modulates the invasive capacity.

Cell-Matrix Adhesion Molecules

CD44

The main form of CD44 (CD44H), a transmembrane 80 kD glycoprotein, is widely expressed in a variety of lymphoid and epithelial cells and also in malignant tumors. CD44 has many variant forms (at least 45 are known today), which are generated by alternative splicing. During the past years, expression of certain variant CD44 isoforms has been correlated with the degree of tumor differentiation, tumor cell invasion and metastatic potential.[43,44] In model systems interaction of CD44 and hyaluronan was a critical determinant for cell adhesion and transendothelial invasion.[23] Other studies provided evidence for proteolytic cleavage of CD44 on the surface of cancer cells at the extracellular domain through membrane-associated metalloproteases. It has been suggested that CD44 cleavage plays an important role in CD44-mediated tumor cell migration[45] and in efficient cell-detachment from the hyaluronate substrate within extracellular matrix.

Laminin Receptor

The 67 kD laminin receptor belongs to the group of nonintegrin adhesion molecules and interacts with the YIGSR sequence in the β1 chain of laminin. Expression of the receptor correlates with invasive potential of malignant melanoma cells, which can be inhibited by the YIGSR amino acid peptide in vitro.[46,47]

Matrix Molecules

Fibronectin

The 400-500kD homodimeric fibronectin protein represents an important component of extracellular matrix and basement membranes. In vitro, fibronectin is chemoattractive for many different types of cancer cells,[48] and overexpression of fibronectin is frequently detected in invading areas of malignant tumors in vivo.[49] Further, an effect of fibronectin expression on enhanced secretion of proteases, including uPA, was observed in vitro.[50] As this effect was dependent on RGD sequences and β1-integrin expression, cell-matrix interactions seem to trigger enhanced protease secretion. Fibronectin was also suggested to be involved in cohort migration, a phenomenon relating to the observation that carcinoma cells frequently invade the surrounding tissue as coherent clusters or nests of cells.[51]

Laminin

The basement membrane molecule laminin forms a family of proteins. Laminin is a cross shaped multifunctional glycoprotein formed by the multimeric assembly of subunits which result from activation of several genes. In vivo, depending on its location, because of its adhesive properties and multivalent affinities, laminin is in association as a part of supramolecular complexes together with compounds of the plasma, the basement membrane and the cell coat. In the basement membrane laminin has structural and functional roles. It may also be adsorbed on the cell coat or secreted. Laminin-5 was identified as a key protein in the anchoring filaments of the basement membrane. The anchoring filaments connect the basement membrane to the epithelial cells and together form the epithelial adhesion complex. Laminin has been implicated in a number of stages in tumor invasion and metastasis. In addition to its roles in cell adhesion and migration, laminin may be important in mediating interactions of tumor cells with the immune system and have more subtle roles in controlling metastatic behavior. Tumor invasion through the basement membrane has been correlated with a decreased expression of laminin. Laminin stimulates migration of several tumor cells tested[52] and laminin-5 is specifically degraded by MMP-2 to produce a soluble chemotactic fragment.[53]

Proteoglycans

Abnormal expression of proteoglycans, highly glycosylated matrix molecules, has been implicated in cancer and metastasis primarily because these macromolecules are involved in control of cell growth and matrix assembly. Significant upregulation of perlecan expression, a major heparan sulfate proteoglycan of basement membranes and pericellular matrices, was detected in melanomas and other malignant tumors.[54] Perlecan was also shown to be upregulated by the growth factor neutrophin, suggesting a further role in early steps of invasion.

Growth Factors

Extracellular signaling by growth factors has been demonstrated to elicit deregulated cell motility in invasive tumor cells. Recent findings suggest that growth factor receptor-mediated motility is one of the most common alterations in malignant tumors, which is causally involved in acquisition of an invasive cell phenotype and represents a biological phenomenon distinct from adhesion-related haptokinetic and haptotactic migration.[55] In addition, extracellular growth factor signaling was also shown to elicit upregulation of MMP expression in malignant tumors.

EGF and TGF-α

Transforming growth factor alpha (TGF-α) is a polypeptide growth factor sharing significant structural and functional homology with epidermal growth factor (EGF). TGF-α is synthesized as a 160 amino acid transmembrane precursor molecule from which the mature 50 amino acid TGF-α peptide is released following proteolytic cleavage by specific elastase-like enzymes. All biological properties of TGF-α and EGF are thought to be elicited by binding of the proteolytically processed ligands to transmembrane tyrosine kinase-coupled receptors of the EGF receptor family. Thus, the biochemical cascade following activation of EGF receptor tyrosine kinase is likely to be identical in response to either ligand. However, heparin-binding EGF-like growth factor (HB-EGF), an additional EGF family member, represents a much more potent mitogen for smooth muscle cells than either EGF or TGF, despite the fact that all three ligands bind the same receptor. Epidermal growth factor (EGF) was shown to stimulate tumor cell migration in nanomolar concentrations. This effect involved upregulation of MMP-2 and uPA expression,[56,57] which seems to be mediated by a phosphatidylinositol 3'-kinase and phospholipase C-dependent mechanism.[58] Signaling of EGF to migratory pathways appears to be dissociated, at least in part, from the proliferative pathway and therefore, EGF can elicit different biological responses depending on the cellular context and further downstream signal transduction cascades.

HGF

Hepatocyte growth factor (HGF), also referred to as scatter factor (SC) or hepatopoietin, was originally identified as a pleiotropic growth factor from rat platelets. Mature HGF is an 82 kD heterodimeric glycoprotein of 674 amino acids that lacks significant homology with other known growth factors. The transmembrane tyrosine kinase-coupled receptor c-Met functions as the HGF-receptor. HGF is synthesized as a 728 amino acid precursor protein and proteolytically processed into the mature growth factor consisting of a disulfide-linked 69 kD chain and a 34 kD chain. Signaling through the HGF/c-Met system elicits a cellular response linked in several ways to invasive growth behavior. HGF was shown to promote cell adhesion to laminin, fibronectin and vitronectin through a PI3-K-dependent mechanism.[59] Subsequently, increased adhesion induced by HGF is followed by increased invasiveness through these matrix proteins. This phenomenon seems to be regulated by triggering integrin-clustering[60] and also by upregulation of MMP and uPA expression.[61] Further, elevation of tyrosine-phosphorylated ß-catenin after HGF treatment was detected.[62]

TGF-β

The TGF superfamily comprehends a large number of polypeptide growth and differentiation factors, including transforming growth factors beta (TGF-β) 1, 2, 3 and further growth/differentiation factors (GDFs), Mullerian inhibiting substance (MIS), bone morphogenic proteins (BMPs), glial cell line-derived neurotrophic factor (GDNF), inhibins or activins, Lefty and Nodal. Members of the TGF superfamily are involved in embryonic development and adult tissue homeostasis. Growth regulatory factors of the TGF-β family inhibit proliferation of epithelial, endothelial and hematopoietic cells and stimulate synthesis of extracellular matrix components. Recent evidence suggests that acquisition of resistance to TGF-β-conferred growth inhibition plays a significant role in progression of malignant epithelial and hematopoietic tumors. Under these circumstances secretion of TGF-βs by tumor cells may promote invasion rather than inhibit growth.[63] Consistently, upregulation of MMP-2 and MMP-9 by TGF-β1 has been observed in different kinds of cancers.[57]

Wnt

Products of the highly conserved Wnt gene family, including Wnt-1 through Wnt-10, play key roles in regulating cellular growth and differentiation. Wnt-1 is a cysteine-rich, secreted glycoprotein that associates with cell membranes and likely functions as a key regulator of cellular adhesion. Wnt-1, which is essential for normal development of the embryonic nervous system, contributes to hyperplasia and tumorigenic progression when improperly expressed in mammary tissue.[64,65] Wnt-3 is also involved in tumorigenesis and Wnt-2, Wnt-4 and WNT-5a may be associated with abnormal proliferation in human malignancies.[66,67] Wnt-10b has been implicated along with FGF-3 in the development of mouse mammary tumor virus-induced breast cancers.

Cytokines

Similar to their response to growth factors, invasive tumor cells have been demonstrated to reveal deregulated cell motility in response to extracellular signals from cytokines. Since cytokines represent polypeptide, frequently soluble molecules that interact with specific receptors on the cell surface, the distinction between growth factors and cytokines is largely arbitrary and reflects the discovery of cytokines as modulators of immune responses.

Interleukins

Interleukins comprehend a broad family of well characterized cytokines, primarily of hematopoietic cell origin. For example, IL-2 is secreted primarily by mitogen-activated T helper lymphocytes and functions as an autocrine growth factor, driving the clonal expansion of antigen specific cells. In contrast, IL-7 primarily stimulates proliferation of preB cells while IL-13 is a potent regulator of inflammatory and immune responses and acts to ensure the rapid onset of a Th2-like response. IL-15 is produced by a broad range of tissues and cell types and shares many of its biological properties with IL-2. Four distinct interleukins, IL-2, IL-4, IL-7 and IL-15, functionally interact with the common IL-2R receptor subunit. Additional members of the interleukin family include IL-16, IL-17 and IL-18 (also designated IGFI). IL-1 receptor antagonist (IL-1ra) is a cytokine that inhibits IL-1 binding to interleukin receptors.

Cytokines such as interleukin 1 induce the synthesis and surface expression of ICAM-1 and ELAM-1 adhesion molecules and contribute to autocrine and paracrine induction of pro-metastatic genes in cancer.[68] Further studies in vitro revealed that the invasiveness of tumor cells can be enhanced by addition of recombinant IL-1 and reduced by adding inhibitory anti-IL-1 antibodies.[69] Interleukin 8 expression was shown to be enhanced in many different solid tumors including malignant melanomas[70] and to promote tumorigenicity and metastasis in prostate[71] and breast cancer.[72] Specific upregulation of MMP-9 and MMP-2 expression and activation was described as a consequence of IL-8 overexpression resulting in increased invasion through Matrigel.[71,73] Also enhanced levels of interleukin 6 expression correlate with tumor progression[74,75] and upregulation of MMP-2, MMP-9 and TIMP-1.[76] In the case of interleukin 10 a dose-dependent stimulating effect on glioma invasiveness in Boyden chamber assays was measured. However, this effect is mechanistically poorly understood since MMP expression levels remained unchanged after IL-10 treatment, but Marimastat, a synthetic MMP-inhibitor, markedly reduced IL-10 stimulated invasiveness.[77]

TNF-α

The proinflammatory polypeptides tumor necrosis factor alpha (TNF-α) (also designated cachectin) and TNF-β (also designated lymphotoxin) exhibit approximately 30% sequence homology and bind to a common receptor complex. TNF is produced as a precursor, which exists as a membrane-bound form, from which the soluble mature factor is derived by cleavage of the extracellular domain. The biologically active forms of both TNF-α and TNF-β are noncovalently linked trimers. TNF-α may be involved in the enhancement of tumor invasion and metastasis in part by upregulating the proteolytic enzymes such as MMPs and uPA in tumor tissues.[78] Data suggest that TNF-α tightly regulates gelatinase B secretion in glioma cells, an enzyme which is believed to play an important role in the local invasion of brain tissue by tumor cells[79] and further, that TNF-α stimulates 92 kD gelatinase secretion in myeloblastic leukemia cell invasion.[80] The same was shown for several other types of cancer.

Intracellular Molecules of Signal Transduction Pathways

PI3-Kinase

The prototype member of the phosphatidylinositol kinase family, PI3-kinase, is composed of a 110 kD catalytic and an 85 kD regulatory subunit (p85 alpha or p85 beta). The 85 kD subunit lacks catalytic activity and functions as an adapter protein with one SH3 and two SH2 domains. It has been established that PI3-kinase p110 represents a family of related proteins, including four distinct PI3-kinase catalytic subunits: p110alpha, p110beta, p110gamma and p110delta. Related proteins include the catalytic subunit of PI4-kinase, FRAP (FKBP12-rapamycin-associated protein; also named mTOR or RAFT1), ATM (ataxia telangiectasia mutated protein), FRP1 (FRAP-related protein) and DNA-PK (DNA-dependent protein kinase). In contrast, the PI(4)P5-kinases form a separate nonhomologous family of proteins. In addition to the expanding family of PI3-kinase catalytic subunits and related proteins, a new PI3-kinase regulatory subunit, PIK p55, has been described recently. PIK p55 is composed of a unique 30-residue NH2 terminus, followed by a proline-rich motif and two SH2 domains with significant sequence homology to the p85 SH2 domains. Inhibition of PI3-kinase activity by overexpression of a dominant–negative mutant or by incubation with Wortmannin inhibited invasion of breast cancer and hepatoma cells.[81,82] Further evidence for a role of PI3-kinase in HGF/c-Met-mediated invasiveness (VIII.2)[83,84] indicates an important function of PI3-kinase activation for tissue remodeling and invasive growth of cancer cells.

PLC-gamma

Phospholipase C gamma 1 (PLC gamma 1), a phosphatidylinositol-4,5-bisphosphate (PI-4,5-P-2)-specific phosphodiesterase has been shown to play a critical role in the initiation of receptor-mediated signal transduction by generating the second messengers, inositol 1,4,5-trisphosphate and diacylglycerol from phosphatidylinositol-4,5-bisphosphate. So far eight different mammalian PLC isozymes have been identified (PLC beta 1, PLC beta 2, PLC beta 3, PLC beta 4, PLC gamma1, PLC gamma 2, PLC delta 1 and PLC delta2) with molecular weights ranging from 85 to 150 kD. The gamma-type enzymes are unique in that they contain SH2 and SH3 domains. Moreover, the two gamma-type enzymes, but not the beta and delta isozymes, are subject to activation by phosphorylation in response to downstream signaling from a number of different protein tyrosine kinases which interact directly with the PLC gamma SH2 domains. In contrast, activation of PLC beta 1, PLC beta2 and PLC beta 3 is mediated by the alpha units of the G-q class of heterotrimeric G proteins and by certain G protein subunits. Detailed studies using a prostate cancer model have revealed that an EGFR-PLC gamma-mediated motility-associated signaling pathway is rate-limiting for tumor cell invasion in vitro and in vivo.[85] Inhibition of PLC gamma signaling either by specific pharmacological inhibitors or expression of dominant-negative PLC constructs decreased invasion.[86,87]

PKC

Members of the protein kinase C (PKC) family play a key regulatory role in a variety of cellular functions including cell growth and differentiation, gene expression, and hormone responses. PKCs were originally identified as protein serine/threonine kinases whose activities are dependent on free cellular calcium and phospholipids. PKCs can be subdivided into two major classes, including conventional (c) PKC isoforms (alpha, betaI, betaII and gamma) and novel (n) isoforms (sigma, delta, zeta, lambda). Patterns of expression for each PKC isoform differ among tissues and PKC family members exhibit clear differences in their cofactor dependencies. Many studies have shown that activities of protein kinase C (PKC) isoenzymes contribute significantly to progression of malignant cell phenotypes. An important role of PKC isoenzymes is assumed in regulation of cell attachment, matrix degradation and migration. PKC overexpression was shown to lead to enhanced migration and invasion in vitro, whereas inhibition of PKC activity in tumor cells elicited reduced invasiveness.[88-90]

FAK

Integrins are cell surface molecules which promote adhesion of cells to the extracellular matrix. In addition to providing a molecular "glue" essential for tissue organization and survival, integrins are dynamic signaling molecules. They allow cells to sense adhesion to the extracellular matrix and the nature of the specific matrix composition. Thereby, a cell survival signal is provided preventing a particular form of cell death in detached cells which has been referred to as anoikis.[91] An important mediator of this signal is the focal adhesion kinase (FAK), which becomes phosphorylated and activated during integrin-mediated cell adhesion. FAK was initially identified as a major 125 kD substrate for the intrinsic protein tyrosine kinase activity of Src encoded pp60. Visualization of p125 by immunofluorescence revealed its preferred subcellular localization in focal adhesions, leading to its designation as focal adhesion kinase (FAK). FAK is concentrated at the basal edge of focal contacts of cells that are actively migrating and spreading.

FAK was shown to be frequently overexpressed in many different kinds of malignant tumors such as colon, breast and prostate cancers.[92-94] FAK overexpression causes increased cell migration whereas cells derived from FAK -/- mice exhibit reduced migratory capabilities as compared to wild type cells. FAK is thought to exert a dual function with respect to invasion: Overexpression leads to increased cell migration and increased cell survival under conditions of anchorage-independent growth.[95] FAK also seems to be important for enhanced MMP-9 expression after exposure of ovarian cancer cells to fibronectin.[96]

β-catenin

The catenins (alpha, beta and gamma) are ubiquitously expressed, cytoplasmic proteins that associate with E-cadherin at cellular junctions. The p85 kD protein β-catenin was initially described as an E-cadherin associated protein, but subsequent studies have shown that it can also bind P-cadherin and N-cadherin. In addition to its ability to bind to cadherins, β-catenin has been shown to coimmunoprecipitate with APC. β-catenin is a major component of desmosomes, where it is complexed with desmoglein. Increased tyrosine phosphorylation can disrupt catenin-cadherin complexes and thereby attenuate cellular adhesion. β-catenin belongs also to the important downstream effectors of the Wnt signaling pathway[97] and has been implicated in two major biological processes of vertebrates: early embryonic development and tumorigenesis.[98,99] GSK-3β and possibly other, yet unidentified kinases destabilize β-catenin by phosphorylation at serine 33, 37, 45 and threonine 41.[100] Stabilizing mutations of these phosphorylation sites in β–catenin were detected in many tumor cell lines in vitro[101] and subsequently in malignant tumors in vivo.[102] In addition to binding to cadherins, beta-catenin also interacts with transcription factors of the TCF-subfamily of HMG box proteins and regulates their activity. Elevated levels of β-catenin, sometimes concentrated in the invasive front, have been described in several kinds of cancer[103] and β-catenin was also shown to be involved in MMP-7 upregulation in colorectal cancer.[104]

Ras

Ha-, Ki- and N-Ras genes encode three highly homologous 21 kD guanine nucleotide-binding proteins. Ha-Ras and Ki-Ras were first identified as oncogenes in the Harvey and Kirsten retrovirus strains. Subsequently, mutated cellular *Ras* genes have been found in many human tumors, most frequently in pancreatic cancers, providing evidence for a common genetic defect. Ras-encoded proteins bind to GDP and to GTP with high affinity and possess a low level of intrinsic GTPase activity that can be stimulated more than 100-fold by interaction with cytosolic GTPase activating protein (GAP). In mammals, a variety of extracellular growth factors that act through protein tyrosine kinase receptors, such as insulin, platelet-derived growth factor, epidermal growth factor and nerve growth factor, require Ras proteins as part of the cytoplasmic signal transduction pathway linking extracellular signals with regulation of gene expression in the nucleus. Ras p21 in its active GTP binding state binds to Raf-1, resulting in downstream activation of the MAP kinase signaling cascade. An additional member of the Ras family, Rheb, also interacts with Raf-1. This interaction is potentiated by growth factors and agents that increase cAMP levels. Activation of Ras and phosphatidylinositol-3-kinase through the multifunctional docking site is required for receptor-mediated invasive growth. In a number of malignant tumors Ras family members are mutated, amplified or overexpressed. Oncogenically activated members confer transforming, invasive and metastatic properties to normal cells.[105]

Cytoskeleton

Over the past years an increasing number of genes have been identified that modify the cytoskeleton and play an active role in cell motility and invasion. Oncogenic signaling can also be significantly influenced by cytoskeletal regulation. Thus, the cytoskeleton has evolved from a static structure to a dynamic modulator of signal transduction pathways.

Keratins

The cytokeratins are intermediate filament proteins characteristic of epithelial cells. In human cells, some 20 different cytokeratin isotypes have been identified. Epithelial cells express between two and ten cytokeratin isotypes and the consequent profile which reflects both epithelial type and differentiation status may be useful in tumor diagnosis. In some cancers, particularly malignant melanoma and breast carcinoma, there is a strong indication that vimentin and keratin IFs are coexpressed, thus presenting as a dedifferentiated or interconverted (between epithelial and mesenchymal) phenotype.[106-109] These studies have provided direct evidence linking overexpression of keratin IFs in human melanoma and breast carcinoma with increased migratory and invasive activity in vitro,[110] which can be down-regulated by substituting dominant-negative keratin mutants.[111,112] Further, cytokeratin 19 was described to be lost in invasive squamous cell carcinoma.[113]

Rho, Rac

The Ras superfamily of GTPases comprises several subfamilies of small GTP-binding proteins whose functions include control of proliferation, differentiation, apoptosis, but also cytoskeleton organization. As the multistep process of invasion requires coordinate activation of migration, motility and cytoskeletal reorganization, the small GTPases Rho, Cdc42 and Rac (Rho-like GTPases) play a major role in this ability. Rho-like GTPases have been implicated in the orchestration of changes in the actin cytoskeleton in response to receptor stimulation, but have also been shown to be involved in transcriptional activation and cell cycle regulation.[114-116] Moreover, they can induce oncogenic transformation in fibroblast cells. RhoC was recently shown to enhance metastasis when overexpressed in cells[117] whereas a dominant-negative Rho inhibits metastasis. Consistently, members of the Rho family are very frequently overexpressed in breast cancers and malignant melanomas.

Among several proteins isolated as putative target molecules of Rho, the ROCK (ROK) family of Rho-associated serine-threonine protein kinases are thought to participate in the induction of focal adhesions and stress fibers in cultured cells. A specific ROCK inhibitor (Y-27632) blocked both Rho-mediated activation of actomyosin and invasive activity of cells ROCK also seems to play a part in tumor cell invasion.[118]

Recently Identified Genes with Complex Functions

MIA

The melanoma inhibitory activity (MIA) protein was identified within growth-inhibiting activities purified from tissue culture supernatant of the human melanoma cell line HTZ-19.[119] It is translated as a 131 amino acid precursor molecule and processed into a mature 107 amino acid protein after cleavage of a secretion signal. MIA provides a clinically useful parameter in

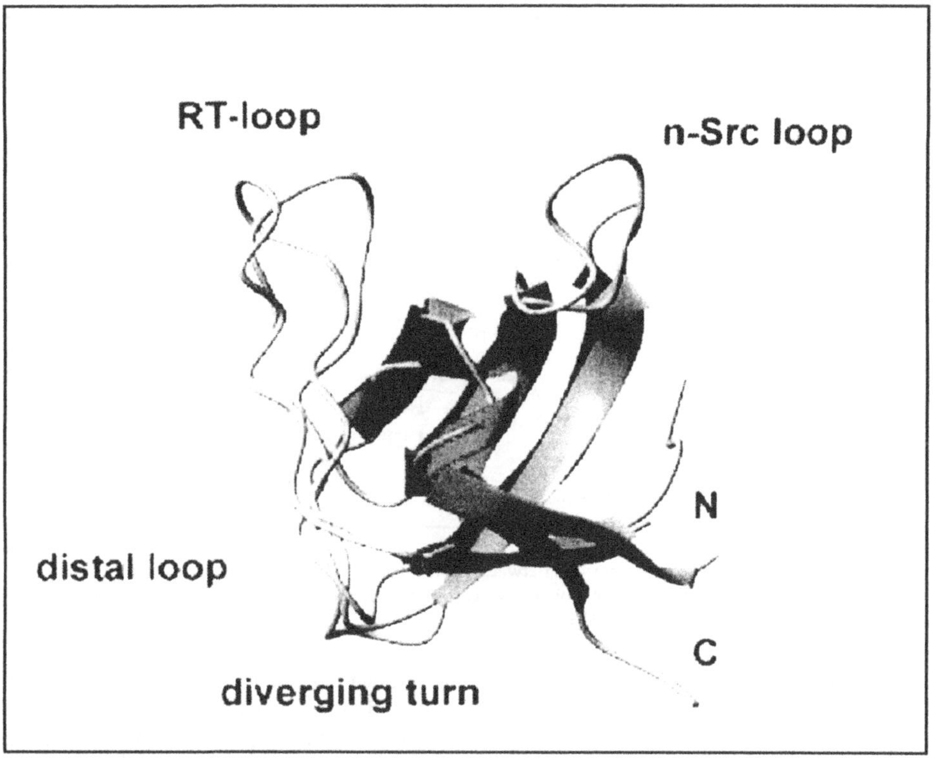

Fig. 2. MIA 3D structure. MIA represents the first extracellular protein adopting a SH3 domain-like structure. The structure is stabilized by two disulfide bonds. [EMBO 20:340-349 (2001), Figure 6. 2; By permission of Oxford University Press.]

patients with metastatic melanoma stages III and IV.[120-123] MIA was described to elicit antitumor activity by inhibiting proliferation of melanoma cell lines in vitro.[119,124] However, further studies have revealed expression patterns inconsistent with a tumor suppressor. Expression of the wild-type MIA protein gene was not detected in normal skin and melanocytes but was associated with progression of melanocytic tumors.[120,125] More recently, it was suggested that the MIA protein specifically inhibits attachment of melanoma cells to fibronectin and laminin, thereby masking the binding site of integrins to these extracellular matrix (ECM) components and promoting invasion and metastasis in vivo.[120,126,127] Thus, the growth-inhibitory activity in vitro reflects the ability of the protein to interfere with the attachment of cell lines to the surface of tissue culture dishes in vitro.[119] The 3D structure of recombinant human MIA in solution was determined recently by multidimensional NMR spectroscopy and revealed that MIA is the first extracellular protein known to adopt an SH3 domain-like fold (Fig. 2).[128] These studies also provided evidence of specific interaction between a binding fold of MIA and a partial fibronectin peptide that has been implicated in integrin binding. It appears that MIA belongs to a growing family of proteins that promote invasion and metastasis by inhibiting specific interactions between integrins and ECM molecules within the local tumor milieu.

TIAM

Studying an in vitro model system of T-cell lymphoma invasion and metastasis identified a novel protein, Tiam1, which shares a Dbl-homology domain with GDP dissociation stimulator (GDS) proteins. In the in vitro system Tiam1 enhanced invasion and metastasis of mouse T-

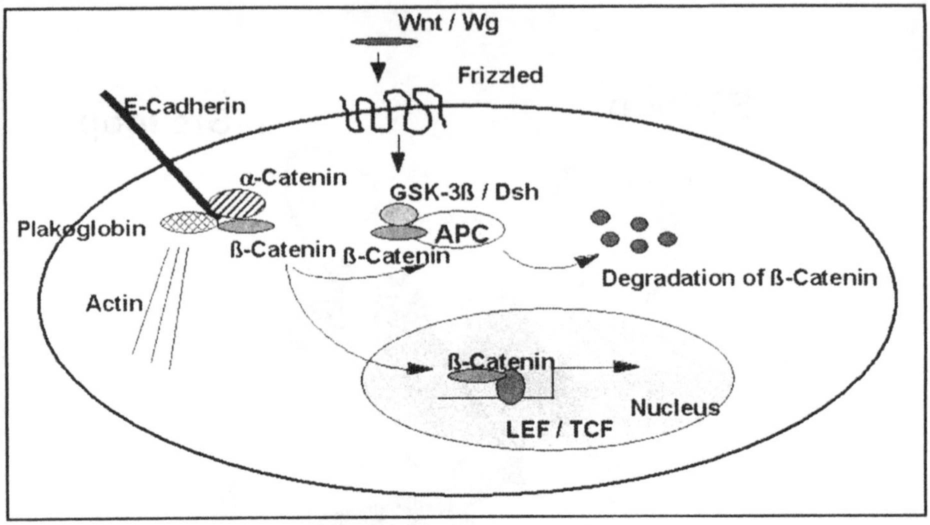

Fig. 3. Schematic overview on interactions between molecules involved in E-cadherin signaling pathways All molecules discussed in this review have to be seen in a close context of interactions. See sections on cadherins, Wnt and β-catenin for details.

cell lymphoma cells.[114] Subsequent investigations revealed that Tiam1 functions as a guanine nucleotide exchange factor for the Rho-like GTPase Rac1, a member of the Ras superfamily of small GTP-binding proteins (see XI.3) and thereby induces Rac1 signaling.[129] Binding to the ankyrin repeat domain of ankyrin activates GDP/GTP exchange on Rho GTPases.[129] Via ankyrin binding and Rac1 activation tumor cell invasion and migration is activated in vitro. However, so far in vivo studies are merely descriptive and show TIAM expression in lymphomas, melanomas, neuroblastomas and breast cancer.[130-132]

SPARC

SPARC (secreted protein, acidic and rich in cysteine, also known as osteonectin or BM40) is a matricellular protein that modulates cell adhesion, migration and growth and is thought to play an important role in tissue remodeling and angiogenesis. Alterations of SPARC expression have been observed in a variety of solid tumors.[133-136] Overexpression in vitro indicated that SPARC contributes to increased invasion and altered adhesion.[137] Mechanistically, binding of SPARC to collagen type IV has been proposed[138] but other investigators have postulated activation of MMP-2.[139]

Amphoterin / RAGE

Amphoterin (HMG1) is a 30-kD heparin-binding protein which is functionally associated with the outgrowth of cytoplasmic processes in developing neurones. Amphoterin has been shown to mediate adhesive and proteolytic interactions at the leading edge of motile cells. Recently it was shown that inhibition of amphoterin interactions with its cell surface receptor (RAGE) suppresses tumor growth and metastasis. Enhanced expression of the high-mobility-group protein amphoterin/HMG-1 correlates with increasing invasive potential and progression of tumors.[140]

RAGE, the receptor for advanced glycation end products, is a multiligand member of the immunoglobulin superfamily of cell surface molecules. RAGE interacts with distinct molecules implicated in homeostasis, development and inflammation. Engagement of RAGE by a ligand triggers activations of key signaling pathways including Ras-signaling to MAP kinases,

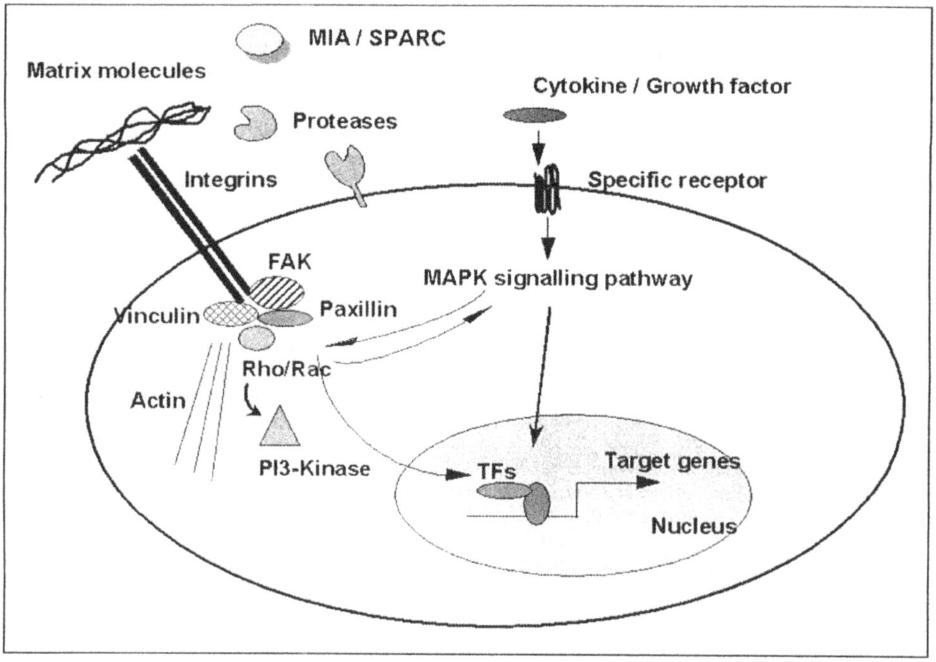

Fig. 4. Schematic overview on relations between molecules discussed in the review. Matrix molecules, cytokines / growth factors, receptors and signaling pathways documented in this review are placed in direct contact. Changes observed in malignant tumors can not be understood without analyzing the effects on the whole system of a cell.

NF-κB and Cdc42/rac activation.[141] RAGE functions also as a cell surface receptor for amphoterin. Inhibition of amphoterin-RAGE interaction suppressed activation of p44/p42, p38 and SAP/JNK MAP kinases and decreased growth of metastases assayed by an in vivo model.[141] Intracellular signaling pathways regulated by RAGE are important molecular effector mechanisms also linked to tumor proliferation, invasion and expression of MMPs.

Discussion

Although many pieces of evidence linking molecular pathways have been gathered much of the information remains isolated and has been extracted from somewhat artificial experimental systems. Nonetheless, a network of interactions emerges linking extracellular and cell surface molecules to cytoplasmic signal transduction pathways and nuclear regulation of gene expression patterns (Figs. 3 and 4).

This network is constantly subject to modification and further detailing by current application of microarrays and other novel techniques which lead to detection of a wide variety of new molecules relevant for invasion. After completion of the human genome project, with the new dimension of proteomics, even further new genes and the respective proteins will be isolated by methods like 2-dimensional electrophoresis, peptide sequencing and mass spectrometry. However, much effort will be needed to understand the biological significance of this novel information and to unravel the precise interactions and molecular pathways. As Jules Henri Poincare said more than a century ago:

> "Science is built up with facts, as a house is with stones. But a collection of facts is no more science than a heap of stones is a house."

Invasion is not a separated tumor cell function but closely integrated in mechanisms as proliferation, immune response, apoptosis, cell-cell contact und others. Therefore, a precise understanding of the metastatic process will lead to fundamental knowledge in tumor cell biology. This will offer new possibilities for therapeutic option in tumor treatment.

Many of the proteins discussed in this review are already used as marker molecules in tumor pathology. In the future they could also serve as prognostic markers to distinguish tumors which are highly likely to metastasize from those that are not. Further, these proteins provide potential targets for tumor therapy.

References

1. Kos J, Werle B, Lah T et al. Cysteine proteinases and their inhibitors in extracellular fluids: Markers for diagnosis and prognosis in cancer. Int J Biol Markers 2000; 15:84-89.
2. Bartenjev I, Rudolf Z, Stabuc B et al. Cathepsin D expression in early cutaneous malignant melanoma. Int J Dermatol 2000; 39:599-602.
3. Dohchin A, Suzuki JI, Seki H et al. Immunostained cathepsins B and L correlate with depth of invasion and different metastatic pathways in early stage gastric carcinoma. Cancer 2000; 89:482-487.
4. Mai J, Waisman DM, Sloane BF. Cell surface complex of cathepsin B/annexin II tetramer in malignant progression. Biochim Biophys Acta 2000; 1477:215-230.
5. Fujii H, Nakajima M, Saiki I et al. Human melanoma invasion and metastasis enhancement by high expression of aminopeptidase N/CD13. Clin Exp Metastasis 1995; 13:337-344.
6. Saiki I, Fujii H, Yoneda J et al. Role of aminopeptidase N (CD13) in tumor-cell invasion and extracellular matrix degradation. Int J Cancer 1993; 54:137-143.
7. Kido A, Krueger S, Haeckel C et al. Possible contribution of aminopeptidase N (APN/CD13) to invasive potential enhanced by interleukin-6 and soluble interleukin-6 receptor in human osteosarcoma cell lines. Clin Exp Metastasis 1999; 17:857-863.
8. Kosir MA, Quinn CC. orting of heparan sulfate proteoglycan into matrix compartments of prostate adenocarcinoma cells. J Surg Res 1995; 58:46-52.
9. Friedmann Y, Vlodavsky I, Aingorn H et al. Expression of heparanase in normal, dysplastic, and neoplastic human colonic mucosa and stroma: Evidence for its role in colonic tumorigenesis [In Process Citation]. Am J Pathol 2000; 157:1167-1175.
10. Vlodavsky I, Elkin M, Pappo O et al. Mammalian heparanase as mediator of tumor metastasis and angiogenesis. Isr Med Assoc J 2000; 45:37-45.
11. Vlodavsky I, Friedmann Y, Elkin M et al. Mammalian heparanase: gene cloning, expression and function in tumor progression and metastasis. Nat Med 1999; 5:793-802.
12. Csoka TB, Frost GI, Stern R. Hyaluronidases in tissue invasion. Invasion Metastasis 1997; 17:297-311.
13. Maleski M, Hockfield S. Glial cells assemble hyaluronan-based pericellular matrices in vitro. Glia 1997; 20:193-202.
14. Madan AK, Pang Y, Wilkiemeyer MB et al. Increased hyaluronidase expression in more aggressive prostate adenocarcinoma. Oncol Rep 1999; 6:1431-1433.
15. Madan AK, Yu K, Dhurandhar N et al. Association of hyaluronidase and breast adenocarcinoma invasiveness. Oncol Rep 1999; 6:607-609.
16. Pham HT, Block NL, Lokeshwar VB. Tumor-derived hyaluronidase: A diagnostic urine marker for high-grade bladder cancer. Cancer Res 1997; 57:778-783.
17. Novak U, Stylli SS, Kaye AH et al. Hyaluronidase-2 overexpression accelerates intracerebral but not subcutaneous tumor formation of murine astrocytoma cells. Cancer Res 1999; 59:6246-6250.
18. Victor R, Chauzy C, Girard N et al. Human breast-cancer metastasis formation in a nude-mouse model: studies of hyaluronidase, hyaluronan and hyaluronan-binding sites in metastatic cells. Int J Cancer 1999; 82:77-83.
19. Liu D, Pearlman E, Diaconu E et al. Expression of hyaluronidase by tumor cells induces angiogenesis in vivo. Proc Natl Acad Sci USA 1996; 93:7832-7837.
20. Saeki Y, Hazeki K, Matsumoto M et al. Correlation between metastatic potency and the down-regulation of E-cadherin in the mouse hepatoma cell lines G-1 and G-5. Oncol Rep 2000; 7:731-735.
21. Mialhe A, Levacher G, Champelovier P et al. Expression of E-, P-, n-cadherins and catenins in human bladder carcinoma cell lines. J Urol 2000; 164:826-835.
22. Luo J, Lubaroff DM, Hendrix MJ. Suppression of prostate cancer invasive potential and matrix metalloproteinase activity by E-cadherin transfection. Cancer Res 1999; 59:3552-3556.
23. Okada T, Hawley RG, Kodaka M et al. Significance of VLA-4-VCAM-1 interaction and CD44 for transendothelial invasion in a bone marrow metastatic myeloma model. Clin Exp Metastasis 1999; 17:623-629.

24. Sun JJ, Zhou XD, Liu YK et al. Invasion and metastasis of liver cancer: Expression of intercellular adhesion molecule 1. J Cancer Res Clin Oncol 1999; 125:28-34.
25. Umansky V, Schirrmacher V, and Rocha MNew insights into tumor-host interactions in lymphoma metastasis. J Mol Med 1996; 74:353-363.
26. Koyama S, Ebihara T, Fukao K. Expression of intercellular adhesion molecule 1 (ICAM-1) during the development of invasion and/or metastasis of gastric carcinoma. J Cancer Res Clin Oncol 1992; 118:609-614.
27. Iliopoulos D, Ernst C, Steplewski Z et al. Inhibition of metastases of a human melanoma xenograft by monoclonal antibody to the GD2/GD3 gangliosides. J Natl Cancer Inst 1989; 81:440-444.
28. Merzak A, Koochekpour S, Pilkington GJ. Adhesion of human glioma cell lines to fibronectin, laminin, vitronectin and collagen I is modulated by gangliosides in vitro. Cell Adhes Commun 1995; 3:27-43.
29. Merzak A, Koochekpour S, McCrea S et al. Gangliosides modulate proliferation, migration, and invasiveness of human brain tumor cells in vitro. Mol Chem Neuropathol 1995; 24:121-135.
30. Laidler P, Litynska A.Tumor cell N-glycans in metastasis. Acta Biochim Pol 1997; 44:343-357.
31. Renkonen J, Makitie A, Paavonen T et al. Sialyl-Lewis(x/a)-decorated selectin ligands in head and neck tumours. J Cancer Res Clin Oncol 1999; 125:569-576.
32. Renkonen J, Paavonen T, Renkonen R. Endothelial and epithelial expression of sialyl Lewis(x) and sialyl Lewis(a) in lesions of breast carcinoma. Int J Cancer 1997; 74:296-300.
33. Majuri ML, Niemela R, Tiisala S et al. Expression and function of alpha 2,3-sialyl- and alpha 1,3/1,4-fucosyltransferases in colon adenocarcinoma cell lines: role in synthesis of E-selectin counter-receptors. Int J Cancer 1995; 63:551-559.
34. Izawa M, Kumamoto K, Mitsuoka C et al. Expression of sialyl 6-sulfo Lewis X is inversely correlated with conventional sialyl Lewis X expression in human colorectal cancer. Cancer Res 2000; 60:1410-1416.
35. Hosono J, Narita T, Kimura N et al. Involvement of adhesion molecules in metastasis of SW1990, human pancreatic cancer cells. J Surg Oncol 1998; 67:77-84.
36. Sato M, Narita T, Kimura N et al. The association of sialyl Lewis(a) antigen with the metastatic potential of human colon cancer cells. Anticancer Res 1997; 17:3505-3511.
37. Kannagi R. Carbohydrate-mediated cell adhesion involved in hematogenous metastasis of cancer. Glycoconj J 1997; 14:577-584.
38. Camby I, Belot N, Rorive S et al. Galectins are differentially expressed in supratentorial pilocytic astrocytomas, astrocytomas, anaplastic astrocytomas and glioblastomas, and significantly modulate tumor astrocyte migration. Brain Pathol 2001; 11:12-26.
39. Kawachi K, Matsushita Y, Yonezawa S et al. Galectin-3 expression in various thyroid neoplasms and its possible role in metastasis formation. Hum Pathol 2000; 31:428-433.
40. Nakamura M, Inufusa H, Adachi T et al. Involvement of galectin-3 expression in colorectal cancer progression and metastasis. Int J Oncol 1999; 15:143-148.
41. Matarresea P, Tinari N, Semeraroa ML et al. Galectin-3 overexpression protects from cell damage and death by influencing mitochondrial homeostasis. FEBS Lett 2000; 473:311-315.
42. Matarrese P, Fusco O, Tinari N et al. Galectin-3 overexpression protects from apoptosis by improving cell adhesion properties. Int J Cancer 2000; 85:545-554.
43. Kuryu M, Ozaki T, Nishida K et al. Expression of CD44 variants in osteosarcoma. J Cancer Res Clin Oncol 1999; 125:646-652.
44. Ladeda V, Aguirre GJ, Bald KJ. Function and expression of CD44 during spreading, migration, and invasion of murine carcinoma cells. Exp Cell Res 1998; 242:515-527.
45. Okamoto I, Kawano Y, Tsuiki H et al. CD44 cleavage induced by a membrane-associated metalloprotease plays a critical role in tumor cell migration. Oncogene 1999; 18:1435-1446.
46. Iwamoto Y, Nomizu M, Yamada Y et al. Inhibition of angiogenesis, tumour growth and experimental metastasis of human fibrosarcoma cells HT1080 by a multimeric form of the laminin sequence Tyr-Ile-Gly-Ser-Arg (YIGSR). Br J Cancer 1996; 73:589-595.
47. Graf J, Iwamoto Y, Sasaki M et al. Identification of an amino acid sequence in laminin mediating cell attachment, chemotaxis, and receptor binding. Cell 1987; 48:989-996.
48. Ito H, Miyazaki M, Nishimura F et al. Secretion of extracellular matrix (fibronectin), growth factor (transforming growth factor beta) and protease (cathepsin D) by hepatoma cells. Oncology 2000; 58:261-270.
49. Ryu S, Jimi S, Eura Y et al. Strong intracellular and negative peripheral expression of fibronectin in tumor cells contribute to invasion and metastasis in papillary thyroid carcinoma. Cancer Lett 1999; 146:103-109.

50. Urtreger AJ, Aguirre GJ, Werbajh SE et al. Involvement of fibronectin in the regulation of urokinase production and binding in murine mammary tumor cells. Int J Cancer 1999; 82:748-753.
51. Shimao Y, Nabeshima K, Inoue T et al. Role of fibroblasts in HGF/SF-induced cohort migration of human colorectal carcinoma cells: Fibroblasts stimulate migration associated with increased fibronectin production via upregulated TGF-beta1. Int J Cancer 1999; 82:449-458.
52. Hunt G, Sherbet GV. Effects of laminin on the attachment of glioma cells to type IV collagen. Clin Exp Metastasis 1989; 7:353-359.
53. Kleiner DE, Stetler-Stevenson WG. Matrix metalloproteinases and metastasis. Cancer Chemother Pharmacol 1999; 43 Suppl:S42-51.
54. Cohen IR, Murdoch AD, Naso MF et al. Abnormal expression of perlecan proteoglycan in metastatic melanomas. Cancer Res 1994; 54:5771-5774.
55. Ueda M, Ueki M, Terai Y et al. Biological implications of growth factors on the mechanism of invasion in gynecological tumor cells. Gynecol Obstet Invest 1999; 48:221-228.
56. Ueda M, Fujii H, Yoshizawa K et al. Effects of EGF and TGF-alpha on invasion and proteinase expression of uterine cervical adenocarcinoma OMC-4 cells. Invasion Metastasis 1998; 18:176-183.
57. Rooprai HK, Rucklidge GJ, Panou C et al. The effects of exogenous growth factors on matrix metalloproteinase secretion by human brain tumour cells. Br J Cancer 2000; 82:52-55.
58. Price JT, Tiganis T, Agarwal A et al. Epidermal growth factor promotes MDA-MB-231 breast cancer cell migration through a phosphatidylinositol 3'-kinase and phospholipase C-dependent mechanism. Cancer Res 1999; 59:5475-5478.
59. Jiang W, Hiscox S, Matsumoto K et al. Hepatocyte growth factor/scatter factor, its molecular, cellular and clinical implications in cancer. Crit Rev Oncol Hematol 1999; 29:209-248.
60. Trusolino L, Cavassa S, Angelini P et al. HGF/scatter factor selectively promotes cell invasion by increasing integrin avidity. FASEB J 2000; 14:1629-1640.
61. Rosenthal EL, Johnson TM, Allen ED et al. Role of the plasminogen activator and matrix metalloproteinase systems in epidermal growth factor- and scatter factor-stimulated invasion of carcinoma cells. Cancer Res 1998; 58:5221-5230.
62. Hiscox S, Jiang WG. Association of the HGF/SF receptor, c-met, with the cell-surface adhesion molecule, E-cadherin, and catenins in human tumor cells. Biochem Biophys Res Commun 1999; 261:406-411.
63. Taipale J, Saharinen J, Keski-Oja J. Extracellular matrix-associated transforming growth factor-beta: role in cancer cell growth and invasion. Adv Cancer Res 1998; 75:87-134.
64. Naylor S, Smalley MJ, Robertson D et al. Retroviral expression of Wnt-1 and Wnt-7b produces different effects in mouse mammary epithelium. J Cell Sci 2000; 113:2129-2138.
65. Vider BZ, Zimber A, Chastre E et al. Evidence for the involvement of the Wnt 2 gene in human colorectal cancer. Oncogene 1996; 12:153-158.
66. Iozzo RV, Eichstetter I, Danielson KG. Aberrant expression of the growth factor Wnt-5A in human malignancy. Cancer Res 1995; 55:3495-3499.
67. Lejeune S, Huguet EL, Hamby A et al. Wnt5a cloning, expression, and up-regulation in human primary breast cancers. Clin Cancer Res 1995; 1:215-222.
68. Nozaki S, Sledge GWJ, Nakshatri H. Cancer cell-derived interleukin 1alpha contributes to autocrine and paracrine induction of pro-metastatic genes in breast cancer. Biochem Biophys Res Commun 2000; 275:60-62.
69. Yasuda T, Matsui H, Kanamori M et al. Effects of tumor cell-derived interleukin 1 alpha on invasiveness of metastatic clones of murine RCT sarcoma through endothelial cells. Tumour Biol 1999; 20:105-116.
70. Bar-Eli M. Role of interleukin-8 in tumor growth and metastasis of human melanoma. Pathobiology 1999; 67:12-18.
71. Inoue K, Slaton JW, Eve BY et al. Interleukin 8 expression regulates tumorigenicity and metastases in androgen-independent prostate cancer. Clin Cancer Res 2000; 6:2104-2119.
72. Youngs SJ, Ali SA, Taub DD et al. Chemokines induce migrational responses in human breast carcinoma cell lines. Int J Cancer 1997; 71:257-266.
73. Luca M, Huang S, Gershenwald JE et al. Expression of interleukin-8 by human melanoma cells up-regulates MMP-2 activity and increases tumor growth and metastasis. Am J Pathol 1997; 151:1105-1113.
74. Nishino H, Miyata M, Kitamura K. The effect of interleukin-6 on enhancing the invasiveness of head and neck cancer cells in vitro. Eur Arch Otorhinolaryngol 1998; 255:468-472.
75. Obata NH, Tamakoshi K, Shibata K et al. Effects of interleukin-6 on in vitro cell attachment, migration and invasion of human ovarian carcinoma. Anticancer Res 1997; 17:337-342.

76. Kossakowska AE, Edwards DR, Prusinkiewicz C et al. Interleukin-6 regulation of matrix metalloproteinase (MMP-2 and MMP-9) and tissue inhibitor of metalloproteinase (TIMP-1) expression in malignant nonHodgkin's lymphomas. Blood 1999; 94:2080-2089.
77. Wagner S, Stegen C, Bouterfa H et al. Expression of matrix metalloproteinases in human glioma cell lines in the presence of IL-10. J Neurooncol 1998; 40:113-122.
78. Wu W, Yamaura T, Murakami K et al. Involvement of TNF-alpha in enhancement of invasion and metastasis of colon 26-L5 carcinoma cells in mice by social isolation stress. Oncol Res 1999; 11:461-469.
79. Esteve PO, Tremblay P, Houde M et al. In vitro expression of MMP-2 and MMP-9 in glioma cells following exposure to inflammatory mediators. Biochim Biophys Acta 1998; 1403:85-96.
80. Kubota S, Takeda K, Yamada T et al. Tumor necrosis factor alpha and lymphotoxin stimulate human myeloblastic leukemia cell (ML-1) invasion through a reconstituted basement membrane (Matrigel) with concomitant induction of 92 kDa gelatinase secretion. Cancer Lett 1996; 98:233-240.
81. Reshkin SJ, Bellizzi A, Albarani V et al. Phosphoinositide 3-kinase is involved in the tumor-specific activation of human breast cancer cell Na(+)/H(+) exchange, motility, and invasion induced by serum deprivation. J Biol Chem 2000; 275:5361-5369.
82. Saeki Y, Seya T, Hazeki K et al. Involvement of phosphoinositide 3-kinase in regulation of adhesive activity of highly metastatic hepatoma cells. J Biochem (Tokyo) 1998; 124:1020-1025.
83. Bardelli A, Longati P, Williams TA et al. A peptide representing the carboxyl-terminal tail of the met receptor inhibits kinase activity and invasive growth. J Biol Chem 1999; 274:29274-29281.
84. Bardelli A, Basile ML, Audero E et al. Concomitant activation of pathways downstream of Grb2 and PI 3-kinase is required for MET-mediated metastasis. Oncogene 1999; 18:1139-1146.
85. Turner T, Epps-Fung MV, Kassis J et al. Molecular inhibition of phospholipase cgamma signaling abrogates DU-145 prostate tumor cell invasion. Clin Cancer Res 1997; 3:2275-2282.
86. Kassis J, Moellinger J, Lo H et al. A role for phospholipase C-gamma-mediated signaling in tumor cell invasion. Clin Cancer Res 1999; 5:2251-2260.
87. Khoshyomn S, Penar PL, Rossi J et al. Inhibition of phospholipase C-gamma1 activation blocks glioma cell motility and invasion of fetal rat brain aggregates. Neurosurgery 1999; 44:568-577.
88. Cho KK, Mikkelsen T, Lee YJ et al. The role of protein kinase Calpha in U-87 glioma invasion. Int J Dev Neurosci 1999; 17:447-461.
89. Shimao Y, Nabeshima K, Inoue T et al. TPA-enhanced motility and invasion in a highly metastatic variant (L-10) of human rectal adenocarcinoma cell line RCM-1: Selective role of PKC-alpha and its inhibition by a combination of PDBu-induced PKC downregulation and antisense oligonucleotides treatment. Clin Exp Metastasis 1999; 17:351-360.
90. Gomez DE, Skilton G, Alonso DF et al. The role of protein kinase C and novel phorbol ester receptors in tumor cell invasion and metastasis (Review). Oncol Rep 1999; 6:1363-1370.
91. Frisch SM and Ruoslahti EIntegrins and anoikis. Curr Opin Cell Biol 1997; 9:701-706.
92. Judson PL, He X, Cance WG et al. Overexpression of focal adhesion kinase, a protein tyrosine kinase, in ovarian carcinoma. Cancer 1999; 86:1551-1556.
93. Owens LV, Xu L, Craven RJ et al. Overexpression of the focal adhesion kinase (p125FAK) in invasive human tumors. Cancer Res 1995; 55:2752-2755.
94. Weiner TM, Liu ET, Craven RJ et al. Expression of focal adhesion kinase gene and invasive cancer. Lancet 1993; 342:1024-1025.
95. Kornberg LJ. Focal adhesion kinase and its potential involvement in tumor invasion and metastasis. Head.Neck 1998; 20:745-752.
96. Shibata K, Kikkawa F, Nawa A et al. Both focal adhesion kinase and c-Ras are required for the enhanced matrix metalloproteinase 9 secretion by fibronectin in ovarian cancer cells. Cancer Res 1998; 58:900-903.
97. Cadigan KM. Nusse R. Wnt signaling: a Common theme in animal development. Genes Dev 1997; 11:3286-3305.
98. Polakis P. The oncogenic activation of beta-catenin. Curr Opin Genet Dev 1999; 9:15-21.
99. Wodarz A, Nusse R. Mechanisms of Wnt signaling in development. Annu Rev Cell Dev Biol 1998; 14:59-88.
100. Yost C, Torres M, Miller JR et al. The axis-inducing activity, stability, and subcellular distribution of beta-catenin is regulated in *Xenopus* embryos by glycogen synthase kinase 3. Genes Dev 1996; 10:1443-1454.
101. Morin PJ, Sparks AB, Korinek V et al. Activation of beta-catenin-Tcf signaling in colon cancer by mutations in beta-catenin or APC. Science 1997; 275:1787-1790.
102. Wijnhoven BP, Nollet F, De Both NJ et al. Genetic alterations involving exon 3 of the beta-catenin gene do not play a role in adenocarcinomas of the esophagus. Int J Cancer 2000; 86:533-537.

103. Brabletz T, Jung A, Dag S et al. beta-catenin regulates the expression of the matrix metalloproteinase-7 in human colorectal cancer. Am J Pathol 1999; 155:1033-1038.
104. Brabletz T, Jung A, Hermann K et al. Nuclear overexpression of the oncoprotein beta-catenin in colorectal cancer is localized predominantly at the invasion front. Pathol Res Pract 1998; 194:701-704.
105. Hernandez-Alcoceba R, del Peso L, Lacal JC. The Ras family of GTPases in cancer cell invasion. Cell Mol Life Sci 2000; 57:65-76.
106. Kirschmann DA, Seftor EA, Nieva DR et al. Differentially expressed genes associated with the metastatic phenotype in breast cancer. Breast Cancer Res Treat 1999; 55:127-136.
107. Shinohara M, Hiraki A, Ikebe T et al. Immunohistochemical study of desmosomes in oral squamous cell carcinoma: Correlation with cytokeratin and E-cadherin staining, and with tumour behaviour. J Pathol 1998; 184:369-381.
108. Nakayama S, Sasaki A, Mese H et al. Establishment of high and low metastasis cell lines derived from a human tongue squamous cell carcinoma. Invasion Metastasis 1998; 18:219-228.
109. Chu YW, Yang PC, Yang SC et al. Selection of invasive and metastatic subpopulations from a human lung adenocarcinoma cell line. Am J Respir Cell Mol Biol 1997; 17:353-360.
110. Hembrough TA, Kralovich KR, Li L et al. Cytokeratin 8 released by breast carcinoma cells in vitro binds plasminogen and tissue-type plasminogen activator and promotes plasminogen activation. Biochem J 1996; 317:763-769.
111. Hendrix MJ, Seftor EA, Seftor RE et al. Experimental coexpression of vimentin and keratin intermediate filaments in human breast cancer cells results in phenotypic interconversion and increased invasive behavior. Am J Pathol 1997; 150:483-495.
112. Hendrix MJ, Seftor EA, Chu YW et al. Role of intermediate filaments in migration, invasion and metastasis. Cancer Metastasis Rev 1996; 15:507-525.
113. Crowe DL, Milo GE, Shuler CF. Keratin 19 downregulation by oral squamous cell carcinoma lines increases invasive potential. J Dent Res 1999; 78:1256-1263.
114. Michiels F, Collard JG. Rho-like GTPases: their role in cell adhesion and invasion. Biochem Soc Symp 1999; 65:125-146.
115. Evers EE, van der Kammen RA, ten Klooster JP et al. Rho-like GTPases in tumor cell invasion. Methods Enzymol 2000; 325:403-415.
116. Evers EE, Zondag GC, Malliri A et al. Rho family proteins in cell adhesion and cell migration. Eur J Cancer 2000; 36:1269-1274.
117. Clark EA, Golub TR, Lander ES et al. Genomic analysis of metastasis reveals an essential role for RhoC. Nature 2000; 406:532-535.
118. Itoh K, Yoshioka K, Akedo H et al. An essential part for Rho-associated kinase in the transcellular invasion of tumor cells. Nat Med 1999; 5:221-225.
119. Blesch A, Bosserhoff AK, Apfel R et al. Cloning of a novel malignant melanoma-derived growth-regulatory protein, MIA. Cancer Res 1994; 54:5695-5701.
120. Bosserhoff AK, Kaufmann M, Kaluza B et al. Melanoma-inhibiting activity, a novel serum marker for progression of malignant melanoma. Cancer Res 1997; 57:3149-3153.
121. Bosserhoff AK, Golob M, Buettner R et al. MIA ("melanoma inhibitory activity"). Biological functions and clinical relevance in malignant melanoma. Hautarzt 1998; 49:762-769.
122. Dreau D, Bosserhoff AK, White RL et al. Melanoma-inhibitory activity protein concentrations in blood of melanoma patients treated with immunotherapy. Oncol Res 1999; 11:55-61.
123. Deichmann M, Benner A, Bock M et al. S100-Beta, melanoma-inhibiting activity, and lactate dehydrogenase discriminate progressive from nonprogressive American Joint Committee on Cancer stage IV melanoma. J Clin Oncol 1999; 17:1891-1896.
124. Bogdahn U, Apfel R, Hahn M et al. Autocrine tumor cell growth-inhibiting activities from human malignant melanoma. Cancer Res 1989; 49:5358-5363.
125. van Groningen JJ, Bloemers HP, Swart GW. Identification of melanoma inhibitory activity and other differentially expressed messenger RNAs in human melanoma cell lines with different metastatic capacity by messenger RNA differential display. Cancer Res 1995; 55:6237-6243.
126. Bosserhoff AK, Moser M, Hein R et al. In situ expression patterns of melanoma-inhibiting activity (MIA) in melanomas and breast cancers. J Pathol 1999; 187:446-454.
127. Guba M, Bosserhoff AK, Steinbauer M et al. Overexpression of melanoma inhibitory activity (MIA) enhances extravasation and metastasis of A-mel 3 melanoma cells in vivo. Br J Cancer 2000; 83:1216-1222.
128. Stoll R, Renner C, Ambrosius D et al. Letter to the editor: Sequence-specific 1H, 13C, and 15N assignment of the human melanoma inhibitory activity (MIA) protein. J Biomol NMR 2000; 17:87-88.

129. Bourguignon LY, Zhu H, Shao L et al. Ankyrin-Tiam1 interaction promotes Rac1 signaling and metastatic breast tumor cell invasion and migration. J Cell Biol 2000; 150:177-191.
130. Habets GG, Scholtes EH, Zuydgeest D et al. Identification of an invasion-inducing gene, TIAM-1, that encodes a protein with homology to GDP-GTP exchangers for Rho-like proteins. Cell 1994; 77:537-549.
131. Michiels F, Habets GG, Stam JC et al. A role for Rac in Tiam1-induced membrane ruffling and invasion. Nature 1995; 375:338-340.
132. Habets GG, van der Kammen RA, Stam JC et al. Sequence of the human invasion-inducing TIAM1 gene, its conservation in evolution and its expression in tumor cell lines of different tissue origin. Oncogene 1995; 10:1371-1376.
133. Paley PJ, Goff BA, Gown AM et al. Alterations in SPARC and VEGF immunoreactivity in epithelial ovarian cancer. Gynecol.Oncol.2000; 78:336-341.
134. Thomas R, True LD, Bassuk JA et al. Differential expression of osteonectin/SPARC during human prostate cancer progression. Clin Cancer Res 2000; 6:1140-1149.
135. Massi D, Franchi A, Borgognoni L et al. Osteonectin expression correlates with clinical outcome in thin cutaneous malignant melanomas. Hum Pathol 1999; 30:339-344.
136. Porte H, Triboulet JP, Kotelevets L et al. Overexpression of stromelysin-3, BM-40/SPARC, and MET genes in human esophageal carcinoma: Implications for prognosis. Clin Cancer Res 1998; 4:1375-1382.
137. Golembieski WA, Ge S, Nelson K et al. Increased SPARC expression promotes U87 glioblastoma invasion in vitro. Int J Dev Neurosci 1999; 17:463-472.
138. Kato Y, Frankenne F, Noel A et al. High production of SPARC/osteonectin/BM-40 in mouse metastatic B16 melanoma cell lines. Pathol Oncol Res 2000; 6:24-26.
139. Gilles C, Bassuk JA, Pulyaeva H et al. SPARC/osteonectin induces matrix metalloproteinase 2 activation in human breast cancer cell lines. Cancer Res 1998; 58:5529-5536.
140. Xiang YY, Wang DY, Tanaka M et al. Expression of high-mobility group-1 mRNA in human gastrointestinal adenocarcinoma and corresponding noncancerous mucosa. Int J Cancer 1997; 74:1-6.
141. Taguchi A, Blood DC, del Toro G et al. Blockade of RAGE-amphoterin signalling suppresses tumour growth and metastases. Nature 2000; 405:354.

Chapter 6

Common Mechanistic Features in Cell-Extracellular Matrix Interactions Regulating Neurite Outgrowth and Tumor Cell Invasion

Henri J. Huttunen and Heikki Rauvala

Summary

Neurite extension in tissues to establish neuronal connections can be envisioned as a form of invasive migration of growth cones, the distal tips of the neurites that interact with the extracellular environment and contain the essential machinery required for motility. In an analogous manner, leading edges of cancer cells are capable of invading tissues, a phenomenon of obvious importance for tumor spread. Therefore, one would expect that similar or even identical mechanisms would exist to explain invasive process extension required for neurite outgrowth and tumor cell migration. In the present report, we discuss the major cell-matrix interactions that regulate neurite outgrowth and point out similarities found in the mechanisms that regulate tumor cell invasion. Our major focus is in transmembrane signaling at the cell-matrix contact and in localized proteolysis required for process extension through the matrix.

Introduction

Cell motility plays an important role in the development, regeneration and plasticity of tissues. One prominent example is the outgrowth of axons and dendrites—collectively referred to as neurites when they are growing—to establish neuronal connections during development, which is a prerequisite for functional neural networks of the adult organism. Neurite outgrowth implies that the distal parts of the processes, the growth cones (for a comprehensive review of neuronal growth cones and their migration, see Ref. 1), sense extracellular cues for navigation. Cell surface adhesion molecules, extracellular matrix proteins and polypeptide growth factors act as cues for growth cone navigation and are therefore of fundamental importance in attempts to understand the formation of neuronal connections on the molecular level (for a review, see Ref. 2). Cell surface proteins interacting with the components of the extracellular matrix represent mainly four protein families: integrins, membrane-attached proteoglycans, immunoglobulin superfamily members and receptor-type phosphatases (Fig. 1).

Growth cones are highly motile distal tips of the neurites that are indispensable for the neurites to be guided to their targets. Therefore, extracellular cues guiding the growth cones have received much attention and form an intensive area of research in neurobiology.[1] In addition to making decisions regarding the direction of movement, it seems plausible that growth cones may have to degrade extracellular matrix for motility. Indeed, proteolytic zones have been shown around the growth cones (see below), which is in favor of the view that growth cone migration can be understood as a form of invasive migration.

Cell Invasion, edited by Jyrki Heino and Veli-Matti Kähäri.

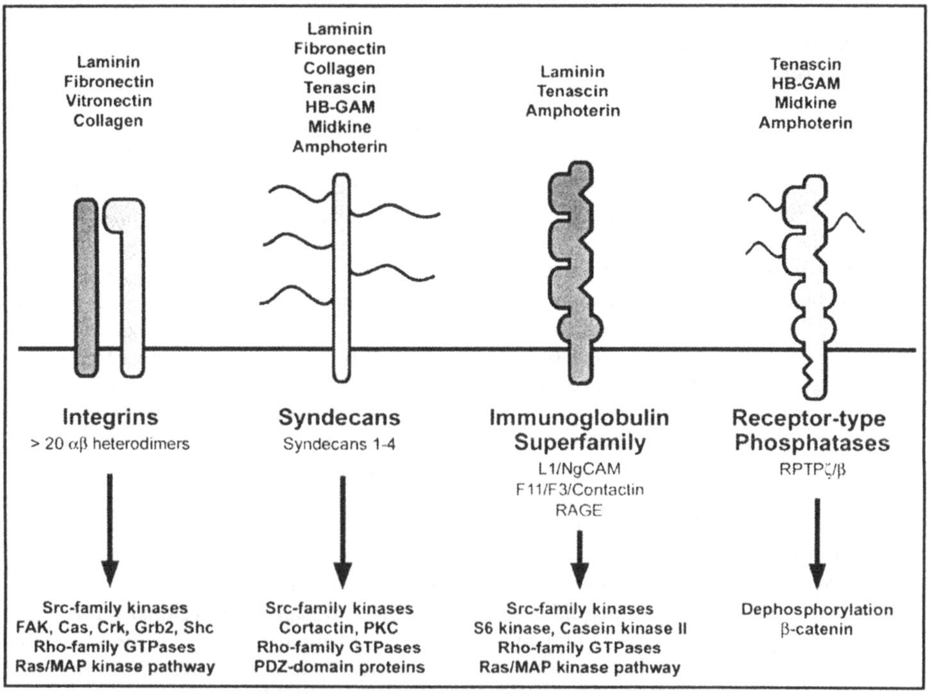

Fig. 1. Summary of receptor families interacting with extracellular matrix. Different ligands and signaling pathways for different family members are indicated above and below the receptors.

In the present report, we will give an overview of cell-extracellular matrix interactions that enhance neurite outgrowth and are accordingly implicated in growth cone migration. In particular, we will highlight mechanisms that appear to be common for neurite outgrowth and tumor cell invasion. We predict that our approach in the interphase of neurobiology and tumor biology may be useful to understand physiological and pathophysiological behaviors of both neuronal and nonneuronal cells.

Integrin-Matrix Interactions in Neurite Outgrowth and Invasive Cell Migration

Normal cells exist in a strictly regulated relationship to their immediate microenvironment formed by components of the extracellular matrix (ECM). Many of the effects of the ECM on cellular morphology, growth control, tissue maintenance and differentiation are mediated by integrins. Integrins are a large family of transmembrane proteins which function as cell adhesion molecules linking macromolecules of the ECM to the cell surface (for a review, see Koistinen and Heino in this book). Integrins are heterodimers formed from α and β subunits. Sixteen α-subunits can associate with eight β-subunits giving more than 20 different heterodimers.

Combinations of different integrins on cell surfaces allow cells to recognize and respond to a variety of different extracellular matrix proteins. Some integrins can bind to the same ligand; others are specific for one ligand. Some integrin receptors recognize the same peptide sequence in different ECM molecules, hence a given integrin may bind to laminin, collagen, and fibronectin. Integrins provide anchorage for cells to the ECM and are clustered into specialized adhesive structures, focal adhesions and focal complexes, in which numerous signaling components are concentrated. Integrins are involved in directed invasion and motility of cells through regulation of the actin cytoskeleton.[3] The loss of integrin-mediated cell–matrix contact induces apoptosis ('anoikis') in certain cell types.[4] Transformed cells often lose contacts with the ECM, have altered integrin expression, demonstrate anchorage-independent growth and display a motile, invasive phenotype.

Laminin is a large, cross-shaped multidomain glycoprotein that is organized in the meshwork of basement membranes. It also occurs in the ECM in sites other than basement membranes at early stages of development, and is localized to specific types of neuronal structures in the central nervous system (CNS) during both embryonic and adult stages. Laminin in the extracellular matrix of the developing CNS localizes to regions where tracts are growing.[5] Laminin is secreted by cells into their extracellular environment where it interacts with receptors at cell surfaces resulting in changes in the behavior of cells such as attachment to a substrate, migration, and neurite outgrowth during embryonic development and regeneration.[6] Laminin affects the velocity and direction of outgrowth by interacting with cell surface receptors at the tips of the advancing growth cones of neurites. Although nonintegrin receptors for laminin exist, β1 integrins are thought to be the major neurite outgrowth receptors for laminin.[5] Different integrin ligands vary in their potential to induce neurite outgrowth. When given a choice between laminin and fibronectin, another glycoprotein of the ECM that has less potential to promote neurite growth for certain types of neurons than laminin, growth cones use filopodial scouts to make choices to move onto one or the other substrate. The choice a growth cone makes does not depend on the degree of adhesivity to the substrate, but rather depends on the nature of substrates contacted.[7]

Integrins play roles in a number of cellular processes that might contribute to the development of tumors, including the regulation of proliferation and apoptosis, cellular motility and invasion, cell surface localization of metalloproteinases, and angiogenesis.[8] The integrin α2β1, a collagen/laminin receptor, has been shown to impart metastatic abilities to some tumor cells.[9] Integrin αvβ3 mediates cellular adhesion to a wide variety of ECM proteins (e.g., vitronectin, fibronectin, fibrinogen, laminin and collagen). Expression of this integrin enhances the ability of a given cell to adhere to, migrate on, or respond to almost any matrix protein it may encounter. The expression of various integrins has been shown to correlate with invasive properties in different types of malignant cells.[10,11]

Adhesive interactions and regulation of cytoskeleton play central roles in motile processes, such as growth cone advance or invasive migration. The regulation of these interactions requires the coordination of a multiplicity of signals, both spatially and temporally (Fig. 2). Ligand binding to integrins and subsequent integrin clustering triggers activation of intracellular protein kinase cascades including focal-adhesion kinase (FAK), Src-family tyrosine kinases, various adapter molecules (e.g., p130Cas, Grb2, Shc, paxillin) and mitogen-activated protein kinases (MAP kinases) (reviewed in Ref. 12). In addition, integrin ligation can modulate cytoskeletal organization through Rho-family GTPases (Rho, Rac and Cdc42).[3] The key downstream components of Rho GTPases are WASP family proteins, adaptor molecules that bind multiple signaling and cytoskeletal proteins, and the Arp2/3 complex, a multi-functional protein complex that nucleates and crosslinks actin filaments.[13] The crucial role of Rho-family GTPases in the regulation of growth cone morphology and guidance has become more evident during the past few years.[14] In addition to downstream signaling molecules, integrins are known to interact laterally with other cell surface molecules, such as urokinase receptor[15] and syndecans.[16] These interactions might play a crucial role in the motile responses elicited by the extracellular matrix molecules.

The recent finding that mRNA localization can be regulated by transmembrane signaling at cell-matrix contacts[17,18] provides another angle to mechanisms regulating cell motility. Ligation of some integrins and some syndecans by a local matrix contact has been shown to localize β-actin mRNA, poly-A RNA and ribosomes near the sites of signal reception. On the other hand, it has been shown that localization of β-actin mRNA close to the leading edge enhances cell motility.[19] Thus, cell contact-induced rapid localization of a cell motility-associated mRNA,

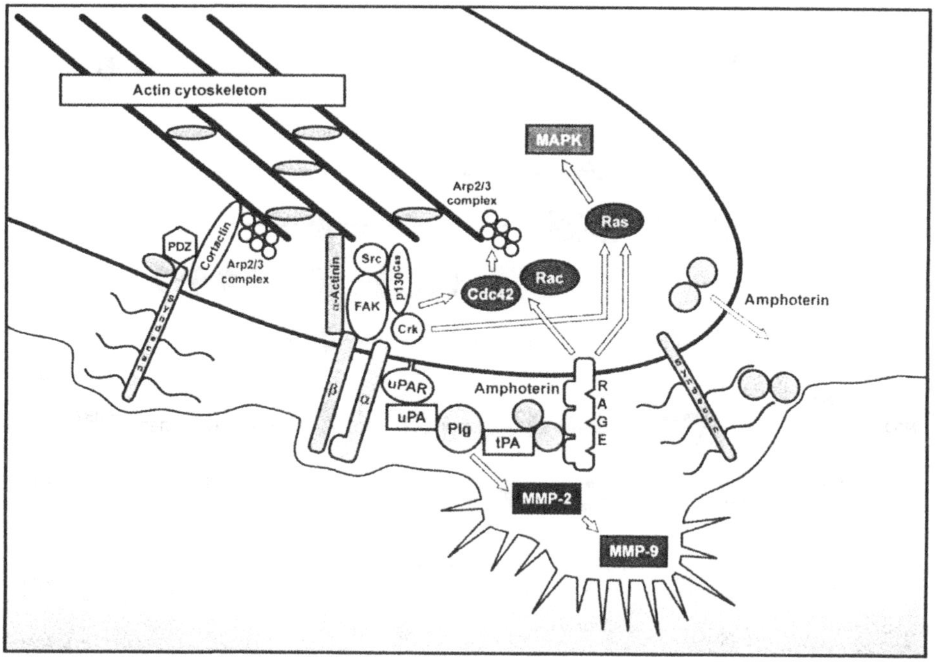

Fig. 2. Hypothetical model summarizing some of the molecular interactions involved in penetration of the leading edge of a neuronal growth cone or a tumor cell through the extracellular matrix. Cell adhesion to extracellular matrix components (e.g., laminin, fibronectin, TSR-family proteins and amphoterin) is mediated by integrins, syndecans and RAGE. On the outer surface of the plasma membrane plasmin and matrix metalloproteinase (MMP) cascades are activated locally through interactions with multimolecular adhesion complexes (e.g., integrin-uPAR-uPA-plasminogen and RAGE-amphoterin-tPA-plasminogen) and cooperate to degrade the surrounding extracellular matrix. Amphoterin is secreted from the leading edge of the cell facilitating this process. On the inner surface of the plasma membrane, the cytoplasmic domains of the receptors interact with signaling molecules (e.g., cortactin-Arp2/3-actin, FAK-Src-p130^{Cas}-Crk, Cdc42/Rac and Ras-MAP kinase pathways) to promote cytoskeletal changes required for protrusion of the process.

like that of β-actin,[17] could provide an efficient way for the cell to regulate local protein expression required for motility.

TSR (Thrombospondin Type 1) Superfamily Proteins

The thrombospondins constitute a family of five extracellular matrix proteins. Thrombospondins 1 and 2 contain a characteristic sequence of about 60 amino acids with conserved cysteine and tryptophan residues, designated as thrombospondin type 1 repeat or TSR. In addition, a wide number of nonthrombospondin extracellular matrix and transmembrane proteins contain TSR repeats (for a partial list of TSR proteins, see Fig. 3; for a review of TSR proteins, see Ref. 20).

Structural studies using heteronuclear NMR have very recently suggested that the TSR sequence motif defines a structure composed of three antiparallel β-strands.[21] This type of structure clearly holds for the four TSR regions of HB-GAM (heparin-binding growth-associated molecule; also known as pleiotrophin) and MK (midkine; see Fig. 3). Further NMR studies and/or solution of crystal structures are required to understand the degree of structural similarity within the TSR regions of different proteins.

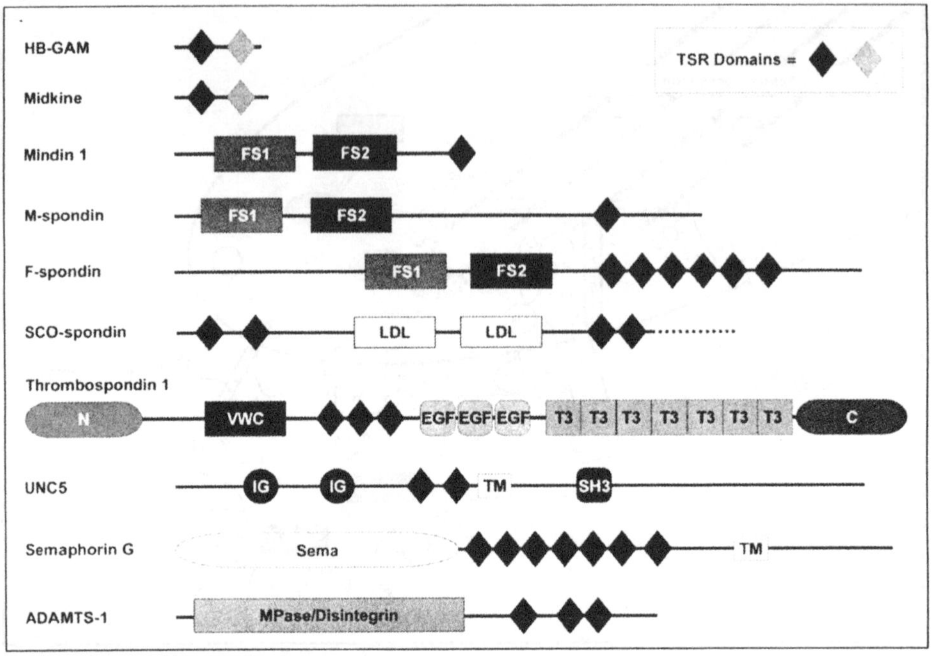

Fig. 3. TSR domains in extracellular matrix and cell surface proteins. The domain designations are as follows: C, thrombospondin C-terminal domain; EGF, epidermal growth factor-like domain; FS1 & FS2, F-spondin homology 1 & 2; LDL, low density lipoprotein-like domain; IG, immunoglobulin domain; N, thrombospondin N-terminal domain; Sema, semaphorin domain; SH3, Src homology 3; T3, thrombospondin type III repeat; TM, transmembrane region; TSR, thrombospondin type I repeat; VWC, von Willebrand factor type C domain; MPase/Disintegrin, metalloproteinase and disintegrin domain. The dotted line in SCOspondin indicates an incomplete sequence.

In general, the TSR module mediates binding to the cell surface or extracellular matrix. In both cases, heparin-type glycans appear to be essential binding partners. This seems clear, e.g., for the TSR regions of HB-GAM (21) and MK,[22] and many other proteins of the superfamily apparently also bind heparin/heparan sulfate through their TSR domains (for review, see Ref. 20). The occurrence of a wide variety of TSR sequences in a multitude of proteins throughout the animal kingdom might create a versatile binding system for the structurally diverse heparan sulfate sequences[23] to regulate biological interactions. However, much remains to be learned about this intriguing protein-carbohydrate interaction.

Extensive literature exists of interactions of TSR proteins with cells in vitro (reviewed in Ref. 20). For both thrombospondin and nonthrombospondin members of the superfamily, cell attachment, motility and guidance form the major line of findings as regarding interactions with various cell types. From the viewpoint that TSR proteins are—with a few exceptions—highly expressed in the developing nervous system, their effects have been widely studied using neuronal cells. In general, TSR proteins promote neurite outgrowth in many types of neurons. In fact, HB-GAM[24] and F-spondin[25] were initially defined by screening for proteins that enhance neurite outgrowth in central neurons, and in both cases the TSR regions are currently known to be indispensable for interactions with the neuron surface. A recently cloned component of Reissner's tract, SCO (subcommissural organ) spondin also promotes neurite outgrowth in a manner that apparently depends on the TSR regions of the protein.[26]

The semaphorin five family proteins also have TSR regions (Fig. 3), but their role in neurite extension has not been studied. From the viewpoint that semaphorins are repulsive for neurites due to their Sema domain but also attractive in some cases, the possible attractive role of the semaphorin TSR regions appears an obvious possibility. However, the role in attracting neurites does not either appear quite straightforward for the TSR domains: although the matrix-bound form of, e.g., HB-GAM, is strongly attractive for neurites, an excess of the protein in solution may also inhibit neurite extension.[27] Another protein family, where the role of the TSR regions is still unclear, is the UNC-5 family of proteins (Fig. 3). UNC-5 acts as a receptor of UNC-6, currently designated as netrin, in neurite growth/guidance in *C. elegans*. As in the case of semaphorins, netrin signaling (reviewed in Ref. 2) also has both attractive and repellant roles in neurite growth and guidance.

A few members of the TSR superfamily proteins have been suggested to play a role in tumor progression. In the case of thrombospondin-1 (TSP-1), the role in tumor formation has been suggested to depend on the effects of TSP-1 on endothelial cell migration and angiogenesis (for review, see Ref. 28). In fact, TSP-1 and many other TSR proteins are highly expressed in both the nervous system and in vascular tissues, and it seems reasonable that they may have similar regulatory roles in controlling motility of neurites in the context of development and regeneration and the motility of endothelial cells required for angiogenesis. In addition to angiogenesis, the TSR region of TSP-1 is suggested to have direct effects on tumor cells,[29] but the whole picture of the role of TSP-1 in cancer is still unclear. A wide literature exists on the roles of HB-GAM/pleiotrophin and MK in tumor progression (for review, see Ref. 30). In general, the roles of HB-GAM and MK in enhancing cancer progression have been ascribed to effects on proliferation of tumor cells or to effects on angiogenesis. From the viewpoint that HB-GAM/MK and the TSR proteins in general have been shown to control motility in many cell systems, the possibility that HB-GAM/MK regulate invasive migration of tumor cells in an autocrine/paracrine manner would be worth studying.

TSR proteins bind to several types of cell surface receptors, and the whole picture of the relevant interactions is still far from being clear. From their distinct binding to heparin/heparan sulfate, it seems probable that syndecans play a role in TSR/cell interactions. Currently, several types of evidence point to an important role of syndecans (see below the Chapter on syndecans and Fig. 1). Syndecans appear relevant transmembrane receptors of TSRs both from the viewpoint of neurite outgrowth and tumor cell migration.

RPTP (receptor-type tyrosine phosphatase) ζ/β, a transmembrane chondroitin sulfate proteoglycan, also binds both HB-GAM and MK[31](for a review, see Ref. 32), and perhaps other TSR proteins as well. RPTP ζ/β mediates migratory responses of neurons[33] (see Fig. 1) and fits well in the postulated role of TSR proteins in cell motility regulation. TSR protein binding to RPTP ζ/β, based on studies using HB-GAM/pleiotrophin, depends at least partially on the glycan chains of the receptor and is equally sensitive to heparin as the HB-GAM binding to syndecan.[31,34]

CD36, also called class B scavenger receptor, acts as the TSR binding site of TSP-1 (for reviews, see Refs. 20 and 35). CD36 is suggested to mediate effects of TSRs on endothelial cells, monocytes and tumor cells, but its possible role in neurons is currently not known.

Very recently, the orphan receptor tyrosine kinase ALK (anaplastic lymphoma kinase) has been reported to bind to HB-GAM/pleiotrophin.[36] In addition, ALK has been shown to be expressed in neurons and to induce neurite outgrowth.[37] Signaling through ALK would therefore be relevant both from the viewpoint of neurite outgrowth and tumor invasion. Further work is clearly warranted to clarify the role of ALK in the signaling mechanism of HB-GAM and other TSR proteins. Up to now, HB-GAM appears to be the first ligand suggested for the ALK kinase.

Syndecans

Syndecans constitute a family of four transmembrane proteoglycans (for reviews, see Refs. 38 and 39). Syndecans act as coreceptors enhancing ligand binding to signaling receptors. Examples of coreceptor functions of syndecans include Wnt-1 (wingless), FGF (fibroblast growth factor) and Shh (sonic hedgehog) signaling. Recent evidence suggests that syndecans play important roles due to their coreceptor function both in development and in tumor formation. For example, syndecan-1 is required for Wnt-1-induced formation of mammary tumors in mice.[40] In addition, syndecans function as cell matrix receptors in collaboration with integrins.[16]

In addition to their coreceptor functions, syndecans also possess independent signaling capacity. In affinity isolation, the cytosolic tail of syndecan-3 (N-syndecan) binds from brain extracts a protein complex in which cortactin and Src-kinase occur as prominent components. Based on this finding and on cell-biological evidence, cortactin/Src kinase signaling has been suggested to mediate neurite outgrowth and brain plasticity upon binding of HB-GAM to syndecan-3.[41,42] Molecular organization of the protein complex binding to the cytosolic tail of syndecan-3 or other syndecans has not been fully clarified, but one component binding directly to the C-terminal end of syndecan is the PDZ-domain protein CASK/LIN-2 that may interact with cortactin/Src-kinase (see Fig. 2).[42,43]

Cortactin appears to be a key regulator of cell motility, both in neurite outgrowth and tumor cell migration. Cortactin is an F-actin-binding protein that was initially defined as a major substrate of Src kinase.[44] Since cortactin undergoes in cells a transient change in phosphorylation upon ligation of syndecan-3 that promotes protrusion of the neurite,[41] a more detailed analysis has been carried out using live cell imaging. These studies have suggested that cortactin is involved both in plasma membrane protrusion and intracellular vesicle movement, and colocalizes with the Arp 2/3 complex[45] that is the key regulator of actin nucleation in the context of cell motility. Very recently, cortactin also has been shown to enhance the activity of Arp 2/3 in a cooperative manner with N-WASP.[46-48]

Although the role of cortactin has not yet been studied using primary neurons, its distinct localization to the advancing plasma membrane at the growth cone, together with the mechanistic studies cited above, make it highly likely that cortactin is a key player in neurite outgrowth as in other forms of cell motility. Interestingly, cortactin is highly expressed in extremely motile and invasive tumor cells, which is due to amplification at chromosome 11q13.[49] It thus appears that tumor cells exploit the same or similar cortactin-based mechanism that is used during development and/or regeneration under conditions that require high motility, of which neuronal growth cone migration is a prominent example. In addition, it is worth noting that migration of osteosarcoma cells appears to use the same HB-GAM/syndecan-3/cortactin/ Src kinase mechanism as the neuronal growth cones.[50] One should also note that the cortactin-dependent mechanism is likely to be used by many transmembrane signaling mechanisms, of which syndecan-mediated motile behavior of the plasma membrane (Fig. 2) is one example.

Interaction of Extracellular Matrix Molecules with the Immunoglobulin Superfamily Proteins

Members of the immunoglobulin superfamily (IgSF) are mainly involved in homophilic and heterophilic intercellular adhesive interactions. However, the interaction of some IgSF members with extracellular matrix components in the development and plasticity of the nervous system adds complexity to the functions of these molecules (reviewed in Refs. 51 and 52; see also Fig. 1). For example, the F11 glycoprotein was found to interact with the extracellular matrix molecules tenascin-R and tenascin-C but has also been shown to bind to axonal adhesion receptors, such as L1/NgCAM (neuron–glia cell adhesion molecule), NrCAM (NgCAM-related cell adhesion molecule), and receptor protein tyrosine phosphatase (RPTP) ζ/β. The most complex interaction pattern has been defined for the axonal glycoprotein L1, which binds to at least seven different proteins in addition to itself: TAG-1/axonin-1, F11, DM-GRASP, laminin, phosphacan, neurocan, and to some integrins.[51]

L1 is involved in the correct patterning of neuronal connections by stimulating selective axon extensions and by stabilizing fasciculated projections. For example, L1 appears to be required for the formation of the corticospinal tract in humans and mice and several mutations in the L1 gene have been linked to human hereditary brain malformations (reviewed in Ref.

53). Although L1 is referred to as a neural cell adhesion molecule, its expression is not restricted to the nervous system: it is also found on several nonneuronal cells, among them melanoma cells and hematopoietic tumor cells.

Members of the immunoglobulin superfamily have been shown to regulate various signaling pathways and cytoskeletal components, such as Src-family kinases, casein kinase II, $p90^{rsk}$ S6-family kinase and ankyrin (reviewed in Refs. 51 and 53; Fig. 1), suggesting direct regulation of the cytoskeletal machinery by the extracellular matrix molecules interacting with these IgSF proteins.

RAGE (Receptor for Advanced Glycation End products) was initially identified as a receptor of AGEs (Advanced Glycation End products) that accumulate in diabetes and during senescence. RAGE belongs to the immunoglobulin superfamily bearing closest homology to NCAM (Neural Cell Adhesion Molecule). Search of additional RAGE-binding molecules has resulted in isolation of several novel RAGE ligands including amyloid-β peptide, S100/calgranulin family proteins and amphoterin (reviewed in Ref. 54). Amphoterin is a protein expressed and secreted by developing neurons and various types of transformed cells.[55,56] RAGE and amphoterin colocalize in developing brain and the ectodomain of RAGE inhibits amphoterin-induced neurite outgrowth in forebrain neurons.[57] The RAGE-mediated process extension depends on the Rho-family GTPases Cdc42 and Rac.[58]

The expression of amphoterin is high in migrating cells but is downregulated upon cell-to-cell contact, suggesting that the level of amphoterin in cells may act as a sensor for cell contact-dependent inhibition of cell migration. Reduction of amphoterin expression by antisense oligonucleotides or inhibition by anti-amphoterin antibodies inhibits haptotactic transfilter migration to laminin.[59] Recently, it was shown blockade of amphoterin-RAGE interaction decreases invasion and growth of both implanted and spontaneously developing tumors by suppressing the activation of MAP kinase pathways and the activity of matrix metalloproteinases (MMPs).[60] A model has been proposed according to which amphoterin is an autocrine/paracrine regulator of invasive migration activating intracellular signals leading to reorganization of the actin cytoskeleton at the same time with local activation of proteolytic cascades on the cell surface (reviewed Ref. 32; see also Fig. 2).[56]

S100 proteins comprise a multigenic family of EF-hand type Ca^{2+}-binding proteins. Some S100 members are released or secreted into the extracellular space and exert trophic or toxic effects depending on their concentration, act as chemoattractants for leukocytes, modulate cell proliferation, or regulate macrophage activation.[61] Interestingly, RAGE has been shown to be a receptor for S100 family of proteins.[62] The secreted S100 proteins appear to have similar effects and overlapping binding to RAGE as amphoterin although S100 proteins are generally not considered as ECM-bound molecules but rather as cytokines or neurotrophic factors. S100B is a well-known promoter of neurite outgrowth and many members of the S100 family have been shown to be overexpressed in tumor cells (e.g., S100A1, S100A4, S100A6, S100A7 and S100B).[61] The expression of S100A4 (also known as metastasis-associated protein Mts1) correlates with high invasiveness in many tumor cells and inhibition of S100A4 function by antisense oligonucleotides suppresses cell motility and in vitro invasiveness.[63] In fact, S100A4 is perhaps the best prognostic marker in breast cancer known to date. Recently, S100A4 was also shown to be a potent promoter of neurite outgrowth.[64] Although other receptors for S100 proteins might exist, RAGE appears to mediate at least some of the cell motility promoting effects of secreted S100 proteins. Recently, some S100 proteins were shown to coregulate neurite outgrowth together with amphoterin through RAGE activation.[65]

Proteolytic Mechanisms in the Cell-Matrix Contacts

Cellular migration is critically dependent on interplay between forces of attachment and detachment. It has long been hypothesized that cells might focus proteinases at their leading edge, where proteolysis can direct migration. The plasmin system driven by urokinase-type plasminogen activator (uPA) and tissue-type plasminogen activator (tPA) operates in such a manner. The uPA receptor (uPAR) has been shown to be spatially and temporally associated with cellular structures that regulate cell adhesion, migration and invasion, colocalizing with integrins in focal contacts and at the leading edge of migrating cells.[66]

The activity of the plasmin system has been suggested to be involved in neurite outgrowth and neuronal migration for a long time. Originally plasminogen activator was found to be released from neuronal growth cones resulting in growth cone proteolytic zones.[67] Although both plasminogen activators are expressed in the developing nervous system during the period of axonal elongation,[68] only recently it has become evident that plasminogen activators play a crucial role regulating motile processes in the developing nervous system (reviewed in Ref. 69). Mice lacking the tPA gene display a markedly retarded rate of neuronal migration through the cerebellar molecular layer by granule neurons.[70] In addition, the tPA-deficient mice display decreased seizure-induced hippocampal mossy fiber outgrowth.[71] On the other hand, uPA has been suggested to regulate cranial neural crest cell migration.[72]

It is generally accepted that the ECM of the nervous system can present to neurons both stimulatory and inhibitory cues for neurite outgrowth. Plasmin can act either directly or through activation of matrix metalloproteinases (MMPs; reviewed in other Chapters in this book) to promote motility by breaking down physical hindrances and inactivating inhibitory matrix components.[73] Neuronal MMP-2 has been shown to promote neurite outgrowth by inactivating an inhibitory chondroitin sulfate proteoglycan and unmasking the neurite-promoting activity of associated laminin in the endoneurial basal lamina.[74] However, none of the mice lacking various MMPs have displayed remarkable alterations in the nervous system. The overall minimal phenotypes observed to date may be due to redundancy suggesting that generation of doubly and multiply MMP deficient mice may be required to unmask full MMP function in the developing nervous system in vivo.

Integrins have been shown to interact directly with MMPs. MMP-2 binds to integrin $\alpha v\beta 3$ and is thus localized, in a proteolytically active form, on the surface of invasive tumor cells or endothelial cells. This localization appears to provide migratory cells with coordinated matrix degradation and cellular motility, thus facilitating cellular invasion processes.[75] Furthermore, an association between ligation of various integrins and an increase in metalloproteinase expression has been documented.[8] The GPI-linked uPA receptor has been shown to interact with $\beta 1$, $\beta 2$ and $\beta 3$ integrins activating key signaling pathways (e.g., the classical MAP kinase pathway; Fig. 2) and modulating integrin function during adhesion.[15] As with tumor cells, the interaction of integrins with the proteolytic systems might be utilized by neuronal growth cones to promote penetration into the matrix. It seems plausible that integrins are part of a multimolecular complex involving matrix molecules and proteolytic components on the outer surface of the cell and cytoskeletal and signaling molecules inside the cell.

The interaction of RAGE and amphoterin provides an unorthodox example of an immunoglobulin superfamily cell surface molecule interacting with a matrix-bound protein coupled to the proteolytic mechanisms acting on the cell surface. Amphoterin binds both plasminogen activators (tPA and uPA) and plasminogen resulting in a ternary complex facilitating plasmin production.[56] The ternary complex appears transient since activation of plasminogen results in hydrolysis of amphoterin thus creating temporary binding sites for the motile processes (Fig. 2). Interestingly, it was shown recently that implanted tumors expressing RAGE have increased activity of MMP-2 and MMP-9 correlating with invasive capacity of tumor cells.[60] However, currently it is not known whether this increase in MMP activity is due to the interaction of cell surface RAGE with amphoterin-plasmin system or because of some other mechanisms.

In addition to matrix degradation, proteolytic processing of the ECM can regulate cellular behavior by releasing matrix-bound growth factors and generating modulatory neo-epitopes from processed matrix molecules. Plasmin can activate or liberate growth factors from extracellular matrix including latent transforming growth factor (TGFβ-1), basic fibroblast growth factor (bFGF) and vascular endothelial growth factor (VEGF).[76] Angiostatin, a proteolytic plasmin fragment, and endostatin, a proteolytic collagen XVIII fragment, inhibit neovascularization whereas elastin fragments may regulate monocyte migration in chronic inflammation.[73] Specific cleavage of laminin-5 (Ln-5) by MMP-2 has been shown to induce migration of breast epithelial cells by exposing a putative cryptic promigratory site on laminin-5 that triggers cell motility. This altered form of laminin-5 is found in tumors and in tissues undergoing remodeling, but not in quiescent tissues.[77]

Adhesion mediated by many cell surface receptors can be regulated by protease-mediated cleavage of the ectodomain of the receptor. Metalloproteases and plasmin can release the L1 ectodomain from the melanoma cell surface leading to downregulation of L1 adhesivity (reviewed in Ref. 78). The release of the syndecan ectodomains by membrane shedding is thought to play a regulatory role in the physiology of syndecans.[38] However, the release mechanism of syndecan shedding is currently not understood.

Conclusions and Future Prospects

Both growth cone migration for the extension of neurites and tumor cell invasion require localized proteolytic activity at the leading membrane, which creates space for membrane processes that advance in tissue. In the case of neuronal development, it now appears clear that local proteolytic activity of the growth cones is required for migration of neuronal precursors, but less is known about the role of proteolytic activity during later stages of development. However, neurite extension for migration and pathway development are generally thought to use the same or similar mechanisms, and it appears likely that local proteolytic activity is also required for the development of neuronal connections in the differentiating nervous system and even in the adult nervous system for remodeling of the connections in the context of neuronal plasticity.

Generation of plasmin activity in the growth cones, like at the leading membrane of tumor cells, has been well documented. In the case of tumor cells, the current research has emphasized the role of matrix metalloproteases (MMPs) in tissue invasion. In the case of neuronal development, only very few studies have addressed the role of MMPs in the extension of neuronal or glial cell processes. However, a role for MMPs in neurite extension also has been suggested, and this also appears likely from the viewpoint that local plasmin activity is known to enhance MMP activation.

Several examples show that transmembrane signaling for neurite outgrowth and tumor cell invasion use similar or analogous mechanisms. One example is the immunoglobulin superfamily member RAGE that has been shown to mediate neurite extension when ligated by matrix-bound amphoterin. Very recently, amphoterin/RAGE signaling has been shown to enhance tumor invasion in several different mouse models. The signaling mechanism of RAGE is clearly relevant to enhance cell motility. An additional attractive feature in amphoterin/RAGE signaling is that amphoterin also binds plasminogen and plasminogen activators. Therefore, besides displaying a relevant cytoskeletal regulation required for motility, amphoterin/RAGE signaling may be linked to local proteolytic activation, as might be expected for a mechanism mediating invasive process extension.

Intracellular mechanisms mediating motility are expectedly similar in many cell types, including neurons and tumor cells. Arp 2/3 complex is currently suggested to be a key regulator of migration in various cell types. Cortactin, a major Src-kinase substrate that binds to F-actin, has been very recently shown to interact with the Arp 2/3 complex and enhance its activity. Interestingly, cortactin is overexpressed in many highly motile and invasive tumor cells and is also a prominent component of neuronal growth cones. It appears that cortactin is one component in the cytosolic signaling network that can be unregulated under conditions requiring a high level of motility, like neurite extension and tumor invasion.

Although it may be useful to test candidate mechanisms for tumor cell invasion based on knowledge of mechanisms mediating neuronal growth cone migration or vice versa, a major challenge will be to find ways for therapeutic interventions. Obviously, therapies based on inhibition of tumor cell invasion may be eventually developed to complement other forms of cancer treatment. Such anti-invasion therapy might be especially valuable in the case of brain tumors that characteristically destroy surrounding tissue by local invasion. In addition to tumors, nervous tissue traumas form a major medical challenge, especially in the case of the central nervous system, where neurite extension to regenerate the connections only occurs to a very limited degree. To date, only limited success has been achieved in attempts to enhance growth of the connections after trauma of the spinal cord. Further research to provide tools to control growth cone migration and related forms of motility in vivo is undoubtedly warranted.

References

1. Gordon-Weeks PR. Neuronal Growth Cones. Cambridge: Cambridge University Press, 2000.
2. Tessier-Lavigne M, Goodman CS. The molecular biology of axon guidance. Science 1996; 274:1123-1133.
3. Keely P, Parise L, Juliano R. Integrins and GTPases in tumour cell growth, motility and invasion. Trends Cell Biol 1998; 8:101-107.
4. Frisch SM, Ruoslahti E. Integrins and anoikis. Curr Opin Cell Biol 1997; 9:701-706.
5. Luckenbill-Edds L. Laminin and the mechanism of neuronal outgrowth. Brain Res Rev 1997; 23:1-27.
6. Beck K, Hunter I, Engel J. Structure and function of laminin: Anatomy of a multidomain glycoprotein. FASEB J 1990; 4:149-160.
7. Gomez TM, Letourneau PC. Filopodia initiate choices made by sensory neuron growth cones at laminin/fibronectin borders in vitro. J Neurosci 1994; 14:5959-5972.
8. Varner JA, Cheresh DA. Integrins and cancer. Curr Opin Cell Biol 1996; 8:724-730.
9. Chan BM, Chan C, Matsuura N et al. In vitro and in vivo consequences of VLA-2 expression on rhabdomyosarcoma cells. Science 1992; 251:1600-1602.
10. Nip J, Shibata H, Loskutoff D et al. Human melanoma cells derived from lymphatic metastases use integrin avb3 to adhere to lymph node vitronectin. J Clin Invest 1992; 90:1406-1413.
11. Morini M, Mottolese M, Ferrari N et al. The $\alpha3\beta1$ integrin is associated with mammary carcinoma cell metastasis, invasion, and gelatinase B (MMP-9) activity. Int J Cancer 2000; 87:336-342.
12. Macieira-Coelho A, ed. Signaling Through the Cell Matrix. Heidelberg: Springer-Verlag, 2000.
13. Mullins RD. How WASP-family proteins and the Arp2/3 complex convert intracellular signals into cytoskeletal structures. Curr Opin Cell Biol 2000; 12:91-96.
14. Dickson BJ. Rho GTPases in growth cone guidance. Curr Opin Neurobiol 2001; 11:103-110.
15. Ossowski L, AguirreGhiso JA. Urokinase receptor and integrin partnership: Coordination of signaling for cell adhesion, migration and growth. Curr Opin Cell Biol 2000; 12:613–620.
16. Couchman JR, Woods A. Syndecan-4 and integrins: Combinatorial signaling in cell adhesion. J Cell Sci 1999; 112:3415-3420.
17. Fages C, Kaksonen M, Kinnunen T et al. Regulation of mRNA localization by transmembrane signalling: Local interaction of HB-GAM (heparin-binding growth-associated molecule) with the cell surface localizes beta-actin mRNA. J Cell Sci 1998; 111:3073-3080.
18. Chicurel M, Singer RH, Meyer CJ et al. integrin binding and mechanical tension induce movement of mRNA and ribosomes to focal adhesions. Nature 1998; 392:730-733.
19. Kislauskis EH, Zhu X, Singer RH. Beta-actin messenger RNA localization and protein synthesis augment cell motility. J Cell Biol 1997; 136:1263-1270.
20. Adams JC, Tucker RP. The thrombospondin type 1 repeat (TSR) superfamily: Diverse proteins with related roles in neuronal development. Dev Dyn 2000; 218:280-299.
21. Kilpelainen I, Kaksonen M, Avikainen H et al. Heparin-binding growth-associated molecule contains two heparin-binding beta-sheet domains that are homologous to the thrombospondin type I repeat. J Biol Chem 2000; 275:13564-13570.
22. Iwasaki W, Nagata K, Hatanaka H et al. Solution structure of midkine, a new heparin-binding growth factor. EMBO J 1997; 16:6936-6946.
23. Lindahl U, Kusche-Gullberg M, Kjellén L. Regulated diversity of heparan sulfate. J Biol Chem 1998; 273:24979-24982.
24. Rauvala H. An 18-kd heparin-binding protein of developing brain that is distinct from fibroblast growth factors. EMBO J 1989; 8:2933-2941.
25. Klar A, Baldassare M, Jessell TM. F-spondin: A gene expressed at high levels in the floor plate encodes a secreted protein that promotes neural cell adhesion and neurite extension. Cell 1992; 69:95-110.
26. Gobron S, Creveaux I, Meiniel R et al. Subcommissural organ/Reissner's fiber complex: Characterization of SCOspondin, a glycoprotein with potent activity on neurite outgrowth. Glia 2000; 32:177-191.

27. Raulo E, Julkunen I, Merenmies J et al. Secretion and biological activities of heparin-binding growth-associated molecule. Neurite outgrowth-promoting and mitogenic actions of the recombinant and tissue-derived protein. J Biol Chem 1992; 267:11408-11416.
28. Bornstein P. Diversity of function is inherent in matricellular proteins: An appraisal of thrombospondin 1. J Cell Biol 1995; 130:503-506.
29. Tuszynski GP, Rothman VL, Papale M et al. Identification and characterization of a tumor cell receptor for CSVTCG, a thrombospondin adhesive domain. J Cell Biol 1993; 120:513-521.
30. Zhang N, Deuel TF. Pleiotrophin and midkine, a family of mitogenic and angiogenic heparin-hinding growth and differentiation factors. Curr Opin Hematol 1999; 6:44-50.
31. Maeda N, Nishiwaki T, Shintani T et al. 6B4 proteoglycan/phosphacan, an extracellular variant of receptor-like protein-tyrosine phosphatase zeta/RPTPbeta, binds pleiotrophin/heparin-binding growth-associated molecule (HB-GAM). J Biol Chem 1996; 271:21446-21452.
32. Rauvala H, Huttunen HJ, Fages C et al. Heparin-binding proteins HB-GAM (pleiotrophin) and amphoterin in the regulation of cell motility. Matrix Biol 2000; 19:377-387.
33. Maeda N, Noda M. Involvement of receptor-like protein tyrosine phosphatase zeta/RPTPbeta and its ligand pleiotrophin/heparin-binding growth-associated molecule (HB-GAM) in neuronal migration. J Cell Biol 1998; 142:203-216.
34. Milev P, Chiba A, Haring M et al. High affinity binding and overlapping localization of neurocan and phosphacan/protein-tyrosine phosphatase-zeta/beta with tenascin-R, amphoterin, and the heparin-binding growth-associated molecule. J Biol Chem 1998; 273:6998-7005.
35. Greenwalt DE, Lipsky RH, Ockenhouse CF et al. Membrane glycoprotein CD36: A review of its roles in adherence, signal transduction, and transfusion medicine. Blood 1992; 80:1105-1115.
36. Stoica GE, Kuo A, Aigner A et al. Identification of ALK (anaplastic lymphoma kinase) as a receptor for the growth factor pleiotrophin. J Biol Chem 2001; 276:16772-16779.
37. Souttou B. Brunet-De Carvalho N, Raulais D et al. Activation of anaplastic lymphoma kinase receptor tyrosine kinase induces neuronal differentiation through the mitogen-activated protein kinase pathway. J Biol Chem 2001; 276:9526-9531.
38. Bernfield M, Gotte M, Park PW et al. Functions of cell surface heparan sulfate proteoglycans. Annu Rev Biochem 1999; 68:729-777.
39. Bandtlow CE, Zimmermann DR. Proteoglycans in the developing brain: New conceptual insights for old proteins. Physiol Rev 2000; 80:1267-1290.
40. Alexander CM, Reichsman F, Hinkes MT et al. Syndecan-1 is required for Wnt-1-induced mammary tumorigenesis in mice. Nat Genet 2000; 25:329-332.
41. Kinnunen T, Kaksonen M, Saarinen J et al. Cortactin-Src kinase signaling pathway is involved in N-syndecan-dependent neurite outgrowth. J Biol Chem 1998; 273:10702-10708.
42. Lauri SE, Kaukinen S, Kinnunen T et al. Regulatory role and molecular interactions of a cell-surface heparan sulfate proteoglycan (N-syndecan) in hippocampal long-term potentiation. J Neurosci 1999; 19:1226-1235.
43. Hsueh YP, Sheng M. Regulated expression and subcellular localization of syndecan heparan sulfate proteoglycans and the syndecan-binding protein CASK/LIN-2 during rat brain development. J Neurosci 1999; 19:7415-7425.
44. Wu H, Reynolds AB, Kanner SB et al. Identification and characterization of a novel cytoskeleton-associated pp60src substrate. Mol Cell Biol 1991; 11:5113-5124.
45. Kaksonen M, Peng HB, Rauvala H. Association of cortactin with dynamic actin in lamellipodia and on endosomal vesicles. J Cell Sci 2000; 113:4421-4426.
46. Uruno T, Liu J, Zhang P et al. Activation of Arp2/3 complex-mediated actin polymerization by cortactin. Nat Cell Biol 2001; 3:259-266.
47. Weaver AM, Karginov AV, Kinley AW et al. Cortactin promotes and stabilizes Arp2/3-induced actin filament network formation. Curr Biol 2001; 11:370-374.
48. Zettl M, Way M. New tricks for an old dog? Nat Cell Biol 2001; 3:E74-5.
49. Lagarkova MA, Boitchenko VE, Mescheryakov AA et al. Human cortactin as putative cancer antigen. Oncogene 2000; 19:5204-5207.
50. Imai S, Kaksonen M, Raulo E et al. Osteoblast recruitment and bone formation enhanced by cell matrix-associated heparin-binding growth-associated molecule (HB-GAM). J Cell Biol 1998; 143:1113-1128.
51. Sonderegger P. Axonin-1 and NgCAM as "recognition" components of the pathway sensor apparatus of growth cones: A synopsis. Cell Tissue Res 1997; 290:429-439.
52. Schachner M. Neural recognition molecules and synaptic plasticity. Curr Opin Cell Biol 1997; 9:627-634.
53. Brümmendorf T, Kenwrick S, Rathjen FG. Neural cell recognition molecule L1: From cell biology to human hereditary brain malformations. Curr Opin Neurobiol 1998; 8:87-97.

54. Schmidt AM, Yan SD, Yan SF et al. The biology of the receptor for advanced glycation end products and its ligands. Biochim Biophys Acta 2000; 1498:99-111.
55. Rauvala H, Pihlaskari R. Isolation and some characteristics of an adhesive factor of brain that enhances neurite outgrowth in central neurons. J Biol Chem 1987; 262:16625-16635.
56. Parkkinen J, Raulo E, Merenmies J et al. Amphoterin, the 30-kDa protein in a family of HMG1-type polypeptides. Enhanced expression in transformed cells, leading edge localization, and interactions with plasminogen activation. J Biol Chem 1993; 268:19726-19738.
57. Hori O, Brett J, Slattery T et al. The receptor for advanced glycation end products (RAGE) is a cellular binding site for amphoterin: Mediation of neurite outgrowth and co-expression of RAGE and amphoterin in the developing nervous system. J Biol Chem 1995; 270:25752-25761.
58. Huttunen HJ, Fages C, Rauvala H. Receptor for advanced glycation end products (RAGE)-mediated neurite outgrowth and activation of NF-KB require the cytoplasmic domain of the receptor but different downstream signaling pathways. J Biol Chem 1999; 274:19919-19924.
59. Fages C, Nolo R, Huttunen HJ et al. Regulation of cell migration by amphoterin. J Cell Sci 2000; 113:611-20.
60. Taguchi A, Blood DC, Del Toro G et al. Blockade of RAGE-amphoterin signalling suppresses tumour growth and metastases. Nature 2000; 405:354-360.
61. Donato R. Functional roles of S100 proteins, calcium-binding proteins of the EF-hand type. Biochim Biophys Acta 1999; 1450:191-231.
62. Hofmann MA, Drury S, Fu C et al. RAGE mediates a novel proinflammatory axis: A central cell surface receptor for S100/calgranulin polypeptides. Cell 1999; 97:889-901.
63. Sherbet GV, Lakshmi MS. S100A4 (MTS1) calcium-binding protein in cancer growth, invasion and metastasis. Anticancer Res 1998; 18:2415-2421.
64. Novitskaya V, Grigorian M, Kriajevska M et al. Oligomeric forms of the metastasis-related Mts1 (S100A4) protein stimulate neuronal differentiation in cultures of rat hippocampal neurons. J Biol Chem 2000; 275:41278-41286.
65. Huttunen HJ, Kuja-Panula J, Sorci G et al. Coregulation of neurite outgrowth and cell survival by amphoterin and S100 proteins through receptor for advanced glycation end products (RAGE) activation. J Biol Chem 2000; 275:40096-40105.
66. Andreasen PA, Kjoller L, Christensen L et al. The urokinase-type plasminogen activator system in cancer metastasis: A review. Int J Cancer 1997; 72:1-22.
67. Krystosek A, Seeds NW. Plasminogen activator release at the neuronal growth cone. Science 1981; 213:1532-1534.
68. Sumi Y, Dent MA, Owen DE et al. The expression of tissue and urokinase-type plasminogen activators in neural development suggests different modes of proteolytic involvement in neuronal growth. Development 1992; 116:625-637.
69. Seeds NW, Siconolfi LB, Haffke SP. Neuronal extracellular matrix proteases facilitate cell migration, axonal growth, and pathfinding. Cell Tissue Res 1997; 290:367-370.
70. Seeds NW, Basham ME, Haffke SP. Neuronal migration is retarded in mice lacking the tissue plasminogen activator gene. Proc Natl Acad Sci USA 1999; 96:14118-14123.
71. Wu YP, Siao CJ, Lu W et al. The tissue plasminogen activator (tPA)/plasmin extracellular proteolytic system regulates seizure-induced hippocampal mossy fiber outgrowth through a proteoglycan substrate. J Cell Biol 2000; 148:1295-1304.
72. Agrawal M, Brauer PR. Urokinase-type plasminogen activator regulates cranial neural crest cell migration in vitro. Dev Dyn 1996; 207:281-290.
73. Shapiro S. Matrix metalloproteinase degradation of extracellular matrix: Biological consequences. Curr Opin Cell Biol 1998; 10:602-608.
74. Zuo J, Ferguson TA, Hernandez YJ et al. Neuronal matrix metalloproteinase-2 degrades and inactivates a neurite-inhibiting chondroitin sulfate proteoglycan. J Neurosci 1998; 18:5203-5211.
75. Brooks PC, Stromblad S, Sanders L et al. Localization of matrix metalloproteinase MMP-2 to the surface of invasive cells by interaction with integrin $\alpha v\beta 3$. Cell 1996; 85:683-693.
76. Saksela O, Rifkin DB. Cell-associated plasminogen activation: Regulation and physiological functions. Annu Rev Cell Biol 1988; 4:93-126.
77. Giannelli G, Falk-Marzillier J, Schiraldi O et al. Induction of cell migration by matrix metalloprotease-2 cleavage of laminin-5. Science 1997; 277:225-228.
78. Kamiguchi H, Lemmon V. IgCAMs: Bidirectional signals underlying neurite growth. Curr Opin Cell Biol 2000; 12:598-605.

Chapter 7

Emmprin (CD147), a Tumor Cell Surface Inducer of Matrix Metalloproteinase Production

Bryan P. Toole

Introduction

Emmprin is a member of the Ig superfamily that plays an essential role in several normal tissues but is particularly enriched on the surface of malignant tumor cells in vitro and in vivo. Tumor cell emmprin stimulates production of several matrix metalloproteinases (MMPs) by fibroblasts and endothelial cells, but it also acts in an autocrine fashion to increase MMP synthesis and invasiveness in tumor cells themselves. In addition, emmprin acts as a docking protein for interstitial collagenase on the surface of tumor cells. Increased expression of emmprin in weakly malignant, human breast cancer cells leads to dramatic augmentation of tumor growth and invasion in vivo.

Several important aspects of tumor progression involve proteolytic modification of the pericellular matrix around tumor cells by matrix metalloproteinases (MMPs). For example, MMPs have been implicated in invasion through basement membranes and interstitial matrices, angiogenesis, and tumor cell growth. Strong support for the involvement of MMPs in tumor invasion in vivo comes from experiments in which natural or synthetic inhibitors of MMPs were shown to prevent metastasis in experimental animal models.[1-3] In this chapter I will discuss the function of emmprin, an important regulator of MMP synthesis, in tumor cell invasion.

Emmprin was initially identified as a factor associated with the surface of tumor cells that stimulates synthesis of matrix metalloproteinases by fibroblasts.[4,5] On cloning of emmprin cDNA,[4] it became apparent that emmprin is a member of the Ig superfamily and that it is identical to human basigin[6] and the M6 antigen present on membranes of leukocytes from patients with arthritis,[7] proteins whose functions were not then known. Emmprin is homologous to proteins independently discovered in a wide variety of systems in other species, e.g., mouse gp42 and basigin,[8,9] rat OX47 and CE9,[10,11] and chick 5A11, HT7 and neurothelin.[12-14] Emmprin and its homologs are now also termed CD147. In addition, it is evident that there is a family of molecules related to emmprin. For example, embigin and basigin are closely related,[6] and rat synaptic membranes contain two major Ig superfamily proteins, gp65 and gp55, that are related but not identical to the rat homolog of emmprin.[15]

Tumor Cell Emmprin Stimulates Fibroblast Production of MMPs

A surprising development with respect to MMP production in tumors was the discovery that most MMPs, e.g., interstitial collagenase (MMP-1), collagenase-3 (MMP-13), gelatinase A (MMP-2), gelatinase B (MMP-9), stromelysin-1 (MMP-3), stromelysin-3 (MMP-11), and membrane type-MMPs (MT-MMPs), are mainly produced by stromal fibroblasts associated with tumors.[16-20] Moreover, these stromal MMPs contribute significantly to tumor progression in vivo.[20-22] However, MMPs are produced both by stromal cells and by tumor cells, possibly depending on the stage of progression of the tumor, and both sources of MMPs are likely to be important.[17,23,24] Matrilysin (MMP-7) appears to be unique in its restriction to epithelial and carcinoma cells.[17,25]

The almost ubiquitous production of MMPs by stromal cells within tumors, but not within most normal adult tissues, implies that tumor cells may exert regulatory effects on the stromal cells, inducing them to express elevated levels of MMPs. Although it is clear that soluble cytokines and growth factors contribute to this process,[26-28] it is also apparent that tumor cell membrane-bound factors are involved. The first systematic investigation of the latter took place in the laboratory of Dr. Chitra Biswas, where initial experiments suggested that tumor cell-secreted or shed factors were responsible for stimulation of synthesis of MMP-1 by fibroblasts.[29,30] However, subsequent experiments in the Biswas lab showed that most of the MMP-1-stimulatory factor produced by B16 murine melanoma and LX-1 human lung carcinoma cells was plasma membrane-derived, and that this factor could act via direct cell-cell interaction or via shedding of the factor from the cell surface.[31,32] An activity-blocking monoclonal antibody was produced against the factor (originally called tumor cell-derived collagenase stimulatory factor or TCSF)[33] which led to its cloning and full characterization as a transmembrane glycoprotein and member of the Ig superfamily.[4,5] It was also shown to be present in normal tissue[34] and to stimulate production of several MMPs by fibroblasts,[35] and was thus renamed emmprin (extracellular matrix metalloproteinase inducer).[4] (Sadly, Chitra Biswas died in 1993, after having completed the molecular characterization of emmprin).

More recent data has revealed that purified emmprin not only stimulates synthesis of MMPs by fibroblasts but also by endothelial cells. Emmprin stimulates production of interstitial collagenase (MMP-1), gelatinase A (MMP-2) and stromelysin-1 (MMP-3) in both cell types (Refs. 5, 35; Zucker S, Ciao J, Rollo EE, Toole BP, unpublished results). Emmprin-mediated stimulation of MMP-1 synthesis in human lung fibroblasts is dependent on the activity of the MAP kinase, p38, but not ERK1/2 or SAPK/JNK.[36] A recent study has shown that emmprin also stimulates synthesis of membrane-type-MMPs (MT-MMPs) in co-cultures of human glioblastoma cells expressing high levels of emmprin with brain tumor-derived fibroblasts.[37] Both MT1- and MT2-MMP were stimulated in this system. Increased activation of MMP-2 by emmprin has also been observed,[35,37] presumably due to the action of MT-MMPs.[38,39] However, it has been noted that different fibroblast populations differ widely in their response to emmprin;[5,35] the basis for this difference has not yet been elucidated.

The effect of emmprin on tumor cell invasion has been examined in co-cultures of oral squamous cell carcinoma cells and peritumor-derived fibroblasts.[40] In this study the tumor cells were plated on a filter coated with reconstituted basement membrane matrix; the fibroblasts were plated in a well beneath the filter. Tumor cell invasion of the matrix was found to be dependent on emmprin and to result from emmprin stimulation of MMP-2 production, presumably by the fibroblasts.[40]

Autocrine Action of Emmprin Promotes Tumor Cell Invasiveness

Recent data suggest that emmprin acts in an autocrine as well as paracrine fashion. Transfection of weakly malignant MB-MDA436 human breast carcinoma cells with emmprin cDNA leads to an increase in MMP-2 and MT-MMP production (Ref. 41; Caudroy S, Polette M, Nawrocki-Raby B, Toole BP, Zucker S, Birembaut P, submitted for publication). These emmprin-transfected cells were found to be more invasive than vector-transfected controls. Similar findings have been made with the more malignant MDA-435 breast carcinoma cell line without transfection, in that MMP-2 production by and invasiveness of these cells were shown to be emmprin-dependent.[42] In the latter study it was also shown that soluble emmprin inhibits endogenous emmprin action,[42] most likely due to interference with homophilic interactions between emmprin molecules.[42-44]

Emmprin Docks MMP-1 on the Tumor Cell Surface

After synthesis and secretion, some MMPs bind back to the tumor cell surface. For example, MMP-2 binds to either $\alpha v\beta 3$ integrin[45] or to a TIMP2-MT-MMP complex; formation of the latter complex leads to activation of MMP-2.[38,39] A similar mechanism of binding and activation involving MT-MMP may occur with collagenase-3.[46] Gelatinase B can bind to the cell surface via interaction with CD44[47] or a component of collagen type IV.[48] Presentation of MMPs at these docking sites has been shown to promote tumor cell invasiveness.[38,45,47] In a recent study we have shown that, in addition to stimulating MMP production, emmprin is a docking protein for MMP-1.[49] We showed by phage display, affinity chromatography and immunocytochemistry that MMP-1 forms a complex with emmprin on the surface of human lung carcinoma cells. Since collagenase activity is essential for invasion of fibrous tissues,[50] localization of MMP-1 on the tumor cell surface would facilitate this process.

Emmprin Promotes Tumor Growth and Invasion In Vivo

Although it is now apparent that many normal embryonic and adult tissues express emmprin, the level of emmprin expression in tumors, especially malignant tumors, is usually much greater than in corresponding normal tissue.[36,51-55] For example, in one study, the relative distribution of emmprin and gelatinase A (MMP-2) mRNAs was compared by in situ hybridization in normal lung tissue vs squamous cell carcinomas of the lung and in benign mammary growths vs ductal carcinomas of the breast.[52] Emmprin mRNA was detected in all breast carcinomas and the majority of lung carcinomas. Both pre-invasive and invasive cancer cells were positive, but tumor stromal cells and peritumoral tissue showed insignificant emmprin mRNA reactivity. Normal and benign epithelia were negative. MMP-2 and MMP-1 mRNAs, on the other hand, were restricted to stromal cells close to tumor clusters.[36,52] The expression of emmprin mRNA was also analyzed by Northern blots which were then densitometrically scanned; the results showed low expression in normal or benign tissues but high levels at all stages of tumor progression in both lung and breast cancers.[52] Analyses of distribution within tumors made by quantitative image cytometry showed that high levels of emmprin mRNA were expressed in pre-invasive and invasive nests of tumor cells versus low amounts in normal or peritumoral tissues.[52] Both normal and tumor epithelia stained with antibody to emmprin, but expression of emmprin was much stronger in tumor tissue.[53] In other studies, emmprin levels were shown to be higher in transitional cell carcinomas of the bladder than in normal bladder epithelium,[51] and in malignant glioblastomas than in benign gliomas and normal brain tissue.[55] Although emmprin is expressed at a moderately high level in normal non-neoplastic keratinocytes,[34] its presence in oral squamous cell carcinoma is associated with MMP production and tumor cell invasion.[40]

Since malignant tumor cells often express emmprin in vivo and in vitro at much higher levels than normal and benign cells, we recently tested whether over-expression of emmprin stimulates tumor progression.[41] We used human breast carcinoma cells that produce slow-growing primary tumors in nude mice and express relatively low levels of emmprin. We transfected the cells with emmprin cDNA and selected stable transfectant clones with increased expression of emmprin. The emmprin transfectants grew at similar rates to vector-transfected controls in monolayer cell culture. The tumor cells were injected into the mammary fat pad of groups of 10 nude mice in three separate in vivo experiments using different transfectant clones. In all three experiments, the mice injected with emmprin transfectants grew large tumors over a 12 week period whereas controls grew small tumors that were primarily detectable only at autopsy. In addition, the emmprin transfectants gave rise to high levels of MMP expression and to extensive invasion into surrounding abdominal wall muscle whereas controls did not. Mouse survival was markedly decreased with the emmprin transfectants compared to controls.[41] We conclude that increased expression of emmprin leads to increased malignant tumor behavior.

The Functions of Emmprin are Diverse

Recently, a knockout mouse has been produced in which basigin, the murine homolog of emmprin, is lacking.[56] The null mutant is in most cases unable to undergo implantation, possibly due to the involvement of MMPs in this process.[57,58] However embryos that successfully implant and survive past birth have deficiencies in spermatogenesis,[56,59] retinal and photoreceptor development and maintenance,[60,61] other sensory functions,[62] and lymphocyte responses.[62] Any relevance of MMP stimulation to these latter processes has not been established.

Structural analyses have demonstrated that the transmembrane and cytoplasmic domains of emmprin are highly conserved among species, suggesting that these regions are of functional importance. The properties of the transmembrane region also suggest that intramembrane interactions with other proteins are likely to occur.[6,7,10] Emmprin interacts with integrins, $\alpha3\beta1$ and $\alpha6\beta1$, within the plasma membrane of HT1080 fibrosarcoma cells.[63] It acts as a chaperone for assembly of lactate transporters in the plasma membrane.[64] It binds to cyclophilin A, facilitating HIV virus entry into cells.[65] These interactions are likely to involve the transmembrane and/or cytoplasmic domains of emmprin. Again, however, it is not known whether proteolytic processes stimulated by emmprin are involved in any of these processes. Rather, it seems likely that emmprin has multiple functions, but the underlying mechanisms are presently unknown.

Conclusions

Increasingly, evidence is appearing that firmly establishes the importance of the stroma in carcinoma progression.[66-69] We propose that interactions of tumor cells and stromal cells lead to synthesis and activation of MMPs that in turn promote tumor invasiveness and that emmprin is a crucial component of these interactions. However, emmprin on the tumor cell surface also appears to be directly involved in tumor cell invasiveness, without stromal interactions, by autocrine stimulation of MMP synthesis and by docking of MMP-1 to the cell surface. It is becoming increasingly apparent that tumor cells create a pericellular environment in which many MMPs and other proteases become concentrated, thereby enhancing the ability of tumor cells to invade extracellular matrices and to process locally precursors of factors that promote tumor progression. Emmprin stimulation of MMP production could play a central role in these processes. However, emmprin is also involved in other pathological and physiological events that may or may not involve regulation of MMP synthesis. Whether or not emmprin serves more than one molecular function in malignant tumor cell behavior remains to be seen.

References

1. Khokha R. Suppression of the tumorigenic and metastatic abilities of murine B16-F10 melanoma cells in vivo by the overexpression of the tissue inhibitor of the metalloproteinases-1. J Natl Cancer Inst 1994; 86: 299-304.
2. Sledge GW, Qulali M, Goulet R, Bone EA, Fife R. Effect of matrix metalloproteinase inhibitor batimastat on breast cancer regrowth and metastasis in athymic mice. J Natl Cancer Inst 1995; 87:1546-1550.
3. Hua J, Muschel R. Inhibition of matrix metalloproteinase 9 expression by a ribozyme blocks metastasis in a rat fibrosarcoma model system. Cancer Res 1996; 56:5279-5284.
4. Biswas C, Zhang Y, DeCastro R, Guo H, Nakamura T et al. The human tumor cell-derived collagenase stimulatory factor (renamed EMMPRIN) is a member of the immunoglobulin superfamily. Cancer Res 1995; 55:434-439.
5. Guo H, Zucker S, Gordon MK, Toole BP, Biswas C. Stimulation of matrix metalloproteinase production by recombinant extracellular matrix metalloproteinase inducer from transfected chinese hamster ovary cells. J Biol Chem 1997; 272:24-27.
6. Miyauchi T, Masuzawa Y, Muramatsu T. The basigin group of the immunoglobulin superfamily: Complete conservation of a segment in and around transmembrane domains of human and mouse basigin and chick HT7 antigen. J Biochem 1991; 110:770-774.
7. Kasinrerk W, Fiebiger E, Steffanova I, Baumruker T, Knapp W et al. Human leukocyte activation antigen M6, a member of the Ig superfamily, is the species homologue of rat OX-47, mouse basigin, and chicken HT7 molecule. J Immunol 1992; 149:847-854.

8. Altruda F, Cervella P, Gaeta ML, Daniele A, Giancotti F et al. Cloning of cDNA for a mouse membrane glycoprotein (gp42): Shared identity to histocompatibility antigens, immunoglobulins and neural-cell adhesion molecules. Gene 1989; 85:445-452.
9. Miyauchi T, Kanekura T, Yamaoka A, Ozawa M, Miyazawa S et al. Basigin, a new, broadly distributed member of the immunoglobulin superfamily, has strong homology with both the immunoglobulin V domain and the β-chain of major histocompatibility complex class II antigen. J Biochem 1990; 107:316-323.
10. Fossum S, Mallett S, Barclay AN. The MRC OX-47 antigen is a member of the immunoglobulin superfamily with an unusual transmembrane sequence. Eur J Immunol 1991; 21:671-679.
11. Nehme CL, Cesario MM, Myles DG, Koppel DE, Bartles JR. Breaching the diffusion barrier that compartmentalizes transmembrane glycoprotein CE9 to the posterior-tail plasma membrane domain of the rat spermatozoon. J Cell Biol 1993; 120:687-694.
12. Fadool JM, Linser PJ. 5A11 antigen is a cell recognition molecule which is involved in neuronal-glial interactions in avian neural retina. Dev Dynamics 1993; 196:252-262.
13. Schlosshauer B, Bauch H, Frank R. Neurothelin: Amino acid sequence, cell surface dynamics and actin colocalization. Europ J Cell Biol 1995; 68:159-166.
14. Seulberger H, Unger CM, Risau W. HT7, neurothelin, basigin, gp42 and OX-47—Many names for one developmentally regulated immunoglobulin-like surface glycoprotein on blood-brain endothelium, epithelial tisuue barriers and neurons. Neurosci Lett 1992; 140:93-97.
15. Langnaese K, Beesley PW, Gundelfinger ED. Synaptic membrane glycoproteins gp65 and gp55 are new members of the immunoglobulin superfamily. J Biol Chem 1997; 272:821-827.
16. DeClerck YA. Interactions between tumour cells and stromal cells and proteolytic modification of the extracellular matrix by metalloproteinases in cancer. Europ J Cancer 2000; 36:1258-1268.
17. Heppner KJ, Matrisian LM, Jensen RA, Rodgers WH. Expression of most matrix metalloproteinase family members in breast cancer represents a tumor-induced host response. Amer J Pathol 1996; 149:273-282.
18. Johnsen M, Lund LR, Romer J, Almholt K, Dano K. Cancer invasion and tissue remodeling: Common themes in proteolytic matrix degradation. Current Opin Cell Biol 1998; 10:667-671.
19. Okada A, Bellocq JP, Rouyer N, Chenard MP, Rio MC et al. Membrane-type matrix metalloproteinase (MT-MMP) gene is expressed in stromal cells of human colon, breast, and head and neck carcinomas. Proc Nat Acad Sci USA 1995; 92:2730-2734.
20. Masson R, Lefebvre O, Noel A, El Fahime M, Chenard MP et al. In vivo evidence that the stromelysin-3 metalloproteinase contributes in a paracrine manner to epithelial cell malignancy. J Cell Biol 1998; 140:1535-1541.
21. Itoh T, Tanioka M, Yoshida H, Yoshioka T, Nishimoto H et al. Reduced angiogenesis and tumor progression in gelatinase A-deficient mice. Cancer Res 1998; 58:1048-1051.
22. Sternlicht MD, Lochter A, Sympson CJ, Huey B, Rougier JP et al. The stromal proteinase MMP3/stromelysin-1 promotes mammary carcinogenesis. Cell 1999; 98:137-146.
23. Wright JH, McDonnell S, Portella G, Bowden GT, Balmain A et al. A switch from stromal to tumor cell expression of stromelysin-1 mRNA associated with the conversion of squamous to spindle carcinomas during mouse skin tumor progression. Mol Carcinogenesis 1994; 10:207-215.
24. Sehgal G, Hua J, Bernhard EJ, Sehgal I, Thompson TC et al. Requirement for matrix metalloproteinase-9 (gelatinase B) expression in metastasis by murine prostrate carcinoma. Amer J Pathol 1998; 152:591-596.
25. Wilson CL, Heppner KJ, Rudolph LA, Matrisian LM. The metalloproteinase matrilysin is preferentially expressed by epithelial cells in a tissue-restricted pattern in the mouse. Mol Biol Cell 1995; 6:851-869.
26. Ito A, Nakajima S, Sasaguri Y, Nagase H, Mori Y. Co-culture of human breast adenocarcinoma MCF-7 cells and human dermal fibroblasts enhances the production of matrix metalloproteinases 1,2 and 3 in fibroblasts. Brit J Cancer 1995; 71:1039-1045.
27. Uria JA, Stahle-Backdal M, Seiki M, Fueyo A, Lopez-Otin C. Regulation of *collagenase-3* expression in human breast carcinomas is mediated by stromal-epithelial cell interactions. Cancer Res 1997; 57:4882-4888.
28. Westermarck J, Li S, Jaakkola P, Kallunki T, Grenman R et al. Activation of fibroblast collagenase-1 expression by tumor cells of squamous cell carcinomas is mediated by p38 mitogen-activated protein kinase and c-jun NH_2-terminal kinase-2. Cancer Res 2000; 60:7156-7162.
29. Biswas C. Tumor cell stimulation of collagenase production by fibroblasts. Biochem Biophys Res Commun 1982; 109:1026-1034.
30. Biswas C. Collagenase stimulation in cocultures of human fibroblasts and human tumor cells. Cancer Letters 1984; 24:201-207.

31. Biswas C, Nugent MA. Membrane association of collagenase stimulatory factor(s) from B-16 melanoma cells. J Cell Biochem 1987; 35:247-258.
32. Nabeshima K, Lane WS, Biswas C. Partial sequencing and characterization of the tumor cell-derived collagenase stimulatory factor. Arch Biochem Biophys 1991; 285:90-96.
33. Ellis SM, Nabeshima KN, Biswas C. Monoclonal antibody preparation and purification of a tumor cell collagenase-stimulatory factor. Cancer Res 1989; 49:3385-3391.
34. DeCastro R, Zhang Y, Guo H, Kataoka H, Gordon MK et al. Human keratinocytes express EMMPRIN, an extracellular matrix metalloproteinase inducer. J Invest Dermatol 1996; 106:1260-1265.
35. Kataoka H, DeCastro R, Zucker S, Biswas C. The tumor cell-derived collagenase stimulatory factor, TCSF, increases expression of interstitial collagenase, stromelysin and 72-kDa gelatinase. Cancer Res 1993; 53:3154-3158.
36. Lim M, Martinez T, Jablons D, Cameron R, Guo H et al. Tumor-derived EMMPRIN (extracellular matrix metalloproteinase inducer) stimulates collagenase transcription through MAPK p38. FEBS Lett 1998; 441:88-92.
37. Sameshima T, Nabeshima K, Toole BP, Yokogami K, Okada Y et al. Glioma cell extracellular matrix metalloproteinase inducer (EMMPRIN) (CD147) stimulates production of membrane-type matrix metalloproteinases and activated gelatinase A in co-cultures with brain-derived fibroblasts. Cancer Lett 2000; 157:177-184.
38. Nakahara H, Howard L, Thompson EW, Sato H, Seiki M et al. Transmembrane/cytoplasmic domain-mediated membrane type 1-matrix metalloprotease docking to invadopodia is required for invasion. Proc Natl Acad Sci USA 1997; 94:7959-7964.
39. Zucker S, Drews M, Conner C, Foda HD, DeClerck YA et al. Tissue inhibitor of metalloproteinase-2 (TIMP-2) binds to the catalytic domain of the cell surface receptor, membrane type 1-matrix metalloproteinase 1 (MT1-MMP). J Biol Chem 1998; 273:1216-1222.
40. Bordador LC, Li X, Toole BP, Chen B, Regezi J et al. Expression of emmprin by oral squamous cell carcinoma. Int J Cancer 2000; 85:347-352.
41. Zucker S, Hymowitz M, Rollo EE, Mann R, Conner CE et al. Tumorigenic potential of extracellular matrix metalloproteinase inducer (EMMPRIN). Amer J Pathol; in press.
42. Sun J, Hemler ME. Regulation of MMP-1 and MMP-2 production through CD147/extracellular matrix metalloproteinase inducer interactions. Cancer Res 2001; 61:2276-2281.
43. Fadool JM, Linser PJ. Evidence for the formation of multimeric forms of the 5A11/HT7 antigen. Biochem Biophys Res Commun 1996; 229:280-286.
44. Yoshida S, Shibata M, Yamamoto S, Hagihara M, Asai N et al. Homo-oligomer formation by basigin, an immunoglobulin superfamily member, via its N-terminal immunoglobulin domain. Europ J Biochem 2000; 267:4372-4380.
45. Brooks PC, Stromblad S, Sanders LC, von Schalscha TL, Aimes RT et al. Localization of matrix metalloproteinase MMP-2 to the surface of invasive cells by interaction with integrin alpha V beta 3. Cell 1996; 85:683-693.
46. Knauper V, Will H, Lopez-Otin C, Smith B, Atkinson SJ et al. Cellular mechanisms for human procollagenase-3 (MMP-13) activation. Evidence that MT1-MMP (MMP-14) and gelatinase A (MMP-2) are able to generate active enzyme. J Biol Chem 1996; 271:17124-17131.
47. Yu Q, Stamenkovic I. Cell surface-localized matrix metalloproteinase-9 proteolytically activates TGF-beta and promotes tumor invasion and angiogenesis. Genes Dev 2000; 14:163-176.
48. Olson MW, Toth M, Gervasi DC, Sado Y, Ninomiya Y et al. High affinity binding of latent matrix metalloproteinase-9 to the alpha2(IV) chain of collagen IV. J Biol Chem 1998; 273:10672-10681.
49. Guo H, Li R, Zucker S, Toole BP. EMMPRIN (CD147), an inducer of matrix metalloproteinase synthesis, also binds interstitial collagenase to the tumor cell surface. Cancer Res 2000; 60:888-891.
50. Benbow U, Schoenermark MP, Mitchell TI, Rutter JL, Shimokawa K et al. A novel host/tumor cell interaction activates matrix metalloproteinase 1 and mediates invasion through type 1 collagen. J Biol Chem 1999; 274:25371-25378.
51. Muraoka K, Nabeshima K, Murayama T, Biswas C, Koono M. Enhanced expression of a tumor cell-derived collagenase-stimulatory factor in urothelial carcinoma: Its usefulness as a tumor marker for bladder cancers. Int J Cancer 1993; 55:19-26.
52. Polette M, Gilles C, Marchand V, Lorenzato M, Toole B et al. Tumor collagenase stimulatory factor (TCSF) expression and localization in human lung and breast cancers. J Histochem Cytochem 1997; 45:703-710.
53. Caudroy S, Polette M, Tournier JM, Burlet H, Toole B et al. Expression of the extracellular matrix metalloproteinase inducer (EMMPRIN) and the matrix metalloproteinase-2 in bronchopulmonary and breast lesions. J Histochem Cytochem 1999; 47:1575-1580.

54. Dalberg K, Eriksson E, Enberg U, Kjellman M, Backdahl M. Gelatinase A, membrane type 1 matrix metalloproteinase, and extracellular matrix metalloproteinase inducer mRNA expression: Correlation with invasive growth of breast cancer. World J Surgery 2000; 24:334-340.
55. Sameshima T, Nabeshima K, Toole BP, Yokogami K, Okada Y et al. Expression of EMMPRIN (CD147), a cell surface inducer of matrix metalloproteinases, in normal human brain and gliomas. Int J Cancer 2000; 88:21-27.
56. Igakura T, Kadomatsu K, Kaname T, Muramatsu H, Fan QW et al. A null mutation in basigin, an immunoglobulin superfamily member, indicates its important roles in peri-implantation development and spermatogenesis. Dev Biol 1998; 194:152-165.
57. Alexander CM, Hansell EJ, Behrendtsen O, Flannery ML, Kishnani NS et al. Expression and function of matrix metalloproteinases and their inhibitors at the maternal-embryonic boundary during mouse embryo implantation. Development 1996; 122:1723-1736.
58. Vu TH, Werb Z. Matrix metalloproteinases: effectors of development and normal physiology. Genes Develop 2000; 14:2123-2133.
59. Toyama Y, Maekawa M, Kadomatsu K, Miyauchi T, Muramatsu T et al. Histological characterization of defective spermatogenesis in mice lacking the basigin gene. Anat Histol Embryol 1999; 28:205-213.
60. Hori K, Katayama N, Kachi S, Kondo M, Kadomatsu K et al. Retinal dysfunction in basigin deficiency. Invest Ophthalmol Vis Sci 2000; 41:3128-3133.
61. Ochrietor JD, Moroz TM, Kadomatsu K, Muramatsu T, Linser PJ. Retinal degeneration following failed photoreceptor maturation in 5A11/basigin null mice. Exp Eye Res 2001; 72:467-477.
62. Igakura T, Kadomatsu K, Taguchi O, Muramatsu H, Kaname T et al. Roles of basigin, a member of the immunoglobulin superfamily, in behavior as to an irritating odor, lymphocyte response and blood-brain barrier. Biochem Biophys Res Commun 1996; 224:33-36.
63. Berditchevski F, Chang S, Bodorova J, Hemler ME. Generation of monoclonal antibodies to integrin-associated proteins: Evidence that a3b1 complexes with EMMPRIN/basigin/OX47/M6. J Biol Chem 1997; 272:29174-29180.
64. Kirk P, Wilson MC, Heddle C, Brown MH, Barclay AN et al. CD147 is tightly associated with lactate transporters MCT1 and MCT4 and facilitates their cell surface expression. EMBO J 2000; 19:3896-3904.
65. Pushkarsky T, Zybath G, Dubrovsky L, Yurchenko V, Guo H et al. CD147 facilitates HIV-1 infection by interacting with virus-associated cyclophilin A. Proc Nat Acad Sci USA, 2001; 98:6360-6365.
66. Olumi AF, Grossfeld GD, Hayward SW, Carroll PR, Tlsty TD et al. Carcinoma-associated fibroblasts direct tumor progression of initiated human prostatic epithelium. Cancer Res 1999; 59:5002-5011.
67. Park CC, Bissell MJ, Barcellos-Hoff MH. The influence of the microenvironment on the malignant phenotype. Mol Med Today 2000; 6:324-329.
68. Moinfar F, Man YG, Arnould L, Bratthauer GL, Ratschek M et al. Concurrent and independent genetic alterations in the stromal and epithelial cells of mammary carcinoma: Implications for tumorigenesis. Cancer Res 2000; 60:2562-2566.
69. Shekhar MP, Werdell J, Santner SJ, Pauley RJ, Tait L. Breast stroma plays a dominant regulatory role in breast epithelial growth and differentiation: Implications for tumor development and progression. Cancer Res 2001; 61:1320-1326.

Chapter 8

The Plasminogen Activation System in Cell Invasion

M. Patrizia Stoppelli

Abstract

The plasminogen activator/plasmin system is an enzymatic cascade involved in the control of fibrin degradation, matrix turnover and cell invasion.

Extracellular conversion of the ubiquitous inactive plasminogen to the broad spectrum serine protease plasmin results in the recruitment of an enormous reservoir of potential proteolytic activity. Plasmin is, in turn, able to degrade fibronectin, laminin, vitronectin, proteoglycans, as well as fibrin and activate latent collagenases. Plasminogen activation is catalyzed by urinary (uPA) or tissue-type (tPA) plasminogen activators (PAs), which are subjected to time and space-dependent regulation. In particular, uPA is regarded as the critical trigger for plasmin generation during cell migration and invasion, under physiological and pathological conditions (such as cancer metastasis), whereas tPA is likely to play an important role in the control of intravascular fibrin degradation. The system includes specific plasminogen activator inhibitors (PAIs) which counteract the activity of PAs, thereby restricting the generation of plasmin for extracellular matrix (ECM) as well as for intravascular fibrin degradation.

Like the proteases of the blood coagulation system, plasminogen and PAs are complex molecules bearing large noncatalytic regions, which mediate their regulatory interactions with matrix or cell-associated proteins. Membrane receptors for all components of the PA system ensure plasminogen activation at the cell surface, thus focusing proteolysis to the immediate pericellular environment.

In addition to its role in regulating the localization of proteolytic activity, the well characterized receptor for uPA (uPAR) has the ability to bind matrix vitronectin and to transduce signals upon ligation with catalytically inactive uPA. By these means, receptor-bound uPA elicits biological outcomes, such as cell adhesion, migration and growth.

Relevant information provided by the analysis of mice with specific gene deficiencies has revealed that this system has a causal role in tissue regeneration, wound healing as well as in tissue involution, immune response, angiogenesis and cancer invasion. The causal link between the multiple effects observed in vivo and the complex molecular interactions of the PA/plasmin system with metalloproteases, integrin receptors, endocytosis receptors and growth factors is the subject of current investigation. Moreover, the clearcut involvement of PAs in cancer metastasis render the whole system an attractive subject for prognostic, diagnostic and therapeutic studies.

Introduction

Cell invasion requires that motile cells cross tissue barriers through the degradation of basement membranes and extracellular matrices (ECM). This is achieved through the cooperation of tightly regulated extracellular proteolytic cascades.[1]

Cell Invasion, edited by Jyrki Heino and Veli-Matti Kähäri. ©2002 Eurekah.com.

The controlled generation of plasmin from plasminogen provides an efficient proteolytic activity directed to the degradation of fibronectin, laminin, vitronectin and proteoglycans.[2] In addition, plasmin may represent a physiological activator of latent metalloproteases (MMPs), thereby switching on collagen degradation. Thus, the uPA-plasmin system may play a pivotal role in the general control of matrix degradation.[3,4]

Further activities of plasmin include the increased availability of active basic fibroblast growth factor (bFGF), as a consequence of ECM degradation, as well as the direct activation of latent transforming growth factor-β (TGF-β).[5] Therefore, the role of plasmin extends beyond ECM degradation to the control of cell growth and differentiation through growth factor activation.

In vivo, the multiple effects of plasmin activity emerge from the analysis of plasminogen-deficient mice which survive embryonic development but are impaired in growth, fertility and survival.[6] In the same mice, skin wound healing is severely impaired, suggesting that plasmin activity is also required for tissue regeneration.[7] Functional cooperation between the plasminogen/plasmin and the metalloprotease cascades is suggested by the finding that complete inhibition of the healing process requires both plasminogen deficiency and metalloprotease inhibition.[8]

The two known plasminogen activators, namely urinary-type (uPA) and tissue-type (tPA) have a restricted substrate specificity, as they recognize and cleave the R560-V561 peptide bond in plasminogen. However, uPA can also activate the scatter factor pro-HGF, which exhibits extensive homologies with plasminogen, and the macrophage stimulating protein (MSP), thereby controlling cell proliferation, invasion of extracellular matrix and prevention of apoptosis.[9]

Unlike the simple digestive proteases, in which only a signaling peptide and a short activation domain are attached to the catalytic region, the fibrinolytic proteases bear large noncatalytic regions.[10] These regions contain functionally autonomous modules, such as "kringle", "growth factor-like" or "EGF-like" and "finger" domains, which also occur outside the serine protease family. Phylogenetic trees constructed for the "kringle" and protease domains strongly suggest an independent domain evolution through exon shuffling.[11]

A graphic representation of the domain arrangement in plasminogen, uPA and tPA is shown in Figure 1. Some of these noncatalytic regions, which are essential for the biological activity of these proteases, may occur in vivo as degradation products with new functional properties. A typical case is that of angiostatin, which is an internal fragment of plasminogen comprising kringle domains 1-4 and part of kringle 5. Interestingly, this fragment can be generated by elastase activity or by plasmin autocatalysis.[12] When given systemically, angiostatin strongly inhibits tumor growth and maintains tumor cells in a dormant state defined by a balance of proliferation and apoptosis.[13]

Plasminogen and Plasminogen Activators

Plasminogen

Human plasminogen is a 791 aminoacid zymogen which can be activated to the broad-spectrum two-chain plasmin by a single proteolytic cleavage of Arg560-Val562 peptide bond. The large amino-terminal region of plasminogen includes a "finger" module and five "kringle" domains, involved in regulatory interaction with fibrin and matrix proteins; the carboxy-terminal domain is responsible for the serine protease activity (Fig. 1). Plasminogen occurs in blood plasma at the concentration of 1-2 μM and is also largely present in tissues. This enormous reservoir of proteolytic activity is controlled by urokinase-type and tissue-type plasminogen activators, which were originally identified in urine and tissue extracts, respectively.[2]

On the other hand, no significant additional pathways for physiological plasminogen activation can be envisaged in mice, as the phenotype of plasminogen-deficient mice is similar to that of mice carrying both uPA and tPA deficiencies.[6]

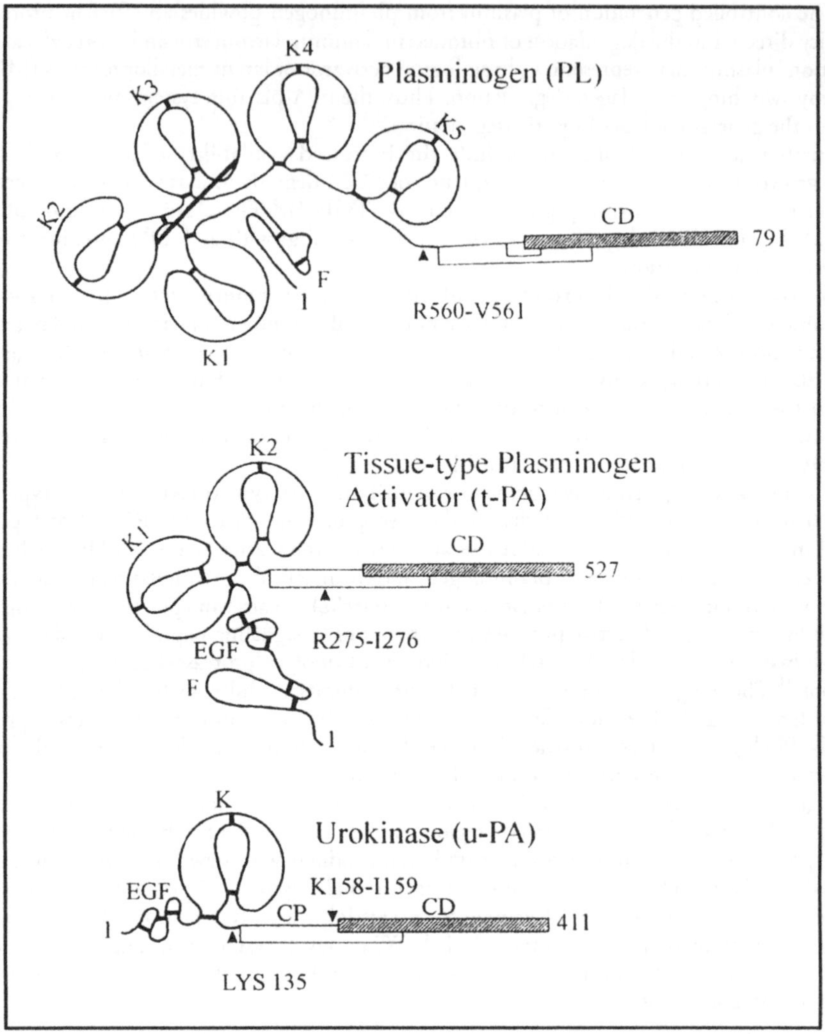

Fig. 1. Graphic representation of the protein domains of plasminogen (PL), tissue-type plasminogen activator (tPA) and urokinase (uPA) proenzymes. The kringle (K), Finger (F), EGF-like (EGF) and catalytic (CD) domains are marked. The activation sites are indicated by an arrow.

The Urokinase-Type Plasminogen Activator (uPA)

The single chain pro-urokinase (pro-uPA) is secreted as a 411 aminoacids zymogen form and becomes activated by plasmin cleavage of K158-I159 peptide bridge, thus generating uPA, a two-chain molecule held together by a single disulfide bond. Pro-uPA can be also cleaved by cathepsin B or kallikrein, which may indirectly control plasmin generation. In vivo, pro-uPA may be activated by glandular kallikrein mGK-6, as shown in the urine from plasminogen-deficient mice.[14]

A further cleavage at K135-K136 releases pro-uPA amino-terminal domain (ATF, aminoacids 1-135), which includes an "EGF-like" and a "kringle" domain. The remaining carboxy-terminal region (LMW urokinase, aminoacids 135-411) retains full catalytic activity (Fig. 1). Pro-uPA undergoes several post-translational modifications, such as glycosylation of Asn 302, phosphorylation on Ser138/303 and fucosylation of Thr18, located in the EGF-like domain.[15-17]

The 6.4 kb pro-uPA gene has been isolated and characterized. The analysis of the exon-intron organization relative to the protease domains strongly suggests that exons encoding single functional domains were exchanged between different genes by recombination events.[18] A variety of biological, chemical and physical agents affect the rate of pro-uPA transcription, thus modulating mRNA and protein levels.[19] The uPA promoter region is highly responsive to phorbol esters, growth factors, steroid hormones and cytoskeletal changes through the activity of a transcriptional enhancer located about 2 kb upstream of the cap site. Recent data assign to the Ets-1 and Ets-2 transcription factors the link connecting epidermal growth factor stimuli with activation of uPA and 92 kDa type IV collagenase (MMP-9) promoters. This mechanism is likely to facilitate invasion and metastasis in breast cancer.[20] Constitutive expression of uPA leading to an highly metastatic phenotype is achieved by increasing mRNA stability in MDA-MB-231 breast carcinoma cells.[21]

The Tissue-Type Plasminogen Activator (tPA)

Unlike uPA, the 527 amino acid single-chain human tPA is proteolytically active. Plasmin cleavage of Arg275-Ile276 peptide bond yields a disulfide-linked two-chain molecule, which exhibits a 10-50 fold increased activity with respect to the uncleaved form. The N-terminal region contains a "fibronectin-type II" domain, an "EGF-like" domain and two "kringle" domains (Fig. 1). Fucosylation of Thr61, which may be relevant to the tPA clearance process in the liver, has also been reported.[22] Similar to uPA, a large catalytic domain is located at the carboxy-terminus.

It is generally assumed that the major role of tPA is to degrade fibrin in blood vessels. The activity of tPA is, in fact, greatly stimulated by fibrin which interacts with kringle 2, finger and EGF-like domains. Accordingly, tPA synthesis is induced under ischemic conditions. Also, in tPA-deficient mice, clot lysis is strongly impaired whereas, in uPA-deficient mice, there is an occasional fibrin deposition.[23]

However, other roles for tPA are emerging from the analysis of nervous system development in mice. In tPA/- mice, cerebellar granule neurons migrate significantly slower than granule neurons from wild-type mice; as a consequence, late arriving neurons are impaired in their synaptic interactions.[24]

Plasminogen Activator Inhibitors

Plasminogen Activator Inhibitor-Type 1 (PAI-1)

Specific inhibitors of plasminogen activation belong to the serine protease inhibitor superfamily (SERPINS).

The 379 amino acid plasminogen activator inhibitor-type 1 or PAI-1 is the primary inhibitor of plasminogen activators in plasma and in the pericellular matrices. Covalent complex formation occurs between the reactive center loop of PAI-1 and the active site serine of uPA or tPA through an ester bond.[25] Pro-uPA does not react with PAI-1, whereas two-chain active uPA rapidly associates with this inhibitor. However, when two-chain uPA is phosphorylated on Ser138/303, its capability to form complexes with PAI-1 is strongly reduced, although the catalytic ability to activate plasminogen is unchanged.[26]

The physiologically relevant and stable form of PAI-1 is in complex with vitronectin, a glycoprotein occurring in plasma and in the extracellular matrices. This association, which stabilizes the inhibitor in a reactive conformation, involves the N-terminal somatomedin B domain of vitronectin (aminoacids 1-44). Interestingly, the latter domain overlaps with the binding site of vitronectin to the $\alpha v\beta 3$ integrin. Therefore, PAI-1 may prevent vitronectin binding to vitronectin receptor, thereby interfering with vitronectin receptor-dependent effects, such as smooth muscle cell migration.[27] Because of its ability to associate with vitronectin, PAI-1 may also prevent vitronectin binding to the urokinase receptor, thereby inducing cell release from vitronectin substrates.[28]

These previously unrecognized interactions are independent of PAI-1 capability to function as a protease inhibitor. However, it is conceivable that active uPA may regulate integrin-dependent effects by complexing PAI-1, thereby releasing vitronectin which can bind to vitronectin receptor. As a consequence, the level of tree PAI-1 can affect the biological outcome/s of vitronectin/ vitronectin receptor in a particular context.

Plasminogen Activator Inhibitor-Type 2 (PAI-2)

Plasminogen activator inhibitor 2 or PAI-2 is a 47 kDa single chain protein. This inhibitor blocks uPA and, less efficiently, tPA, although not as fast as PAI-1, raising the possibility that PAI-2 may have some yet unidentified roles. The physical localization of PAI-2 is peculiar, as this molecule exists in a 60 kDa glycosylated secreted and in a 47 kDa cytosolic forms, both consisting in 415 amino acids devoid of an N-terminal signal peptide sequence. These two N-glycosylated forms of PAI-2 are generated by facultative translocation.[29]

The role of intracellular PAI-2, which accounts for most of the produced inhibitor, still remains to be elucidated. Recent data show that it may play a role in macrophage protection from TNF-α mediated apoptosis. Although PAI-2 ability to inhibit uPA and tPA is poorer than that exhibited by PAI-1, extracellular PAI-2 is considered an inhibitor of plasmin generation both in blood vessels as well as in the ECM. Like PAI-1, extracellular PAI-2 binds to receptor-bound uPA; however, uPA-PAI-2 complexes are rapidly cleaved into a 70 kDa fragment, which is endocytosed or released and a 22 kDa species which remains on the cell surface preventing the binding of intact uPA.[30]

Plasminogen Activator Receptors

Receptors for Plasminogen and for tPA

The study of the plasminogen activator/plasmin system has revealed that the proteolytic components interact with cell surface through residues not directly involved in the catalytic mechanism.

Plasminogen binds to cells with low affinity and high capacity via its "kringle" domains which recognize carboxy-terminal lysines of proteins exposed on cell surface. Following binding to membrane receptors, plasminogen is activated more efficiently; bound plasmin exhibits an increased enzymatic activity and is protected from inhibition by a2-antiplasmin and α2-macroglobulin.[31,32]

Functional evidence for specific and saturable binding sites for tPA, involved in cell surface plasmin generation has been recently obtained in endothelial cells. High affinity binding of tPA on vascular smooth muscle cells occurs by a novel mechanism involving the serine protease domain of tPA and leading to the stimulation of cell-associated plasminogen activation.[33] Saturable high affinity binding sites for tPA on rat and human hepatoma cells which mediate internalization and degradation of the bound protease by a PAI-1- and mannose receptor-independent mechanism. Clearance of tPA occurs through low density lipoprotein receptor protein (LRP) which plays an important role in the clearance of circulating tPA, thus regulating plasma fibrinolytic activity.[34] The role of specific noncatalytic regions in this recognition/internalization process is documented by the inhibition of rat hepatocytes clearance by the recombinant tPA "finger"/"EGF-like" regions.[22]

The Urokinase Receptor (uPAR)

The interaction of uPA with cell surface has received considerable attention over the past 15 years. The initial studies showed that uPA binds specifically and with high affinity to human blood monocytes and to monocyte-like U937 cells.[35] Urokinase ability to become membrane-associated is retained by the purified amino-terminal fragment of uPA (ATF, residues 1-135) of uPA. Interestingly, uPA receptors (uPARs) are subjected to a 10-20 fold differentiation-dependent increase in U937 myelomonocytic cells.[36] Binding to uPAR occurs through the "EGF-like" domain of uPA, with a Kd in the nanomolar range.[37] In particular, Lys23, Tyr24, Phe25, Ile28, and Trp30 in the B loop are important determinants for uPA binding.[38]

The 313 residue uPAR molecule becomes membrane-associated with a concurrent removal of a 21 amino acid signal peptide and is processed to add a glycosyl-phosphatidylinositol(glycosyl-PtdIns) (GPI) anchor to the carboxy-terminal region which targets it to cell membrane.[39] Site-directed mutagenesis of the uPAR carboxy-terminal region indicated that Gly283 is the likely attachment site.[40] The uPAR includes 28 cysteine residues arranged in three homologous repeats and can be assigned to the Ly-6 superfamily which includes CD59 and a variety of elapid snake venom toxins.[41] The uPAR exhibits a three domain structure: the amino-terminal D1 domain which directly contacts uPA, the linker D2 domain and the carboxy-terminal D3 domain which maintains the ligand-receptor high affinity interaction. Upon binding to uPA, uPAR undergoes a conformational change that uncovers the linker region between D1 and 2 that has a potent chemotactic activity. This conformational change can be mimicked in vitro by enzymatic processing of a soluble uPAR with chimotrypsin, that exposes a chemotactic epitope (residues 88-92, SRSRY). The cleaved suPAR and the isolated chemotactic epitope exhibit a strong chemotactic activity in the sub-nanomolar range.[42,43]

D2 and D3 domains are involved in a high affinity interaction with the matrix protein vitronectin which is promoted by simultaneous binding of uPA or ATF.[44] However, efficient binding to vitronectin only occurs with intact uPAR.[45] The ability of uPAR to form ternary complexes with ATF and also vitronectin has been observed in cell lines, such as MCF-7 breast carcinoma or HT1080 fibrosarcoma cells and in membranes from breast ductal carcinoma specimens.[46] Recent findings uncovered the possibility that uPAR may interact with a cleaved form of kininogen in a zinc-dependent manner, through D2 and D3 domains. This interaction may underlie uPAR ability to promote kallikrein-dependent cell surface plasmin generation.[47]

In the GPI-anchored form, unoccupied uPAR is relatively mobile. Upon ligation with uPA, uPARs stably cluster at focal adhesions and cell-cell contacts.[48] This property is likely to play an important role in concentrating cell-surface proteolysis at the leading edge of migrating cells, as shown in monocytes.[49]

By virtue of its N-terminal signal peptide, anchorless uPAR can be secreted in the extracellular milieu. Several mechanisms may account for the lack of uPAR anchoring. Alternative splicing generates a protein lacking the carboxy-terminal aminoacidic sequence for GPI anchor attachment. This mRNA variant is expressed in different human cell lines and tissues and upregulated by phorbol ester in A549 cells.[50]

An anchorless uPAR variant is secreted from blood leukocytes affected by the stem cell disorder, paroxysmal nocturnal hemoglobinuria (PNH), due to a defect in the synthesis of GPI anchors. Unlike normal leukocytes, the PNH-affected cells do not express surface uPARs, although they contain apparently normal levels of uPAR-specific mRNA. In this context, uPAR is found in plasma as well as in the conditioned medium from cultured PNH leukocytes.[51] A full-length soluble uPAR missing the GPI region is also found in ascites and serum of patients with ovarian carcinoma.[52] However, in this case uPAR release is likely to be catalyzed by a cellular phospholipase D, which cleaves GPI anchors.[53] Short forms of soluble uPAR also exist in vivo as a result of limited proteolytic cleavage. Fragments corresponding to the uPAR domains D1 and D2+D3 have recently been isolated from human urine.[54]

Cell-Surface Associated Plasminogen Activation

The concept of pericellular proteolysis through the specific association of plasminogen activation components to the cell surface has received extensive experimental support. The uPAR is a high affinity site for pro-uPA, which is the major form of the enzyme in cells, tissues, and body fluids. After secretion, pro-uPA may become associated to uPARs expressed by the same cell, in an autocrine manner.[55] Alternatively, a paracrine interaction would involve a cooperation between uPA-producing and uPAR-bearing cells. This model has been confirmed in vivo: a four-fold stimulation of chick embryo chorioallantoic membrane invasion by uPAR expressing mouse L-cells is observed following co-cultures with uPA-producing L-cells.[56]

Regardless of the producing cell, membrane-bound pro-uPA may be converted by plasmin or, possibly, other proteolytic enzymes to two-chain uPA, not significantly affecting the binding parameters and not inducing uPA release.[57] Moreover, while bound to uPAR through its amino-terminal moiety, the carboxy-terminal catalytic domain of uPA retains full catalytic activity. In particular, kinetic studies have shown that receptor-bound uPA exhibits a 40-fold lower Km than soluble uPA and a concurrent 6-fold reduction in kcat, thus resulting in an overall increase of catalytic efficiency.[58] Since plasminogen can become membrane-bound, the occurrence of receptors for plasminogen and uPA on the same cell results in the formation of surface-associated plasmin.[59] This machinery generates broad-spectrum proteolytic activity which is restricted to cell surface and protected by circulating inhibitors, such as α2-antiplasmin.

Unlike bound plasmin, uPAR-bound uPA can still interact with the inhibitor PAI-1, which is therefore able to inhibit plasmin formation. Then, the uPA-PAI-1 complex bound to uPAR is internalized and degraded.[60] This clearance process occurs through direct binding of uPAR D3 domain to low density lipoprotein receptor (LRP) and can be blocked by purified recombinant D3 domain.[61]

All these data, taken together, suggest the existence of an uPA cycle that can be summarized as follows: after synthesis pro-uPA is secreted, bound to the receptor and activated to two-chain uPA. On the membrane, uPA can activate surface bound plasminogen to produce surface-associated plasmin. However, in the presence of PAI-1, uPA activity is inhibited and plasmin generation interrupted, while the uPA-PAI-1 complexes are internalized and degraded. Therefore, PAI-1 blocks uPA activity and also causes its degradation.

It is beyond doubt that in culture, cell surface is the site of a powerful proteolytic activity. In vivo, cell surface uPA is proven to be relevant, at least in one case. Mice overexpressing either uPA or uPAR in basal epidermis and hair follicles do not exhibit detectable alterations. However, the combined overexpression of both uPA and uPAR resulted in an extensive alopecia, as a consequence of hair follicles' involution, epidermal thickening and sub-epidermal blisters. These findings show that uPA and uPAR act synergistically in promoting pathogenic extracellular proteolysis in vivo and confirm the importance of uPAR in directing surface-associated proteolysis.[62]

Biological Role of the uPA/uPA Receptor System

uPAR and Cell Adhesion

Early evidence pointed to a role of uPAR in the enhancement of myelomonocytic cell adhesion. If anchorage-independent, nondifferentiated U937 cells are incubated for 20 h with transforming growth factor type β-1 (TGF-β), 1,25-(OH)2 vitamin D3 and, subsequently, with nanomolar concentrations of diisopropyl fluorophosphate-inactivated urokinase (DFP-uPA), they become adherent within minutes.[63]

The acquisition of uPA-induced adherence is accompanied by dramatic cytoskeletal changes and by a rapid inhibition of p56/59(hck) and p55(fgr) tyrosine kinases. This transient p56/59(hck) downregulation is required for adherence, as this process is inhibited by the expression of a constitutively active p56/59(hck) variant. Ligand-independent adherence can also result from the expression of a kinase-defective p56/59(hck) variant, confirming that p56/59(hck) down-modulation increases adherence and that this switch is controlled by uPA.[64]

Exposure of myelomonocytic cells and other cell types, such as melanoma cells, to uPA causes a specific increase of cell adhesion to vitronectin.[65,66] Although the underlying mechanisms of uPA-induced adhesion have not been fully elucidated, a reasonable possibility is that engaged uPAR may directly bind matrix vitronectin, as shown with recombinant molecules in vitro.[45] This hypothesis is also supported by the occurrence of ternary complexes with uPAR, vitronectin and uPA in breast cancer cell membranes.[46] Recent data also indicate that vitronectin binding to uPAR initiates a p130Cas/Rac-dependent signaling pathway controlling actin polymerization state.[67] However, direct binding of uPAR to vitronectin does not provide a satisfactory explanation of the uPA ability to enhance general cell adhesion. For example, in freshly isolated monocytes, urokinase receptors (CD87) form complexes with the β2-integrin complement receptor 3 (CR3, CD11b/CD18) and this association increases adhesion of CR3 to fibrinogen. Conversely, adhesion of monocytes to fibrinogen promotes CR3-uPAR association at the monocyte/macrophage ventral surface.[68]

Recent evidence suggests that uPAR may be a dynamic regulator of integrin function, possibly, through "lateral" interactions. First of all, uPAR is associated in large molecular complexes with various integrins.[69] Secondly, the physical interaction uPAR/activated integrin can be reproduced in vitro and disrupted in vivo by a specific peptide homologous to integrin sequences.[70] Functional consequences of this dissociation include the inhibition of integrin-dependent spreading and migration.[71] Finally, the concept that uPAR signals through integrin activation is fully supported in a variety of systems, in which anti-integrin antibodies inhibit uPA-dependent signaling. Among the most significant examples linking uPAR association with integrins to the biological outcomes is the activation of α5β1-integrin by uPAR in Hep3 human carcinoma cells which generates a persistently high level of active extracellular signal-regulated kinase-1 (ERK-1) necessary for tumor growth in vivo. Disruption of uPAR-α5β1 complexes with peptides or antibodies inhibits the fibronectin-dependent ERK-1 activation, thereby reducing tumorigenicity.[72] A further level of complexity is added by the recent finding that uPAR is a true integrin ligand in both soluble and GPI-anchored forms. In fact, it specifically binds to integrins on adjacent cells, suggesting that uPAR-integrin binding may also mediate cell-cell interaction.[73]

uPAR and Cell Migration

As shown over a decade ago, both uPA native molecule and the ATF are true chemoattractants in the Boyden system, a two-compartment device for testing random and directional cell migration. The involvement of uPAR is supported by the finding that ATF-induced endothelial cell translocation is impaired by antibodies which inhibit ligand-receptor interaction.[74] Inactivation of uPA does not affect its chemotactic ability, whereas the reduction of uPAR expression with an antisense oligonucleotide strongly inhibits chemotaxis, showing that uPAR is required for migration.[75]

Since uPAR expression is widespread, a variety of cell lines and primary cells respond to nanomolar concentrations of uPA, either DFP-inactivated or lacking the catalytic domain, by increasing their motility.[76] In vivo, the important role of uPAR in directing cell locomotion emerges from the analysis of uPAR-deficient mice which exhibit reduced neutrophil recruitment in response to *P. aeruginosa pneumonia* as compared to control mice.[77] Interestingly, the impaired neutrophil migration in uPAR-deficient mice is not due to the disruption of uPAR-mediated proteolysis, but to uPAR occupancy.[78]

Although uPA-directed chemotaxis is dependent on uPAR, the underlying events may be far more complex than expected. For example, uPA phosphorylated on Ser138/303 binds to uPAR with unchanged affinity, but it is unable to elicit cell migration. This suggests that binding to uPAR is not sufficient for uPA-dependent cell mobilization.[16] It is beyond any doubt that cell response to uPA involves the functional and physical association of uPAR with integrin receptors. In breast carcinomas, uPAR physically associates with αvβ5 vitronectin receptor and this association leads to a functional interaction of these receptors. In breast cancer cells, anti-αvβ5 antibodies inhibit uPA-dependent migration, showing that vitronectin receptor is required for uPA signaling.[79] In rat smooth muscle cells, migration and cytoskeletal rearrangements are inhibited by anti-uPAR or anti-αvβ3 vitronectin receptor antibodies, suggesting that a functional association between these receptors is required to elicit both responses.[80]

Migration is a complex process which involves the dynamic participation of cytoskeletal structural and regulatory components. THP-1 macrophage-like cells migrating along a chemotactic ATF gradient undergo a transient activation of the p56/p59hck Src family tyrosine kinase. Under similar conditions, uPAR physically associates with p56/p59hck.[42] A general role of p56/p59hck in migration can be envisaged, as expression of a constitutively active p56/p59hck variant enhances U937 monocyte-like cell migration in a ligand-independent manner.[64]

ATF-dependent stimulation of MCF-7 breast carcinoma cell migration results in the activation of ERK-1 and ERK-2 kinases. Responses to uPA and ATF are blocked when the cells are pretreated with PD-098059, an inhibitor of mitogen-activated protein kinase kinase.[81] A p130Cas/Rac-dependent cascade leading to actin reorganization and increased cell motility is activated by vitronectin binding to uPAR and may play a role at sites where vitronectin and uPAR are co-expressed.[67]

uPAR and Cell Growth

The biological outcomes of the uPA/uPAR interaction are not just related to cell migration, adhesion and actin polymerization state. In the human melanoma cell line GUBSB, inhibition of receptor-bound uPA by specific anti-uPA antibodies reduces cell proliferation, suggesting that cell growth is constantly stimulated by uPAR occupancy, in an autocrine fashion.[82] Additional information is provided by a recent report confirming that ATF is a mitogen for melanoma cells in culture through a yet unidentified membrane-associated mediator of uPA-dependent signal transduction.[83] In vivo, it has been recognized that uPA favors the formation and growth of melanomas, as described later in this chapter. Growth stimulatory effects have been observed also in human SaOS-2 osteosarcoma cells exposed to ATF. In this case, fucosylation of Thr18 within the "EGF-like" domain seems to be required for eliciting this response.[17]

In vivo, a 70% reduction of the uPAR level in the human carcinoma HEp3 cells inoculated into chicken chorioallantoic membrane, while not affecting growth in culture, induces a state of tumor dormancy. The observed G(0)/G(1) arrest may be due to scarce uPA/uPAR/ $\alpha v\beta 5$ complex formation and, consequently, insufficient ERK-1 activity needed for tumor growth in vivo.[84]

uPAR Signaling Mediators

It is generally assumed that GPI-anchored proteins initiate intracellular signaling through the interaction with transmembrane receptors. In this respect, receptor lateral mobility may be essential to the uPAR signaling mechanism.[48] Further evidence of the uPAR ability to move to specific cell sites has been obtained by autoradiographic detection of receptor-bound uPA at the leading edge of migrating human monocytes.[49]

A relationship between uPAR aggregation and signaling initiation is suggested by the increase in Ca^{2+} influx and induction of adherence in polymorphonuclear neutrophils cross-linked to anti-uPAR antibodies, which induce uPAR clustering.[85] Circumstantial evidence that ligand-dependent uPAR aggregation is required for signaling emerges from the study of uPA phosphorylated on Ser138/303 or carrying Glu138/303, which are neither chemotactic nor proadhesive. These nonsignaling uPA forms are also unable to induce uPAR clustering in U937 monocyte-like cells, as shown by confocal microscopy.[16]

In most cases, GPI-anchored proteins are localized in lipid-enriched membrane microdomains known as lipid rafts which contains various proteins, including caveolin.[86,87] In particular, the latter protein is known to enhance the extent of uPAR-mediated cell responses and is also found in complexes with uPAR, $\beta 1$ integrins and Src family kinases.[88] Similar complexes have been detected in leukocytes, in which uPAR associates with p60fyn, p53/56lyn, p58/64hck, and p59fgr tyrosine kinases, as well as with the integrins LFA-1 and CD11b/CD18.[69] As discussed earlier, uPAR capability to associate directly to integrin receptors has received broad experimental support. These findings are consistent with the inhibition of uPAR signaling by anti-integrin blocking antibodies in a variety of cells. Interestingly, unlike vitronectin-dependent signaling, uPA-dependent signaling requires protein kinase C in both MCF-7 and HT1080 cell lines. This finding, together with the evidence that anti-$\alpha v\beta 5$ antibodies block uPA-dependent signaling suggests that uPA directs cytoskeletal rearrangements and cell migration by altering $\alpha v\beta 5$ signaling specificity.[79]

All these observations support a model in which mobile uPAR complexed with caveolin and signaling molecules dynamically associates to ligand-clustered integrins, thereby activating signaling. The complex array of uPAR partners and signaling mediators may be tissue-specific or unique to a particular cell differentiation stage (Fig. 2).

Novel associations between uPAR and other membrane receptors leading to functional effects have been described. For example, L-selectin mediates uPAR-dependent Ca^{2+} mobilization in polymorphonuclear neutrophils.[89] The existence of a link between G-proteins and uPAR signaling is suggested by the finding that the proadhesive ability of pertussis toxin on differentiating U937 cells is prevented by anti-uPAR antibodies.[90] In rat smooth muscle cells, uPA-dependent chemotaxis correlates with a dramatic reorganization of actin cytoskeleton and adhesion plaques: these effects are sensitive to pertussis toxin, thereby supporting the involvement of G proteins.[91]

Consistently with uPAR ability to direct cytoskeletal rearrangements and cell migration, focal adhesion kinase (pp125FAK) and paxillin are tyrosine phosphorylated in bovine aortic endothelial cells exposed to uPA.[92] Mobilization of MCF-7 breast cancer cells initiated by single-chain uPA or ATF requires the activation of a signaling cascade which includes Ras, MEK, ERK-1 and myosin light chain kinase (MLCK).[93]

Other effects include activation of the Janus kinases Jak1 and Tyk2 which mediate uPA-induced activation of transcription in human vascular smooth muscle cells. In particular, Tyk2 triggers a signaling cascade leading to phosphatidylinositol 3-kinase (PI3-K) activation. Inhibition of Tyk2 or PI3-K prevents uPA-dependent migration.[94] Exposure of HT1080 cells to catalytically inactive uPA results in c-fos mRNA stimulation and PAI-2 induction, possibly through AP-1 transcriptional activator.[95] Finally, in breast carcinoma cells uPAR engagement activates a signaling cascade resulting in a rapid upregulation of the transcriptional factor Sp1 binding activity, which, in turn, may upregulate uPAR levels.[96]

Plasminogen Activators and Tissue Remodeling

In the emerging picture, plasminogen activators participate in a wide spectrum of biological events, including the remodeling of the normal surrounding tissue induced by cancer cells, as well as the non-neoplastic tissue involution and regeneration processes. Urokinase-type plasminogen activator expression is induced in the mouse mammary gland during development and post-lactational involution. Results confirming the important role of plasmin have been obtained in plasminogen-deficient mice which are impaired in the lactational differentiation and mammary gland remodeling during involution.[97] Increasing evidence shows that plasmin activity is not just involved in the degradation processes but is also required for tissue regeneration. In acute liver injury, induced by carbon tetrachloride intoxication, mice with targeted disruption of uPA gene exhibit a remarkable delay in the repair process. Interestingly, tPA-deficient mice are slightly defective in the hepatic regeneration. In vivo experiments have assessed that the accumulation of fibrin, fibronectin and necrotic cells within injured areas are responsible for the delayed regeneration process.[98] A severe regeneration defect is observed in uPA deficient mice with experimentally damaged skeletal muscle.[99] Like the previous case, tPA deficient mice are indistinguishable from controls, suggesting that uPA is selectively involved in the regeneration processes. In the remodeling processes, the plasminogen/plasmin system acts together with the metalloprotease system: this possibility is supported by the requirement for both plasminogen deficiency and metalloprotease inhibition to prevent the wound healing process.[8]

Plasminogen Activators and Invasion in Animal Models

A reasonable hypothesis is that permanent alterations in the proteolytic balance of malignant cells may contribute to tumor dissemination. Early evidence showed that inhibition of plasminogen activation may prevent tumor metastasis in animal models. In the first report, human carcinoma HEp3 cell were allowed to grow on the chorioallantoic membrane and metastasize to the chicken embryo. Inoculation of anti-uPA antibodies delayed the onset of pulmonary metastases.[100] Similar results were obtained in mice inoculated with B16 melanoma cells which

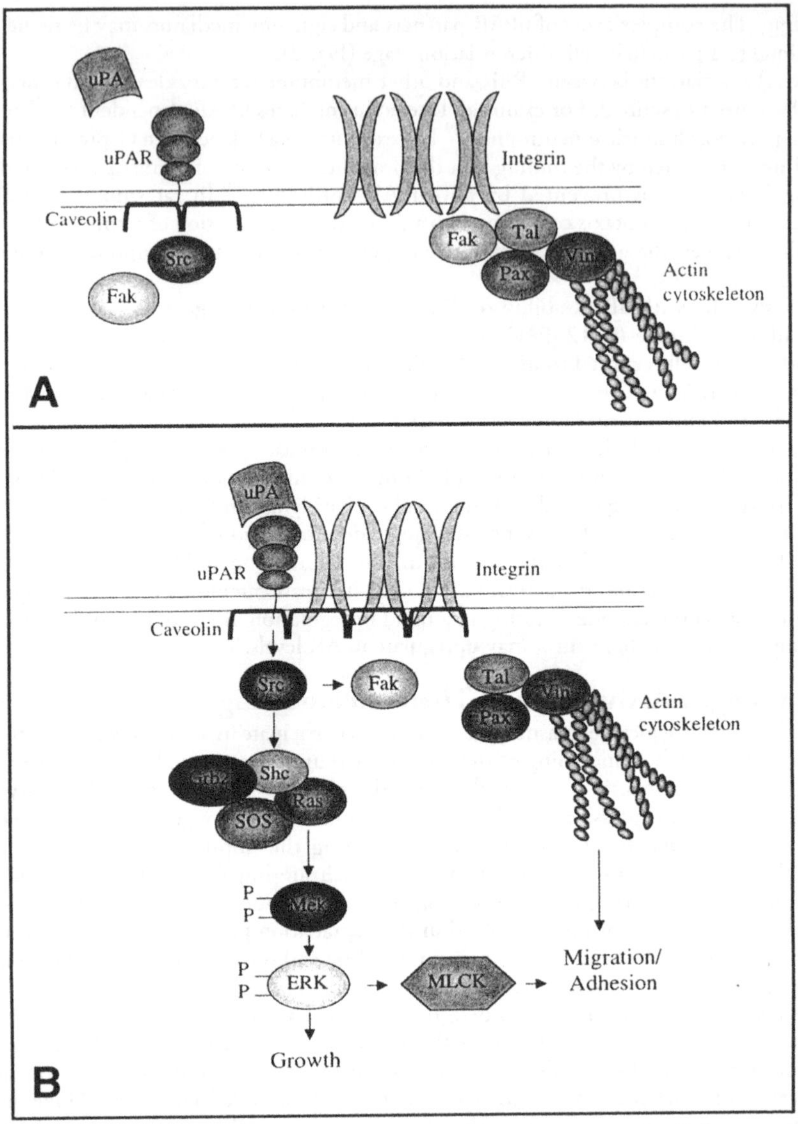

Fig. 2. Model for the urokinase-activated, catalytic-independent signaling cascade leading to cell migration, adhesion and growth. A. The nonengaged uPAR is located in lipid-enriched membrane microdomains together with caveolin, Src kinases, focal adhesion kinase (FAK) and, possibly, other components; integrins are physically connected to cytoskeleton through a variety of molecules, such as talin (Tal), paxillin (Pax), and vinculin (Vin), a marker of adhesion plaques. B. Following uPA binding to uPAR, the GPI-anchored receptor is mobilized and becomes associated to integrins: this is likely to result in the formation of a peculiar signaling initiation complex.

are impaired in their ability to metastasize following preincubation with inhibitory anti-uPA immunoglobulins.[101] More recently, the analysis of tumor growth in mice with targeted gene disruptions has provided new information on the role of individual plasminogen activation/plasmin components. In plasminogen-deficient mice, expressing polyoma middle T antigen, the occurrence and the number of pulmonary metastases are significantly reduced as compared to wild type controls, indicating that tumor dissemination is enhanced by plasmin-dependent matrix degradation.[102] Less obvious is the effect of uPA deficiency on tumor growth and dissemination: chemically induced melanocytic neoplasms in wild type mice progress to melanomas, whereas the uPA-/- mice are not susceptible to melanoma induction. This suggests that uPA contributes to malignant progression, possibly by decreasing the liberation and availability of growth factors such as basic fibroblast growth factor.[103]

The absence of uPAR also negatively affects tumor growth, as shown by the reduced in vivo tumorigenesis of human carcinoma (HEp3) cells bearing a 70% reduced surface uPARs. If these receptor-deficient cells are inoculated in the chorioallantoic membrane of the chick embryo, they enter a state of dormancy, characterized by survival without progressive growth. This seems to be attained through regulation of the balance between the mitogenic extracellular regulated kinase ERK-1 and the apoptotic/growth suppressive stress-activated protein kinase 2, p38(MAPK).[72] An antisense approach to reduce surface uPARs of the highly metastatic HCT116 colon carcinoma cells resulted in an impairment of cell ability to degrade the surrounding matrix. In mice, the artificial downregulation of uPAR strongly reduces the number of pulmonary metastases following intravenous injection of HCT116 cells.[104]

The idea that PAI-1 may not be a rate limiting step in tumor progression was first suggested by PAI-1 overexpression in human malignant tumors and by its correlation with a poor prognosis. Also, in PAI-1-deficient mice, local invasion and tumor vascularization of transplanted keratinocytes is impaired; intravenous injection of an adenoviral vector expressing human PAI-1 restores both invasion and angiogenesis.[105,106] Although it is known that PAI-1, beyond its inhibitory role, binds to vitronectin, thus causing its dissociation from vitronectin receptor and interfering with cell migration/adhesion, the resulting balance of these effects in vivo is unpredictable; in any event, the data obtained in knock-out animals suggest that PAI-1 is required for tumor growth. Some molecular insights have been provided by the analysis of tumor growth in mice bearing single and combined deficiencies of uPA, tPA, uPAR, vitronectin and plasminogen. Interestingly, the data indicate that PAI-1 promotes tumor growth by supporting angiogenesis: it is noteworthy that these effects are related to the PAI-1 inhibition of proteolytic activity, suggesting that excessive plasmin proteolysis prevents tumor vessel formation.[107] Similar to PAI-1, the relationship between PAI-2 and tumor invasion is not obvious: mice overexpressing PAI-2 are more susceptible to skin carcinogenesis and develop epidermal papillomas. Interestingly, overexpressed PAI-2 accumulated predominantly in cells and failed to inhibit extracellular uPA, suggesting that the observed stimulation of tumor progression by PAI-2 is independent of uPA inhibition.[108]

Plasminogen Activators and Human Tumors

Tissue Distribution

Cancer invasion and metastasis is a process requiring efficient ECM degradation which is accomplished through the activity of several proteolytic cascades, including the plasminogen/plasmin system. Based on these considerations, a positive correlation between malignancy and elevated levels of plasminogen activation/plasmin components would be expected.

Consistent with this possibility, the expression level of uPA and uPAR in the malignant tumors are generally higher than in the normal counterparts; according to the results of a quantitative study, breast carcinomas contain five times more uPAR and 19 times more uPA than benign breast lesions.[109] In malignant astrocytomas, especially in glioblastomas, uPA activity is significantly higher with respect to normal brain tissues or low-grade gliomas.[110] However, quantitative studies do not provide any information on the site of PAs production. Studies on the tissue distribution of uPA and uPAR have been carried out by immunocytochemistry and in situ hybridization on colon and breast tumor sections. Receptor-bound uPA is detected on the epithelial breast cancer cell membrane, suggesting that surface proteolytic activity contributes to the invasive phenotype.[111] In human colon cancer, uPA protein and mRNA are expressed in tumor-infiltrating fibroblast-like cells at the invasive foci whereas uPAR is expressed in cancer cells. Thus, surrounding stromal cells actively contribute to tumor invasion, as in tissue remodeling events.[112] In breast carcinomas, uPA may be expressed by the epithelial component or, more frequently, by fibroblast-like stromal cells.[113] Interestingly, fibroblastic rather than epithelial expression of uPA, uPAR and PAI-1 is positively correlated with tumor size and predicts a poor prognosis.[114] The latter studies confirm the complex interplay occurring between cancer and surrounding stromal cells, thus supporting an important role of fibroblast-like cells in the generation and regulation of pericellular proteolysis.

Prognosis and Diagnosis

In the treatment of cancer there is a need to select patients at high risk of recurrence for adjuvant therapy. In many cases, lymph node status is insufficient to predict whether these patients will experience a relapse and specific markers would be needed. Because of the well established correlation between the plasminogen/plasmin system and tumor invasion, the molecular components of this system are potentially useful markers of the tumor metastatic potential. Many studies directed to assess the prognostic impact of the plasminogen/plasmin components have been conducted, mostly based on antigen level quantitation in tissue extracts from surgically removed tumors. These values have been subsequently correlated with prognosis in several types of cancers.[115] In breast cancer, high levels of uPA were associated to a high risk of recurrence and short survival, suggesting that uPA is a more reliable marker than axillary node status, tumor size and estradiol receptor.[116] In a variety of neoplastic conditions, high levels of u-PA and u-PAR are associated to a poor patient prognosis.[117]

Assessment of tPA expression resulted in the identification of patients with a low probability to relapse in melanomas, as lesions with 51-100% tPA-positive tumor cells are associated with the best prognosis, whereas lesions with 6-50% tPA-positive tumor cells with the worst.[118]

An accurate measurement of preformed uPA-PAI-1 complexes with respect to total uPA and total PAI-1 has been performed in tissue extracts from breast cancer patients. Surprisingly, high PAI-1 levels are bad prognostic indicators, whereas uPA-PAI-1 complex levels predict long recurrence-free survival and overall survival.[119] The focus of another study was to assess the level of uPA, uPAR and PAI-1, as well as the expressing cells and their relationship with tumor clinical and pathological data. Interestingly, patients with a strong expression of uPA, uPAR and PAI-1 in fibroblast-like stromal rather than in tumor cells have a high probability to experience a relapse.[114]

From a clinical point of view, quantitation of these molecules in blood or in urine rather than in tissue samples is highly desirable both for patient prognosis as well as for follow-up purposes. This seems to be a realistic possibility, as plasma levels of uPA are, in fact, elevated in breast, prostate, head and colon cancer patients, as compared to control plasma samples from healthy donors.[120] Furthermore, in serum and ascites of patients with ovarian carcinoma the full-length soluble uPAR lacking GPI-anchor is present. Interestingly, high preoperative levels of suPAR predict a poor outcome. In addition to two-chain and LMW uPA, urine samples from healthy volunteers contain measurable amounts of soluble uPAR (suPAR). Urinary suPAR levels are elevated in patients with different types of cancer. Interestingly, part of the suPAR from urine and tumor tissue extracts is present in a cleaved form.[121]

In acute myeloid leukemia patients, a high level of plasma suPAR is found, which correlates with a poor response to chemotherapy. Also, suPAR concentration in plasma and urine decreases during chemotherapy treatment and depends on the number of circulating tumor cells.[122]

Therapy and Antagonists

If cancer dissemination is indeed promoted by unrestrained matrix degradation, one obvious therapeutic approach is to design specific protease inhibitors to be employed as anti-metastatic agents.[123,124] To this purpose, a novel X-ray crystallography-driven screening technique has been recently employed for the discovery and optimization of a new orally available class of uPA inhibitors for cancer treatment.[125]

New approaches include specific targeting of the PA components on tumor cell surface for tumor-selective cytotoxins. Considering that most tumor cells overexpress uPAR, they could be killed by a fusion protein between urokinase and saporin (a ribosome-inactivating protein) which is subjected to a PAI-1-mediated internalization.[126] If tumor cells bear receptor-bound uPA, a mutated anthrax toxin protective antigen in which the furin cleavage site is replaced by sequences cleaved by uPA may be selectively activated on cell surface. This causes internalization of the recombinant cytotoxin, thereby killing the uPAR-expressing tumor cells.[127]

Although the control of excess proteolytic activity in tumors is desirable, novel therapeutic approaches should take into account the nonproteolytic roles of PAs, as well as uPAR physical and functional association with integrins, leading to cell mobilization. Inhibition of uPAR interaction with integrins by the aid of specific peptides has been successfully attempted in model systems.[71,123]

Another approach to control malignancy is to reduce uPAR expression levels, which results in tumor dormancy, a novel promising anticancer strategy in which tumor cell proliferation is balanced by apoptosis.[76] This effect has already been obtained in vivo by the angiogenesis inhibitor angiostatin.[13] The knowledge of angiostatin biology is now being transferred to clinical practice: phase I clinical trials with recombinant angiostatin (kringles 1-3), as well as with an "angiostatin-cocktail" which induces the conversion of plasminogen to kringle 1-4 and part of kringle 5 are currently ongoing.[128]

Naturally occurring inhibitors of uPAR signaling have also been described. As mentioned earlier, serine phosphorylated uPA lacks the motogen ability, although it retains uPAR binding. The formation of such a receptor competitive antagonist is dependent on protein kinase C activity, which regulates the in vivo phosphorylation state of uPA.[129]

A novel inhibitor of tumor growth and invasion for which the mechanism of action is unknown has recently been described. An 8-mer capped peptide (A6) corresponding to the amino acids 136-143 of uPA inhibited breast cancer cell invasion and endothelial cell migration in a dose-dependent manner in vitro without altering cell doubling time. Intraperitoneal administration of A6 results in a significant inhibition of tumor growth and lymph node metastases development, in several models of breast cancer cell growth and metastasis.[130] The combination of A6 and cisplatin efficiently inhibits malignant glioma growth and significantly reduces neovascularization, suggesting a mechanism involving A6-mediated inhibition of endothelial cell motility.[131]

Conclusions and Perspectives

New concepts are emerging in this field based on the growing molecular knowledge of the plasminogen/plasmin components and their plasmin-independent interactions. However, the in vivo relevance as well as the biological outcome of such molecular events is difficult to predict. Recently, mice with targeted disruption of specific genes highlighted the role of plasminogen/plasmin system components in tumor growth and dissemination, wound healing, tissue regeneration and involution. These apparently contradictory roles of PAs and PAIs can be reconciled with the reasonable assumption that time- and space-regulated proteolysis may be required in all cases.

On the other hand, the unexpected finding that the level of PAI-1 in tumors is associated with a poor prognosis together with its ability to promote angiogenesis by inhibiting plasminogen activation in mouse models is not unique to the plasminogen activation/plasmin system. Interestingly, high plasma levels of the metalloprotease inhibitor TIMP-1 are correlated to advanced disease in colorectal cancer patients.[132] Useful information for the interpretation of these findings and the design of new therapeutic strategies may be provided by the effects of the metalloprotease inhibitor, batimastat, which stimulates the outgrowth of capillary structures by human foreskin microvascular endothelial cells in a 3-dimensional fibrin matrix. It has been proposed that batimastat prevents the cleavage of uPAR D1 domain by MMP-12, thereby increasing the number of functional uPARs on endothelial cells and stimulating capillary growth.[133]

These findings show that it is possible to counteract protease activity in vivo but also suggest that the uncontrolled use of proteolytic inhibitors in pathological conditions, such as cancer, may lead to undesired effects.

Acknowledgments

I am grateful to F. Blasi, M.V. Carriero and P. Ragno for stimulating discussions on this subject and critically reviewing the manuscript. Most of the work from this group was supported by AIRC (Associazione Italiana per la Ricerca sul Cancro).

References

1. Blasi F, Stoppelli MP. Proteases and cancer invasion: from belief to certainty. AACR meeting on proteases and protease inhibitors in cancer. Biochim Biophys Acta 1999; 1423:R35-44.
2. Dano K, Andreasen PA, Grondahl-Hansen J et al. Plasminogen activators, tissue degradation, and cancer. Adv Cancer Res 1985; 44:139-266.
3. Mazzieri R, Masiero L, Zanetta L et al. Control of type IV collagenase activity by components of the urokinase-plasmin system: a regulatory mechanism with cell-bound reactants. EMBO J 1997; 16:2319-32.
4. Ramos-DeSimone N, Hahn-Dantona E, Sipley J et al. Activation of matrix metalloproteinase-9 (MMP-9) via a converging plasmin/stromelysin-1 cascade enhances tumor cell invasion. J Biol Chem 1999; 274:13066-76.
5. Rifkin DB, Mazzieri R, Munger JS et al. Proteolytic control of growth factor avaialbility. APMIS 1999; 107:80-85.
6. Ploplis VA, Carmeliet P, Vazirzadeh S et al. Effects of disruption of the plasminogen gene on thrombosis, growth, and health in mice. Circulation 1995; 92:2585-93.
7. Romer J, Bugge TH, Pyke C et al. Impaired wound healing in mice with a disrupted plasminogen gene. Nat Med 1996; 2:287-92.
8. Lund LR, Romer J, Bugge TH et al. Functional overlap between two classes of matrix-degrading proteases in wound healing. EMBO J 1999; 18:4645-56.
9. van der Voort R, Taher TE, Derksen PW et al. The hepatocyte growth factor/Met pathway in development, tumorigenesis, and B-cell differentiation. Adv Cancer Res 2000; 79:39-90.
10. Patthy L. Evolution of proteases of blood coagulation and fibrinolysis by assembly from modules. Cell 1985; 41:657-663.
11. Ikeo K, Takahashi K, Gojobori T. Different evolutionary histories of kringle and protease domains in serine proteases: a typical example of domain evolution. J Mol Evol 1995; 40:331-6.
12. Westphal JR, Van't Hullenaar R, Geurts-Moespot A et al. Angiostatin generation by human tumor cell lines: involvement of plasminogen activators. Int J Cancer 2000; 86:760-7.
13. O'Reilly MS, Holmgren L, Chen C et al. Angiostatin induces and sustains dormancy of human primary tumors in mice. Nat Med 1996; 2:689-92
14. List K, Jensen ON, Bugge TH et al. Plasminogen-independent initiation of the pro-urokinase activation cascade in vivo. Activation of pro-urokinase by glandular kallikrein (mGK-6) in plasminogen-deficient mice. Biochemistry 2000; 39:508-15.
15. Mastronicola MR, Stoppelli MP, Migliaccio A et al. Serine phosphorylation of biosynthetic pro-urokinase from human tumor cells. FEBS Lett 1990; 266:109-14.
16. Franco P, Iaccarino C, Chiaradonna F et al. Phosphorylation of pro-urokinase at Ser138/303 impairs its receptor-dependent ability to promote myelomonocytic adhesion and motility. J Cell Biology 1997; 137:779-791.
17. Rabbani SA, Mazar AP, Bernier SM et al. Structural requirements for the growth factor activity of the amino-terminal domain of urokinase. J Biol Chem 1992; 267:14151-6.
18. Blasi F, Verde P. Urokinase-dependent cell surface proteolysis and cancer. Semin Cancer Biol 1990; 1:117-126.
19. Irigoyen JP, Munoz-Canoves P, Montero L et al. The plasminogen activator system: biology and regulation. Cell Mol Life Sci 1999; 56:104-32.
20. Watabe T, Yoshida K, Shindoh M et al. Fujinaga The Ets-1 and Ets-2 transcription factors activate the promoters for invasion-associated urokinase and collagenase genes in response to epidermal growth factor. Int J Cancer 1998; 77:128-37.
21. Nanbu R, Montero L, D'Orazio D et al. Enhanced stability of urokinase-type plasminogen activator mRNA in metastatic breast cancer MDA-MB-231 cells and LLC-PK1 cells down-regulated for protein kinase C-correlation with cytoplasmic heterogeneous nuclear ribonucleoprotein C. Eur J Biochem 1997; 247:169-74.

22. Camani C, Gavin O, Bertossa C et al. Studies on the effect of fucosylated and non-fucosylated finger/growth-factor constructs on the clearance of tissue-type plasminogen activator mediated by the low-density-lipoprotein-receptor-related protein. Eur J Biochem 1998; 251:804-11.
23. Carmeliet P, Schoonjans L, Kieckens L et al. Physiological consequences of loss of plasminogen activator gene function in mice. Nature 1994; 368:419-24.
24. Seeds NW, Basham ME, Haffke SP. Neuronal migration is retarded in mice lacking the tissue plasminogen activator gene. Proc Natl Acad Sci USA 1999; 96:14118-23.
25. Andreasen PA, Egelund R, Petersen HH. The plasminogen activation system in tumor growth, invasion, and metastasis. Cell Mol Life Sci 2000; 57:25-40.
26. Franco P, Mastronicola MR, De Cesare D et al. Separation and characterization of nonphosphorylated and serine-phosphorylated urokinase. Catalytic properties and sensitivity to plasminogen activator inhibitor type 1. J Biol Chem 1992; 267:19369-72.
27. Stefansson S and Lawrence DA. The serpin PAI-1 inhibits smooth muscle cell migration by blocking integrin binding to vitronectin. Nature 1996; 383:441-443.
28. Deng G, Curriden SA, Wang S et al. Is plasminogen activator inhibitor-1 the molecular switch that governs urokinase receptor-mediated cell adhesion and release? J Cell Biol 1996; 134:1563-71.
29. Belin D, Bost S, Vassalli JD et al. A two-step recognition of signal sequences determines the translocation efficiency of proteins. EMBO J 1996; 15:468-78.
30. Ragno P, Montuori N, Vassalli JD et al. Processing of complex between urokinase and its type-2 inhibitor on the cell surface. A possible regulatory mechanism of urokinase activity. FEBS Lett 1993; 323:279-84.
31. Felez J, Miles LA, Fabregas P et al. Characterization of cellular binding sites and interactive regions within reactants required for enhancement of plasminogen activation by tPA on the surface of leukocytic cells. Thromb Haemost 1996; 76:577-84.
32. Hall SW, Humphries JE, Gonias SL. Inhibition of cell surface receptor-bound plasmin by α2-antiplasmin and α2-macroglobulin. J Biol Chem 1991; 266:12329-36.
33. Werner F, Razzaq TM, Ellis V. Tissue plasminogen activator binds to human vascular smooth muscle cells by a novel mechanism. Evidence for a reciprocal linkage between inhibition of catalytic activity and cellular binding. J Biol Chem 1999; 274:21555-61.
34. Bu G, Maksymovitch EA, Schwartz AL. Receptor-mediated endocytosis of tissue-type plasminogen activator by low density lipoprotein receptor-related protein on human hepatoma HepG2 cells. J Biol Chem 1993; 268:13002-9.
35. Vassalli JD, Baccino D, Belin D. A cellular binding site for the Mr 55,000 form of the human plasminogen activator, urokinase. J Cell Biol 1985; 100:86-92.
36. Stoppelli MP, Corti A, Soffientini A et al. Differentiation-enhanced binding of the amino-terminal fragment of human urokinase plasminogen activator to a specific receptor on U937 monocytes. Proc Natl Acad Sci USA. 1985; 82:4939-43.
37. Appella E, Robinson E, Ulrich S et al. The receptor binding sequence of urokinase: biological function for the growth factor module of proteases. J Biol Chem. 1987; 262:4437-4440.
38. Magdolen V, Rettenberger P, Koppitz M et al. Systematic mutational analysis of the receptor-binding region of the human urokinase-type plasminogen activator. Eur J Biochem 1996; 237:743-51.
39. Roldan AL, Cubellis MV, Masucci MT et al. Cloning and expression of the receptor for human urokinase plasminogen activator, a central molecule in cell surface, plasmin dependent proteolysis. EMBO J 1990; 9:467-74.
40. Ploug M, Ronne E, Behrendt N et al. Cellular receptor for urokinase plasminogen activator. Carboxyl-terminal processing and membrane anchoring by glycosyl-phosphatidylinositol. J Biol Chem 1991; 266:1926-33.
41. Casey JR, Petranka JG, Kottra J et al. The structure of the urokinase-type plasminogen activator receptor gene. Blood 1994; 84:1151-6.
42. Resnati M, Guttinger M, Valcamonica S et al. Proteolytic cleavage of the urokinase receptor substitutes for the agonist-induced chemotactic effect. EMBO J 1996; 15:1572-82.
43. Fazioli F, Resnati M, Sidenius N et al. A urokinase-sensitive region of the human urokinase receptor is responsible for its chemotactic activity. EMBO J 1997; 16:7279-86.
44. Wei Y, Waltz DA, Rao N et al. Identification of the urokinase receptor as an adhesion receptor for vitronectin. J Biol Chem 1994; 269:32380-8.
45. Hoyer-Hansen G, Behrendt N, Ploug M et al. The intact urokinase receptor is required for efficient vitronectin binding: receptor cleavage prevents ligand interaction. FEBS Lett 1997; 420:79-85.
46. Carriero MV, Del Vecchio S, Franco P et al. Vitronectin binding to urokinase receptor in human breast cancer. Clin Cancer Res 1997; 3:1299-308.

47. Colman RW, Pixley RA, Najamunnisa S et al. Binding of high molecular weight kininogen to human endothelial cells is mediated via a site within domains 2 and 3 of the urokinase receptor. J Clin Invest 1997; 100:1481-7.
48. Myohanen HT, Stephens RW, Hedman K et al. Distribution and lateral mobility of the urokinase-receptor complex at the cell surface. J Histochem Cytochem 1993; 41:1291-301.
49. Vassalli JD, Sappino AP, Belin D. The plasminogen activator/plasmin system. J Clin Invest 1991; 88:1067-1072.
50. Pyke C, Eriksen J, Solberg H et al. An alternatively spliced variant of mRNA for the human receptor for urokinase plasminogen activator. FEBS Lett 1993; 326:69-74.
51. Ploug M, Eriksen J, Plesner T et al. A soluble form of the glycolipid-anchored receptor for urokinase-type plasminogen activator is secreted from peripheral blood leukocytes from patients with paroxysmal nocturnal hemoglobinuria. Eur J Biochem 1992; 208:397-404.
52. Sier CF, Stephens R, Bizik J et al. The level of urokinase-type plasminogen activator receptor is increased in serum of ovarian cancer patients. Cancer Res 1998; 58:1843-9.
53. Wilhelm OG, Wilhelm S, Escott GM et al. Cellular glycosylphosphatidylinositol-specific phospholipase D regulates urokinase receptor shedding and cell surface expression. J Cell Physiol 1999; 180:225-35.
54. Sidenius N, Sier CF, Blasi F. Shedding and cleavage of the urokinase receptor (uPAR): identification and characterisation of uPAR fragments in vitro and in vivo. FEBS Lett 2000; 475:52-6.
55. Stoppelli MP, Tacchetti C, Cubellis MV et al. Autocrine saturation of pro-urokinase receptors on human A431 cells. Cell 1986; 45:675-84.
56. Ossowski L, Clunie G, Masucci MT et al. In vivo paracrine interaction between urokinase and its receptor: effect on tumor cell invasion. J Cell Biol 1991; 115:1107-12.
57. Cubellis MV, Nolli ML, Cassani G et al. Binding of single-chain prourokinase to the urokinase receptor of human U937 cells. J Biol Chem 1986; 261:15819-22.
58. Ellis V, Behrendt N, Dano K. Plasminogen activation by receptor-bound urokinase. A kinetic study with both cell-associated and isolated receptor. J Biol Chem 1991; 266:12752-8.
59. Stephens RW, Pollanen J, Tapiovaara H et al. Activation of pro-urokinase and plasminogen on human sarcoma cells: a proteolytic system with surface-bound reactants. J Cell Biol 1989; 108:1987-95.
60. Cubellis MV, Wun TC, Blasi F. Receptor-mediated internalization and degradation of urokinase is caused by its specific inhibitor PAI-1. EMBO J 1990; 9:1079-85.
61. Czekay RP, Kuemmel TA, Orlando RA et al. Direct Binding of Occupied Urokinase receptor (uPAR) to LDL Receptor-related Protein Is Required for Endocytosis of uPAR and Regulation of Cell Surface Urokinase Activity. Mol Biol Cell 2001; 12:1467-1479.
62. Zhou HM, Nichols A, Meda P et al. Urokinase-type plasminogen activator and its receptor synergize to promote pathogenic proteolysis. EMBO J 2000; 19:4817-26.
63. Waltz DA, Sailor LZ, Chapman HA. Cytokines induce urokinase-dependent adhesion of human myeloid cells. A regulatory role for plasminogen activator inhibitors. J Clin Invest 1993; 91:1541-52.
64. Chiaradonna F, Fontana L, Iavarone C et al. Urokinase receptor-dependent and -independent p56/59(hck) activation state is a molecular switch between myelomonocytic cell motility and adherence. EMBO J 1999; 18:3013-23.
65. Waltz DA, Chapman HA. Reversible cellular adhesion to vitronectin linked to urokinase receptor occupancy. J Biol Chem 1994; 269:14746-50.
66. Stahl A, Mueller BM. Melanoma cell migration on vitronectin: regulation by components of the plasminogen activation system. Int J Cancer 1997; 71:116-22.
67. Kjoller L, Hall A. Rac mediates cytoskeletal rearrangements and increased cell motility induced by urokinase-type plasminogen activator receptor binding to vitronectin. J Cell Biol 2001; 152:1145-57.
68. Sitrin RG, Todd RF 3rd, Albrecht E et al. The urokinase receptor (CD87) facilitates CD11b/CD18-mediated adhesion of human monocytes. J Clin Invest 1996; 97:1942-51.
69. Bohuslav J, Horejsi V, Hansmann C et al. Urokinase plasminogen activator receptor, beta 2-integrins, and Src-kinases within a single receptor complex of human monocytes. J Exp Med 1995; 181:1381-90.
70. Wei Y, Lukashev M, Simon DI et al. Regulation of integrin function by the urokinase receptor. Science 1996; 273:1551-1555.
71. Simon DI, Wei Y, Zhang L et al. Identification of a urokinase receptor-integrin interaction site. Promiscuous regulator of integrin function. J Biol Chem 2000; 275:10228-34.
72. Aguirre-Ghiso JA, Liu D, Mignatti A et al. Urokinase receptor and Fibronectin Regulate the ERK(MAPK) to p38(MAPK) Activity Ratios That Determine Carcinoma Cell Proliferation or Dormancy In Vivo. Mol Biol Cell 2001; 12:863-79.
73. Tarui T, Mazar AP, Cines DB et al. Urokinase-type plasminogen activator receptor (CD87) is a ligand for integrins and mediates cell-cell interaction. J Biol Chem 2001; 276:3983-90.

74. Fibbi G, Ziche M, Morbidelli L et al. Interaction of urokinase with specific receptors stimulates mobilization of bovine adrenal capillary endothelial cells. Exp Cell Res 1988; 179:385-95.
75. Gyetko MR, Todd RF 3rd, Wilkinson CC et al. The urokinase receptor is required for human monocyte chemotaxis in vitro. J Clin Invest 1994; 93:1380-7.
76. Ossowski L, Aguirre-Ghiso JA. Urokinase receptor and integrin partnership: coordination of signaling for cell adhesion, migration and growth. Curr Opin Cell Biol 2000; 12:613-620.
77. Gyetko MR, Sud S, Kendall T et al. Urokinase receptor-deficient mice have impaired neutrophil recruitment in response to pulmonary Pseudomonas aeruginosa infection. J Immunol 2000; 165:1513-9.
78. Waltz DA, Fujita RM, Yang X et al. Nonproteolytic role for the urokinase receptor in cellular migration in vivo. Am J Respir Cell Mol Biol 2000; 22:316-22.
79. Carriero MV, Del Vecchio S, Capozzoli M et al. Urokinase receptor interacts with alpha(v)beta5 vitronectin receptor, promoting urokinase-dependent cell migration in breast cancer. Cancer Res 1999; 59:5307-14.
80. Degryse B, Orlando S, Resnati M et al. Urokinase/urokinase receptor and vitronectin/alpha(v)beta(3) integrin induce chemotaxis and cytoskeleton reorganization through different signaling pathways. Oncogene 2001; 20:2032-43.
81. Nguyen DH, Hussaini IM, Gonias SL. Binding of urokinase-type plasminogen activator to its receptor in MCF-7 cells activates extracellular signal-regulated kinase 1 and 2 which is required for increased cellular motility. J Biol Chem 1998; 273:8502-7.
82. Kirchheimer JC, Wojta J, Christ G et al. Functional inhibition of endogenously produced urokinase decreases cell proliferation in a human melanoma cell line. Proc Natl Acad Sci USA. 1989; 86:5424-8.
83. Koopman JL, Slomp J, de Bart AC et al. Mitogenic effects of urokinase on melanoma cells are independent of high affinity binding to the urokinase receptor. J Biol Chem 1998; 273:33267-72.
84. Aguirre Ghiso JA, Kovalski K, Ossowski L. Tumor dormancy induced by downregulation of urokinase receptor in human carcinoma involves integrin and MAPK signaling. J Cell Biol 1999; 147:89-104.
85. Sitrin RG, Pan PM, Harper HA et al. Clustering of urokinase receptors (uPAR; CD87) induces proinflammatory signaling in human polymorphonuclear neutrophils. J Immunol 2000; 165:3341-9.
86. Harder T, Simons K. Caveolae, DIGs, and the dynamics of sphingolipid-cholesterol microdomains. Curr Opin Cell Biol 1997; 9:534-42.
87. Stahl A, Mueller BM. The urokinase-type plasminogen activator receptor, a GPI-linked protein, is localized in caveolae. J Cell Biol 1995; 129:335-44.
88. Wei Y, Yang X, Liu Q et al. A role for caveolin and the urokinase receptor in integrin-mediated adhesion and signaling. J Cell Biol 1999; 144:1285-94.
89. Sitrin RG, Pan PM, Blackwood RA et al. Cutting edge: evidence for a signaling partnership between urokinase receptors (cd87) and l-selectin (cd62l) in human polymorphonuclear neutrophils. J Immunol 2001; 166:4822-5.
90. Wong WS, Simon DI, Rosoff PM et al. Mechanisms of pertussis toxin-induced myelomonocytic cell adhesion: role of Mac-1(CD11b/CD18) and urokinase receptor (CD87). Immunology 1996; 88:90-7.
91. Degryse B, Resnati M, Rabbani SA et al. Src-dependence and pertussis-toxin sensitivity of urokinase receptor-dependent chemotaxis and cytoskeleton reorganization in rat smooth muscle cells. Blood 1999; 94:649-62.
92. Tang H, Kerins DM, Hao Q et al. The urokinase-type plasminogen activator receptor mediates tyrosine phosphorylation of focal adhesion proteins and activation of mitogen-activated protein kinase in cultured endothelial cells. J Biol Chem 1998; 273:18268-72.
93. Nguyen DH, Catling AD, Webb DJ et al. Myosin light chain kinase functions downstream of Ras/ERK to promote migration of urokinase-type plasminogen activator-stimulated cells in an integrin-selective manner. J Cell Biol 1999; 146:149-64.
94. Kusch A, Tkachuk S, Haller H et al. Urokinase stimulates human vascular smooth muscle cell migration via a phosphatidylinositol 3-kinase-Tyk2 interaction. J Biol Chem 2000; 275:39466-73.
95. Konakova M, Hucho F, Schleuning WD. Downstream targets of urokinase-type plasminogen-activator-mediated signal transduction. Eur J Biochem 1998; 253:421-9.
96. Zannetti A, Del Vecchio S, Carriero MV et al. Coordinate up-regulation of Sp1 DNA-binding activity and urokinase receptor expression in breast carcinoma. Cancer Res 2000; 60:1546-51.
97. Lund LR, Bjorn SF, Sternlicht MD et al. Lactational competence and involution of the mouse mammary gland require plasminogen. Development 2000; 127:4481-92.
98. Bezerra JA, Currier AR, Melin-Aldana H et al. Plasminogen activators direct reorganization of the liver lobule after acute injury. Am J Pathol 2001; 158:921-9.
99. Lluis F, Roma J, Suelves M et al. Urokinase-dependent plasminogen activation is required for efficient skeletal muscle regeneration in vivo. Blood 2001; 97:1703-11.
100. Ossowski L, Reich E. Antibodies to plasminogen activator inhibit human tumor metastasis. Cell 1983; 35:611-9.

101. Hearing VJ, Law LW, Corti A et al. Modulation of metastatic potential by cell surface urokinase of murine melanoma cells. Cancer Res 1988; 48:1270-8.
102. Bugge TH, Lund LR, Kombrinck KK et al. Reduced metastasis of Polyoma virus middle T antigen-induced mammary cancer in plasminogen-deficient mice. Oncogene 1998; 16:3097-104.
103. Shapiro RL, Duquette JG, Roses DF et al. Induction of primary cutaneous melanocytic neoplasms in urokinase-type plasminogen activator (uPA)-deficient and wild-type mice: cellular blue nevi invade but do not progress to malignant melanoma in uPA-deficient animals. Cancer Res 1996; 56:3597-604.
104. Wang Y, Liang X, Wu S et al. Inhibition of colon cancer metastasis by a 3'- end antisense urokinase receptor mRNA in a nude mouse model. Int J Cancer 2001; 92:257-262.
105. Eitzman DT, Krauss JC, Shen T et al. Lack of plasminogen activator inhibitor-1 effect in a transgenic mouse model of metastatic melanoma. Blood 1996; 87:4718-22.
106. Bajou K, Noel A, Gerard RD et al. Absence of host plasminogen activator inhibitor 1 prevents cancer invasion and vascularization. Nat Med 1998; 4:923-8.
107. Bajou K, Masson V, Gerard RD et al. The Plasminogen activator inhibitor PAI-1 Controls In Vivo Tumor Vascularization by Interaction with Proteases, Not Vitronectin. Implications for antiangiogenic strategies. J Cell Biol 2001; 152:777-84.
108. Zhou HM, Bolon I, Nichols A et al. Overexpression of plasminogen activator inhibitor type 2 in basal keratinocytes enhances papilloma formation in transgenic mice. Cancer Res 2001; 61:970-6.
109. Del Vecchio S, Stoppelli MP, Carriero MV et al. Human urokinase receptor concentration in malignant and benign breast tumors by in vitro quantitative autoradiography: comparison with urokinase levels. Cancer Res 1993; 53:3198-206.
110. Yamamoto M, Sawaya R, Mohanam S et al. Expression and localization of urokinase-type plasminogen activator in human astrocytomas in vivo. Cancer Res 1994; 54:3656-61.
111. Carriero MV, Franco P, Del Vecchio S et al. Tissue distribution of soluble and receptor-bound urokinase in human breast cancer using a panel of monoclonal antibodies. Cancer Res 1994; 54:5445-54.
112. Johnsen M, Lund LR, Romer J et al. Cancer invasion and tissue remodeling: common themes in proteolytic matrix degradation. Curr Opin Cell Biol 1998; 10:667-71.
113. Kennedy S, Duffy MJ, Duggan C et al. Semi-quantitation of urokinase plasminogen activator and its receptor in breast carcinomas by immunocytochemistry. Br J Cancer 1998; 77:1638-41.
114. Dublin E, Hanby A, Patel NK et al. Immunohistochemical expression of uPA, uPAR, and PAI-1 in breast carcinoma. Fibroblastic expression has strong associations with tumor pathology. Am J Pathol 2000; 157:1219-27.
115. Foekens JA, Peters HA, Look MP et al. The urokinase system of plasminogen activation and prognosis in 2780 breast cancer patients. Cancer Res 2000 60:636-43.
116. Duffy MJ, Reilly D, O'Sullivan C et al. Urokinase-plasminogen activator, a new and independent prognostic marker in breast cancer. Cancer Res 1990; 50:6827-9.
117. Andreasen PA, Kjoller L, Christensen L et al. The urokinase-type plasminogen activator system in cancer metastasis: a review. Int J Cancer 1997; 72:1-22.
118. Ferrier CM, Suciu S, van Geloof WL et al. High tPA-expression in primary melanoma of the limb correlates with good prognosis. Br J Cancer 2000, 83:1351-9.
119. Pedersen AN, Christensen IJ, Stephens RW et al. The complex between urokinase and its type-1 inhibitor in primary breast cancer: relation to survival. Cancer Res 2000; 60:6927-34.
120. Stephens RW, Brunner N, Janicke F et al. The urokinase plasminogen activator system as a target for prognostic studies in breast cancer. Breast Cancer Res Treat 1998; 52:99-111.
121. Sier CF, Sidenius N, Mariani A et al. Presence of urokinase-type plasminogen activator receptor in urine of cancer patients and its possible clinical relevance. Lab Invest 1999; 79:717-22.
122. Mustjoki S, Sidenius N, Sier CF et al. Soluble urokinase receptor levels correlate with number of circulating tumor cells in acute myeloid leukemia and decrease rapidly during chemotherapy. Cancer Res 2000; 60:7126-32.
123. Tressler RJ, Pitot PA, Stratton JR et al. Urokinase receptor antagonists: discovery and application to in vivo models of tumor growth. APMIS 1999; 107:168-73.
124. Magdolen V, Arroyo de Prada N, Sperl S et al. Natural and synthetic inhibitors of the tumor-associated serine protease urokinase-type plasminogen activator. Adv Exp Med Biol 2000; 477:331-41.
125. Nienaber VL, Richardson PL, Klighofer V, Bouska JJ, Giranda VL, Greer J. Discovering novel ligands for macromolecules using X-ray crystallographic screening. Nat Biotechnol 2000; 18:1105-8.
126. Cavallaro U, del Vecchio A, Lappi DA et al. A conjugate between human urokinase and saporin, a type-1 ribosome-inactivating protein, is selectively cytotoxic to urokinase receptor-expressing cells. J Biol Chem 1993; 268:23186-90.
127. Liu S, Bugge TH, Leppla SH. Targeting of tumor cells by cell surface urokinase plasminogen activator-dependent anthrax toxin. J Biol Chem 2001; 276:17976-84.

128. Soff GA. Angiostatin and angiostatin-related proteins. Cancer Metastasis Rev 2000; 19:97-107.103.
129. Franco P, Massa O, Garcia-Rocha M et al. Protein kinase C-dependent in vivo phosphorylation of prourokinase leads to the formation of a receptor competitive antagonist. J Biol Chem 1998; 273:27734-40.
130. Guo Y, Higazi AA, Arakelian A et al. A peptide derived from the nonreceptor binding region of urokinase plasminogen activator (uPA) inhibits tumor progression and angiogenesis and induces tumor cell death in vivo. FASEB J 2000; 14:1400-10.
131. Mishima K, Mazar AP, Gown A et al. A peptide derived from the non-receptor-binding region of urokinase plasminogen activator inhibits glioblastoma growth and angiogenesis in vivo in combination with cisplatin. Proc Natl Acad Sci USA 2000; 97:8484-9.
132. Holten-Andersen MN, Stephens RW, Nielsen HJ et al. High preoperative plasma tissue inhibitor of metalloproteinase-1 levels are associated with short survival of patients with colorectal cancer. Clin Cancer Res 2000; 6:4292-9.
133. Koolwijk P, Sidenius N, Peters E et al. Proteolysis of the urokinase-type plasminogen activator receptor by metalloproteinase-12: implication for angiogenesis in fibrin matrices. Blood 2001; 97:3123-31.

CHAPTER 9

Influence of Cell-Extracellular Matrix Interactions on Keratinocyte Behavior During Repair

Brian K. Pilcher, Jonathan C.R. Jones and William C. Parks

Introduction

The extracellular matrix (ECM), composed of both proteinaceous and nonproteinaceous components, was once thought to merely serve a scaffolding function that provided structural integrity and resiliency to tissues. While this is an important function of the ECM, it is also a key regulator of cell behavior during morphogenic and repair events. Some proteins, such as elastin and many collagen types, are deposited as large, insoluble, protease-resistant fibers. Additionally, several different matrix glycoproteins, such as laminins, fibronectin, thrombospondins, fibrillins, entactin, and many others, form aggregates of various degrees of complexity, providing substrata for cell adhesion and sites for protein-protein interactions. Proteoglycans, in which the mass of carbohydrate exceeds the mass of protein, function in tissue hydration and in a variety of processes related to growth factor/chemokine activity, proteolysis, integrin activation, and more. In addition, extracellular molecules can be organized into specialized structures, such as ligaments, bone, and basement membrane, the matrix upon which epithelia and endothelial cells reside.

Essentially all matrix proteins have been shown to have some role in affecting cell phenotype and behavior. The ability of the ECM to mediate changes in cell behavior is most prominent, however, when cells undergo alterations in their environment that exposes them to a molecularly distinct neo-matrix. Furthermore, the specificity of a cell's response to newly exposed ECM is tightly coordinated by binding and signaling through cell surface receptors, such as integrins. Once activated, these heterodimeric integral membrane proteins affect diverse cellular processes, such as migration and proliferation, which are controlled by intermediate signaling pathways including protein kinase C, mitogen activated protein kinase, and focal adhesion kinase.[1] Thus, a dynamic interplay between cell receptor expression and exposure to diverse ECM components is established, resulting in gene activation and modification of cellular function during morphogenic events.

The epidermal response to wounding is a well-studied example of how changes in cell-ECM interactions regulate cell behavior. Unwounded epidermis consists of a multilayered epithelial sheet that provides a physical barrier against the outside environment. During tissue homeostasis, basal keratinocytes reside on a basement membrane composed of laminins -5, -6, -7 and -10, types -IV and -VII collagen, entactins and heparan sulfate proteoglycans.[2-6] This extracellular mat of proteins provides a substrate for polarized cell adhesion and physically separates the cells from underlying dermal connective tissue, which is rich in type I collagen, fibrillins, and other proteins not found in the basement membrane. While attached to the basement membrane, keratinocytes proliferate, differentiate, and stratify to resupply the superficial layers of epidermis that are continually sloughed. In addition, specific cell-cell and cell-matrix

Cell Invasion, edited by Jyrki Heino and Veli-Matti Kähäri.

adhesion receptors provide structural support and transduce signals involved in regulating epidermal homeostasis. For example, in keratinocytes the laminin-5 binding α6β4 integrin is localized on the basal surface and is associated with the anchoring filaments of hemidesmosomes, which are specialized, polarized junctional complexes required for tissue stability (reviewed in 7, 8). In contrast, the α2β1 and α3β1 integrins are oriented on the basal-lateral surfaces of keratinocytes where they may function in cell-cell contacts.[7,8] However, there is no known ligand or function for the α2β1 integrin in normal, intact skin.[9] But, as is discussed below, ligation of this integrin with dermal collagen by wound-edge keratinocytes is an important early spatial signal regulating the epidermal response to injury. Because the epidermis is not typically in contact with type I collagen, it is tempting to speculate that the constitutive production of the α2β1 integrin by basal keratinocytes keeps the cells primed and ready to respond to injury. It is reasonable that the epidermis is equipped and programmed to respond rapidly to injury and begin wound closure, rather than relying on signals from later events, such as release of soluble factors from inflammatory cells.

In contrast to cells in intact skin, wound keratinocytes undergo global changes in gene expression, switching from a program of proliferation and differentiation to one that supports rapid, sustained, and directed migration.[10] This activation, which typically occurs 18-24 hours prior to the onset of migration, occurs as wound edge keratinocytes disassemble their cell-cell contacts and interact with newly exposed matrix proteins such as fibrillar type I collagen, fibronectin, and fibrin.[11,12] This shift in cell phenotype stimulates a number of complex processes involved in cell migration, including hemidesmosome retraction, upregulation or reorientation of integrin receptors, reorganization of the actin cytoskeleton, lamellipodia formation, and adherence to and dissociation from ECM ligands.[13-15]

Activated keratinocytes express a distinct pattern of matrix-binding integrins, which interact with wound bed matrix and regulate cell activity through specific signaling mechanisms. For example, the α6β4 integrin is redistributed evenly across the cell surface of migrating keratinocytes, which lack hemidesmosomes.[16-19] Expression of the α5β1 and αvβ5 integrins are induced in migrating keratinocytes and may interact with newly deposited fibronectin and vitronectin in the provisional matrix.[20] The constitutively expressed collagen integrins α2β1 and α3β1 are reoriented towards the fronto-basal surface of migrating keratinocytes.[21-23]

Keratinocyte migration from the wound edge occurs as the cells release their contacts from the basement membrane, dissect under a provisional matrix of fibrin, fibronectin, and vitronectin,[24] and over a viable dermis rich in type I collagen. Importantly, the dermal and provisional matrix macromolecules (e.g., fibronectin and type I collagen) with which the keratinocyte interacts are distinct from those in the basement membrane (e.g., laminins and type IV collagen). These newly established cell-matrix contacts are a critical determinant that induces the activated keratinocyte phenotype by enhancing the transcription of genes required for repair. In the second section of this Chapter, the importance of keratinocyte contact with type I collagen and its role in stimulating collagenase-1, a matrix metalloproteinase required for migration, will be discussed.

As reepithelialization proceeds, keratinocytes begin to synthesize and secrete ECM components, specifically α3 subunit containing laminins (laminins-5, -6 and -7) and, most probably, laminin–10, to form a functional basement membrane.[17,25-27] Contact with the newly formed laminin matrix may provide site-specific cues to signal an end to migration and a shift in gene expression to a program that enhances cell proliferation and differentiation. In the last section of this Chapter, recent evidence demonstrating a role for α3 subunit-containing laminins in mediating the responses associated with downregulation of the wound healing response will be discussed.

Type I Collagen, Collagenase-1, and Keratinocyte Migration

Perhaps one of the most thoroughly studied models of ECM influencing cell activation during cutaneous repair is the role of keratinocyte contact with fibrillar type I collagen following injury and the subsequent induction and function of collagenase-1 expression. The matrix metalloproteinases (MMPs), of which collagenase-1 is a member, comprise a family of enzymes that share several common properties.[28] MMPs, or matrixins, are a subgroup of the much larger metalloproteinase superfamily, which also includes astacin and ADAM proteinases, among others. To date, 25 different MMPs have been characterized, and additional members continue to be identified. Distinct from other MMPs that can degrade multiple ECM macromolecules, collagenase-1 (MMP-1) and collagenase-2 (MMP-8, neutrophil collagenase) have a defined substrate spectrum, being limited to the fibrillar collagens, types I, II, and III. In addition, collagenase-1 and -2 do not degrade collagen, but rather make a single, site-specific cleavage within the triple helix of these abundant matrix components.[29] Once cleaved, the triple helix of collagen relaxes at body temperature, and the partially unwound fibril becomes susceptible to further proteolysis by a variety of other MMPs and proteinases.

Typically MMPs are not expressed in normal, healthy, resting tissues, at least their production and activity, with notable exceptions,[30-35] are maintained at nearly undetectable levels. In contrast, some level of MMP expression is seen in any repair or remodeling process, in any diseased or inflamed tissue, and in essentially any cell type grown in culture. Although the qualitative pattern and quantitative levels of MMPs varies among tissues, diseases, tumors, inflammatory conditions, and cell lines, a reasonably safe generalization is that activated cells express MMPs.

Of the many MMPs expressed in wounds, the function of collagenase-1 in the migrating epidermis is the best understood. The notion that epidermal collagenase-1 plays a role in wound healing was first suggested by Grillo and Gross who reported in 1967 that collagenolytic activity is released from the edge of guinea pig epidermal wounds.[36] Subsequently, other laboratories reported that collagenase-1 protein or activity is present in the wound environment,[37,38] and it was thought that this enzyme was produced by fibroblasts, macrophages, or other cells within the granulation tissue. However, as predicted by the initial reports of Grillo and Gross, the epidermis is the principal source of collagenase-1 activity and production in wounded human skin.

In a thorough examination of multiple epidermal wounds from both humans and animals, it was found that collagenase-1 expression is prominently and invariably expressed by basal keratinocytes at the leading edge of reepithelialization.[39-47] Importantly, collagenase-1 expression is limited to wounds with a disrupted basement membrane; collagenase-1 is not expressed in wounds that form above the basement membrane (e.g., freshly formed bullous pemphigoid blisters and nonulcerative pyogenic granuloma).[40] Furthermore, collagenase-1 is not seen in hyperproliferative keratinocytes just behind the wound front and residing on basement membrane or by suprabasal cells in intact or unwounded skin (Fig. 1). Similar results have been reported in ex vivo wounded human skin transplanted onto SCID mice and in wounded organotypic cultures,[48-50] indicating that induction of this MMP is a common response to injury of mammalian skin. The invariant expression of collagenase-1 in a variety of skin wounds indicates that the activity of this proteinase serves a critical role in reepithelialization.

Because collagenase-1 was detected only in wounds exhibiting a disrupted basement membrane, and not in nonulcerated samples or intact skin, it was hypothesized that alterations in cell-ECM interactions, as occurs when keratinocytes contact dermal and provisional matrix of the wound bed, were a critical determinant for induction of collagenase-1 expression. Two key in vivo observations support this idea: (1) Immunostaining for the basement membrane components type IV collagen or laminin-1 showed that collagenase-1-positive keratinocytes are not in contact with a basement membrane,[20,40,50] but rather in close apposition to collagen fibers of the dermal matrix.[48] (2) Collagenase-1 expression ceases once healing has completed and an intact basement membrane is reestablished.[20,49,50] Together, these findings support the idea that keratinocytes acquire a collagenolytic phenotype upon contact with the dermal matrix.

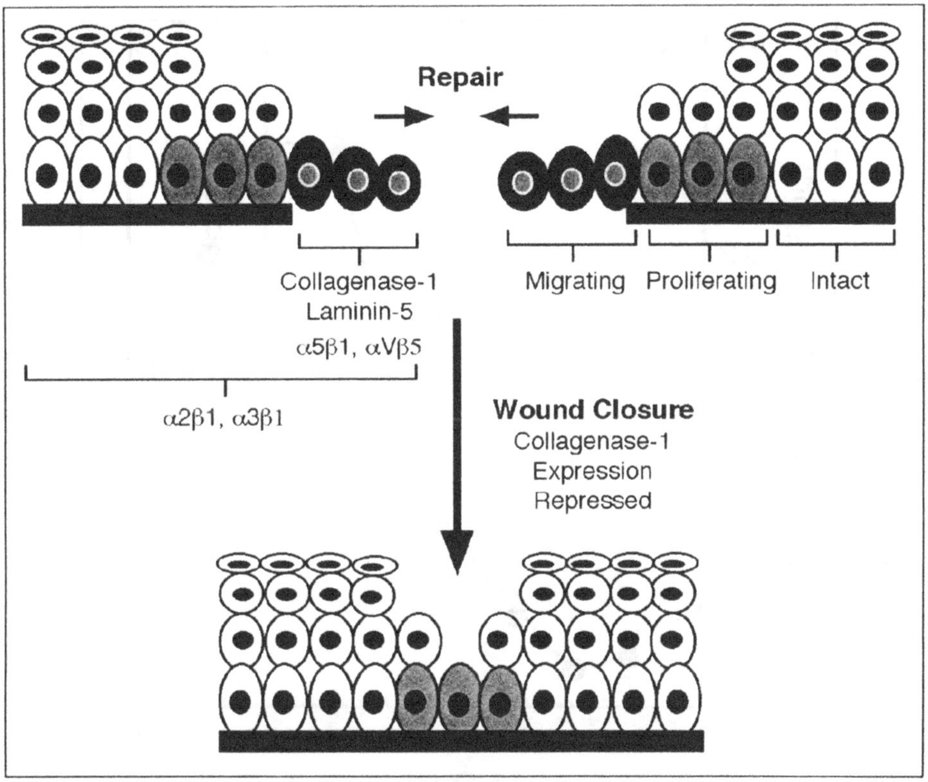

Fig. 1. Spatial patterns of MMP and integrin expression in the epidermis during wound healing. Collagenase-1 is prominently and invariably expressed by migrating basal keratinocytes in all wounds, whether acute or chronic, characterized by disruption of the basement membrane (black line). The integrins $\alpha_5\beta_1$ and $\alpha v\beta 5$ and the basement membrane protein laminin-5 (LM-5) are also induced in the migrating cells. In contrast, the collagen binding integrins $\alpha 2\beta 1$ and $\alpha 3\beta 1$ are constitutively expressed in intact skin and in migrating keratinocytes. At the completion of reepithelialization, metalloproteinase expression is turned off.

The in vivo expression of collagenase-1 by migrating keratinocytes is recapitulated by primary human keratinocytes in vitro. When cultured in high calcium containing medium, keratinocytes form foci of proliferating and differentiating cells surrounded by migrating cells.[48,51] Reflecting the phenotype of basal cells involved in reepithelialization in vivo, collagenase-1 mRNA is expressed only in the migrating keratinocytes away from differentiating colonies and only when the cells are plated on a native type I collagen substrate (Fig. 2).[52] Other dermal or provisional ECM macromolecules that keratinocytes may interact with during the repair process, such as laminin-5, elastin, fibronectin, fibrinogen/fibrin, and type III collagen, do not induce collagenase-1 production.[52] Furthermore, the native, triple helical conformation of collagen is required to mediate this response as plating keratinocytes on gelatin generated by heat denaturation or cleavage with purified collagenase fails to induce enzyme production. Lastly, the kinetics of collagenase-1 induction by collagen mimics that seen in vivo[50] as collagenase-1 mRNA is evident in cells as early as 2 hr following matrix contact.[52] These findings support the idea that contact with native type I collagen is an important, if not the principal, determinant regulating the expression of collagenase-1 by migrating keratinocytes during wound repair. This paradigm implies that the pericellular environment, and the fibrillar ECM in particular, regulates the behavior of migrating keratinocytes by modulating MMP expression.

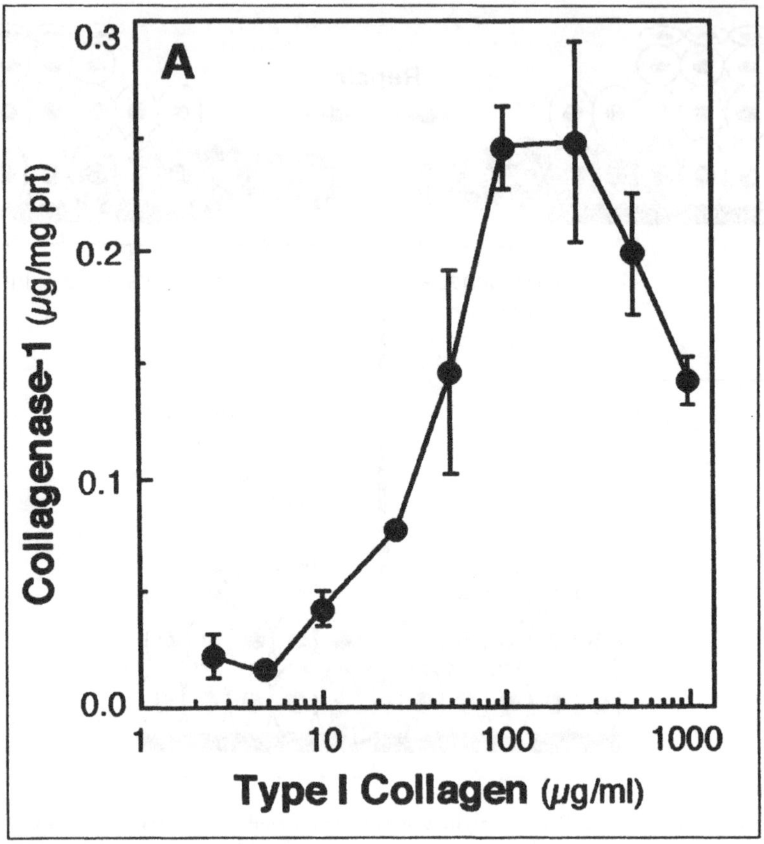

Fig. 2A. Keratinocyte collagenase-1 is induced by native type I collagen and repressed by laminin-1. Primary human keratinocytes were grown on dishes precoated with the indicated concentrations of native type I collagen. Medium was collected 72 h later, and collagenase-1 levels were determined by ELISA and normalized to total protein. Data represent the mean ± SD of triplicate samples (From Ref. 54).

The invariable expression of collagenase-1 by basal keratinocytes in all forms of wounds and the confinement of its expression to periods of active re-epithelialization indicates that this enzyme participates in cell migration. Beyond directly remodeling structural proteins, such as during morphogenesis and tissue resorption, MMPs are thought to breakdown matrix barriers that impede cell migration. Clearly, this is a reasonable role for these proteinases in facilitating cell movement through a three-dimensional matrix, as is seen during blastocyst invasion,[53] angiogenesis,[54] and extravasation and infiltration of inflammatory cells.[55] During normal reepithelialization, however, keratinocytes migrate along a path of least resistance, dissecting underneath the scab while remaining superficial to the underlying viable dermis and wound bed.[56] Thus, epidermal repair involves cell migration over a two-dimensional plane rather than through a three-dimensional matrix-rich environment. Recent accumulated evidence indicates that collagenase-1 is a component of a molecular machine that drives and orients keratinocyte migration.[48] Other key components of this machine are native type I collagen in the dermis and the $\alpha2\beta1$ integrin on keratinocytes.

The requirement of collagenase-1 for keratinocyte movement was demonstrated in various migration assays. In all assays, keratinocyte migration on native collagen was completely blocked

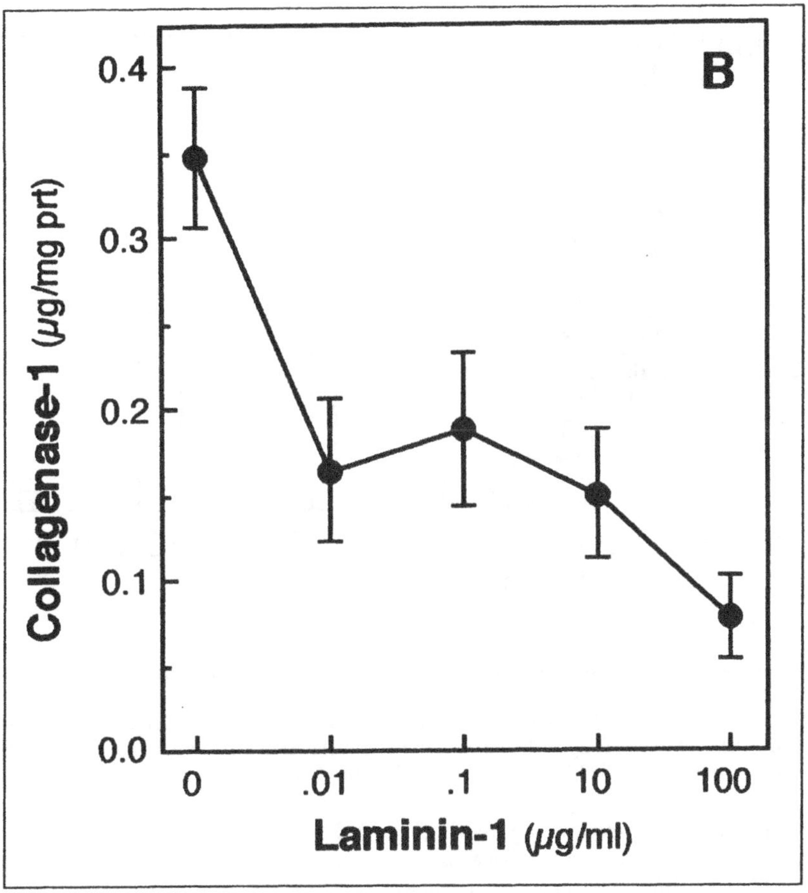

Fig. 2B. Keratinocyte collagenase-1 is induced by native type I collagen and repressed by laminin-1. Primary human keratinocytes were grown on dishes precoated with 100 μg/ml type I collagen containing the indicated concentrations of laminin-1 (B). Medium was collected 72 h later, and collagenase-1 levels were determined by ELISA and normalized to total protein. Data represent the mean ± SD of triplicate samples (From Ref. 54).

by treatment of cells with broad-acting peptide hydroxymate inhibitors (Fig. 3). These compounds, similar to many MMP inhibitors, are substrate-based inhibitors that contain a hydroxamic acid moiety that chelates the active site zinc cation and renders MMPs catalytically inactive.[57] Because these early generation hydroxymates inhibit the activity of all MMPs and because keratinocytes express other MMPs (i.e., stromelysin-2 [MMP-10], gelatinase-B [MMP-9], epilyisn [MMP-28]), these experiments did not demonstrate that the activity of collagenase-1 *alone* is required for keratinocyte migration on collagen. However, keratinocytes plated on collagenase-resistant mutant collagen, in which a double mutation was inserted by homologous recombination in the region of the collagenase cleavage site of the $\alpha1$(I) chain,[58,59] do not migrate, yet the cells express collagenase-1 and adhere equally to those on wild-type collagen.[48]

Because all three human collagenases can cleave type I collagen at the same site, the mutant substrate experiment did not conclusively demonstrate that collagenase-1 is specifically required for keratinocyte migration on a collagen matrix. Reagents that selectively block the activity of collagenase-1 verified that this MMP alone is necessary and sufficient for keratinocyte

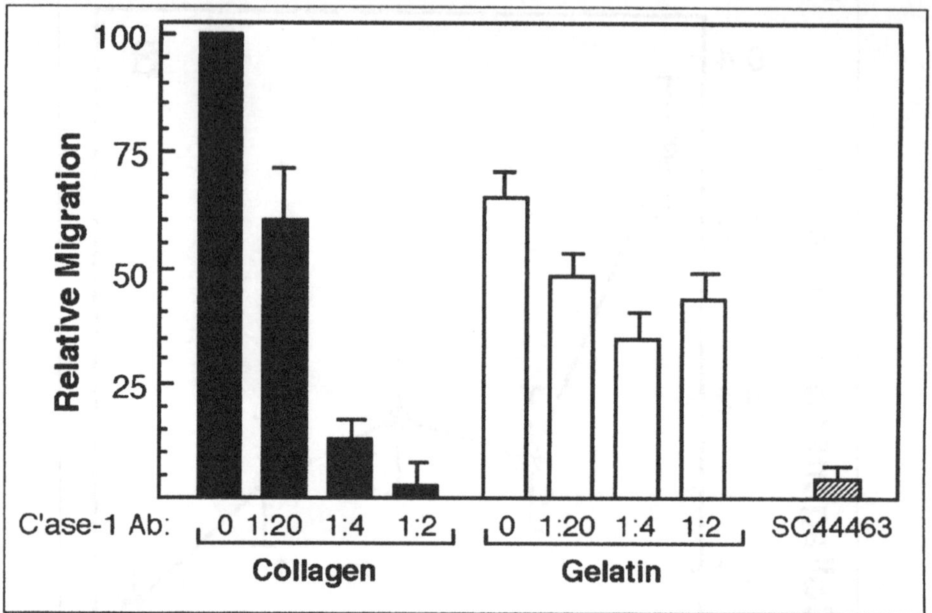

Fig. 3. Keratinocyte migration on type I collagen is collagenase-1-dependent. Primary human keratinocytes were plated on culture slides coated with colloidal gold particles and 100 μg/ml collagen or gelatin and the indicated titers of affinity-purified collagenase-1 antibody or 25 μM SC44463. Keratinocyte migration was quantified 20 h later by measuring the linear phagokinetic tracks created on the gold-collagen substrate. The data shown are the means ± SEM of duplicate samples from four experiments (From Ref. 50).

migration on a collagen-containing matrix (Fig. 3). Keratinocyte migration is completely inhibited by affinity-purified anti-collagenase-1 antibodies, which block the enzyme's catalytic activity.[48] Furthermore, treatment with a newer generation hydroxymate compound, which inhibits all MMPs *except collagenase-1* does not affect keratinocyte migration (BKP and WCP, unpublished observations). Together, these data demonstrate that the proteolytic activity of collagenase-1, and not that of any other MMP, is required for keratinocyte migration on native type I collagen.

Based on the above observations, we proposed that collagenase-1 acting on its principal substrate in the dermis, type I collagen, provides migrating keratinocytes with a mechanism to maintain their course and directionality in the wound environment during reepithelialization (Fig. 4). As stated, basal keratinocytes constitutively express the type I collagen-binding integrin α2β1 on their basal-lateral surfaces.[60,61] In wounds, α2β1 becomes concentrated at the forward-basal tip of migrating keratinocyte,[21] and this redistribution, which is likely regulated by an initial contact with dermal collagen, places the α2β1 integrin in intimate contact with dermal type I collagen. Contact with collagen induces collagenase-1, and this expression is mediated by the α2β1 integrin.[52]

Because the α2β1 integrin binds native collagen with high affinity,[62] clustering this receptor at contact points would tether keratinocytes to the dermis rendering them unable to migrate. Collagenase-1 aids in dissociating keratinocytes from these high affinity attachments by altering the nature of the collagen matrix and, in turn, its affinity with α2β1. As stated, collagenase-1 makes a single, site-specific cleavage through the triple helix of collagen about 3/4 the length from the N-terminus. The resultant TC^A and TC^B fragments are thermally unstable at body temperature and spontaneously unwind into gelatin, which binds the α2β1 integrin with a much lower affinity than does native collagen.[62] Thus, by simply making a single cut through

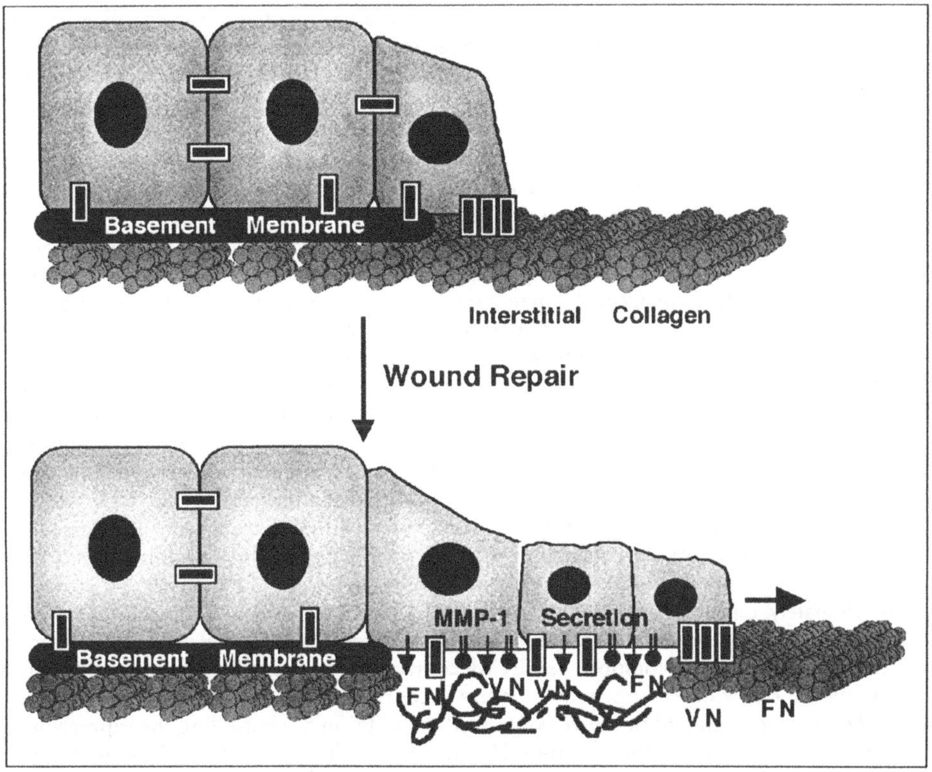

Fig. 4. Collagenase-1-dependent keratinocyte migration during epidermal wound repair. After wounding, basal keratinocytes dislodge from the basement membrane and contact dermal type I collagen, Collagen-binding integrins (black rectangles), such as α2β1, are constitutively expressed on basal keratinocytes and are seen on the basal and lateral surfaces of cells in intact skin, but in wounded skin, these receptors accumulate on the frontobasal surface of migrating keratinocytes. α2β1 binds dermal collagen, and this high affinity interaction may induce collagenase-1 expression as well as that of new integrins, such as α5β1 and αvβ5. High affinity binding to collagen, however, may hinder cell motility. Cleavage of type I collagen by collagenase-1, and its subsequent conversion to gelatin, would reduce the affinity of α2β1's interaction with the matrix, thereby allowing the keratinocytes to use newly expressed integrins to migrate on their ligands, such as fibronectin and vitronectin, which are abundant in the wound bed matrix. The high affinity interaction of α2β1 with dermal collagen, but not with gelatin, provides the migrating cells with a mechanism to control their direction and to remain superficial during reepithelialization.

the type I collagen helix, collagenase-1 effectively mediates the loosening of the tight contacts keratinocytes establish with the dermal matrix, thereby allowing the cells to move forward. This function is distinct from the often-suggested idea that migrating cells use MMPs to remove matrix barriers that may physically impede movement.

Although collagenase-1 facilitates keratinocyte migration by affecting the conformation of type I collagen and, consequently, the avidity with which cells interact with it, one may argue that this is an inherently inefficient mechanism. If activated keratinocytes migrate *over* the viable dermis, rather than *through* matrix, then why do they need to cleave type I collagen? Why would they adhere to the dermis with such high affinity if their objective is to close the wound as quickly as possible? The answer, we believe, is that the process of interacting with and then cleaving type I collagen provides keratinocytes with a mechanism to determine and maintain their directionality during reepithelialization. Because α2β1 binds native collagen much more tightly than it binds gelatin, this integrin, once loosened by the action of collagenase-1, would favor reestablishment of high affinity interactions with native, uncleaved collagen. Thus, by simply making a single cut through the type I collagen helix, MMP-1 effectively loosens the tight contacts established by keratinocytes with the dermal matrix. The cells, in turn, would move forward by forming new contacts with uncleaved collagen on the open, superficial plane of the viable wound bed. By repeatedly establishing tight contacts, then rapidly loosening this hold by the action of collagenase-1, keratinocytes use native type I collagen as a "molecular compass" to guide repair over the open wound surface.

An important observation relevant to the directionality hypothesis is that collagenase-1 production is induced in keratinocytes by native type I collagen but not by denatured forms of the molecule.[52,63] Thus, collagenase-1 acting on collagen creates a mediator that does not support or maintain its own production. The conversion of collagen to gelatin would replace the inductive stimulus with a neutral substrate (gelatin) and, in stationary cells, collagenase-1 expression would decline. Indeed, collagenase-1 expression is rapidly turned off at the completion of reepithelialization.[50] The initial expression of collagenase-1 and cleavage of the collagen substrate would, in effect, neutralize the inductive effect of the underlying matrix. If keratinocytes continue to interact with type I collagen, presumably by migrating, then they would continue to express collagenase-1, which they do throughout reepithelialization.

By which cells and where in the tissue environment an MMP is expressed and released are equally, if not more important considerations when predicting the target of proteolysis than is the affinity of enzyme-substrate interactions. After all, cells do not release proteases indiscriminately, especially enzymes like collagenases with such a defined substrate specificity, but rather they rely on precise cell-ECM interactions to accurately remodel connective tissue in the pericellular space. In addition, cell-ECM contacts can provide an unambiguous signal informing the cell which matrix protein it has encountered and, hence, which proteinase is needed and where it should be delivered. Indeed, in addition to controlling the expression of collagenase-1, collagenase-1 binds to the $\alpha2\beta1$ integrin on the surface of keratinocytes migrating over collagen.[64] Thus, the interaction of collagenase-1 with $\alpha2\beta1$ confines the proteinase to points of cell contact with collagen and the resultant ternary complex of integrin, enzyme, and substrate function together to drive and regulate keratinocyte migration.

Laminin Effects on Keratinocyte Behavior Upon Completion of Reepithelialization

As reepithelialization completes, cell-cell contacts and keratinocyte interactions with basement membrane proteins are restored, and both or either of these events may regulate the switch from a migratory phenotype to one that supports proliferation and differentiation. If contact with the dermal matrix mediates keratinocyte activation, then the phenotype of the resting epidermis may be maintained by distinct cell-ECM interactions.

The literature on laminin involvement in the wound healing process is confusing. Early studies suggested that laminin-1, a heterotrimeric basement membrane protein composed of $\alpha1$, $\beta1$ and $\gamma1$ laminin subunits, is deposited in the newly formed basement membrane just behind the migrating front of epidermis[23] and is a potent inhibitor of keratinocyte migration.[65] Additionally, as reepithelialization progresses a mass of laminin-1 is deposited and accumulates under the previous migrating cells, providing a site-specific mechanism to down-regulate the activated keratinocyte phenotype. Indeed, expression of collagenase-1, a marker of the activated keratinocyte, is not induced in cells plated on a complex mix of basement membrane proteins (EHS tumor matrix) or purified laminin-1.[52] Furthermore, and consistent with this idea, relatively small concentrations of laminin-1 block the inductive effects of type I collagen (Fig. 2).[52] Assuming one binding event per molecule, one laminin-1 molecule effectively blocks the induction of collagenase-1 mediated by approximately 3,000 collagen molecules. Because type I collagen is so abundant in the dermis, it is not surprising that the inhibitory effect of laminin-1 is so potent. However, the relevance of these in vitro studies to the in vivo repair has been questioned since it is now recognized that laminin-1 is a somewhat rare laminin and is most likely either absent from or a minor component of the basement membrane zone of intact and healing epidermis.[66] Moreover, it should be noted that many studies concerning laminins in the skin were performed using poorly characterized laminin antibodies or antibodies that cross-react with more than one laminin subunit (discussed in 68).

The current consensus of opinion is that the major laminins in the basement membrane, i.e., laminins -5, -6 and -7, each possessing an α3 subunit, and laminin-10, which possesses an α5 subunit, not only function to enhance keratinocyte migration over wounds but also to stabilize keratinocyte adhesion to the wound bed.[2,3,6] In particular, recent work has focused on the role of the subunits of laminin-5 in wound healing. This heterotrimeric basement membrane protein is composed of α3, β3 and γ2 laminin chains[6] that assemble into complex arrays in the basement membrane of intact skin and interact with hemidesmosomes and matrix proteins, such as type VII collagen.[67,68] Thus, laminin-5 provides structural stability by forming a critical link between cells and the anchoring fibrils in the dermis.[69,70] In intact skin, adhesion to laminin-5 by keratinocytes is mediated by α6β4 integrin, which interacts with specific sequences in the G domain of the α3 subunit.[71] This laminin-5/α6β4 integrin complex is the backbone of the hemidesmosome and has been proposed to relay signals from the keratinocyte to the basement membrane and vice versa.[72-75]

Laminin-5 is synthesized as a precursor that undergoes rapid proteolytic processing of the α3 chain following secretion into the extracellular space.[76,77] Cleavage of the α3 subunit reduces its molecular weight from 200 to 165 kD while the γ2 subunit is processed from 150 to 105 kD.[76-79] In vitro, plasmin and bone morphogenetic protein-1 (BMP-1) have been reported to cleave the α3 subunit while BMP-1, in addition, processes the γ2 chain.[78,79] Furthermore, it has been suggested that MT1-MMP can cleave the γ2 subunit resulting not only in production of a 105 kD isoform but further proteolysis of the γ2 species giving rise to a 80 kD polypeptide.[80] Preprocessed and truncated laminin-5 forms have distinct effects on the ability of keratinocytes to migrate or form hemidesmosomes in vitro. Laminin-5 containing the preprocessed α3 subunit supports keratinocyte migration but does not induce hemidesmosome assembly, while laminin-5 containing the 165 kD α3 subunit nucleates assembly of hemidesmosomes and retards keratinocyte motility.[78] In contrast, laminin-5 containing an 80 kD γ2 subunit can drive cell migration during periods of active tissue remodeling.[80,81]

Although laminin-5 is a constitutively expressed component of the intact basement membrane, its expression is further upregulated in migrating keratinocytes just behind the wound front.[17,27,78,82,83] Laminin-5 is the first provisional basement membrane protein expressed by wound keratinocytes, preceding the production of laminin 10/11 and type VII collagen.[84] Interestingly, the α3 subunit of laminin-5 produced by migrating keratinocytes is in an unprocessed form that fails to promote hemidesmosome formation.[78] Because this laminin-5 isoform supports migration, it has been hypothesized that in wound healing its main function is to promote keratinocyte migration over the wound bed (Fig. 5). There are, however, certain problems with this idea. For instance, how does a cell lay down the track upon which it subsequently migrates? One possibility is that laminin-5 secreted from keratinocytes diffuses into the wound space where it becomes incorporated into the provisional wound matrix (Model 1, Fig. 5). It would therefore lie in the wound bed ahead of the advancing sheet of keratinocyte cells. There is some support for this possibility as soluble laminin-5 added to culture medium has been shown to integrate into forming and formed basement membranes of tissue explants in vitro.[85] Alternatively, provisional matrix proteins of the wound bed or chemokines in the wound fluid may initiate keratinocyte migration over and/or induce chemotaxis towards the wound. Unprocessed, newly secreted laminin-5 in the matrix that underlies the wound keratinocytes may then enhance the migration or spreading of a cell sheet over the wound bed (Model 2, Fig. 5). That antibodies which functionally perturb cell-laminin-5 interaction and impede keratinocyte motility in vitro provides support for both mechanisms.[26,84] It should be noted that, regardless of its role in cell migration, processed laminin-5 appears to have a major

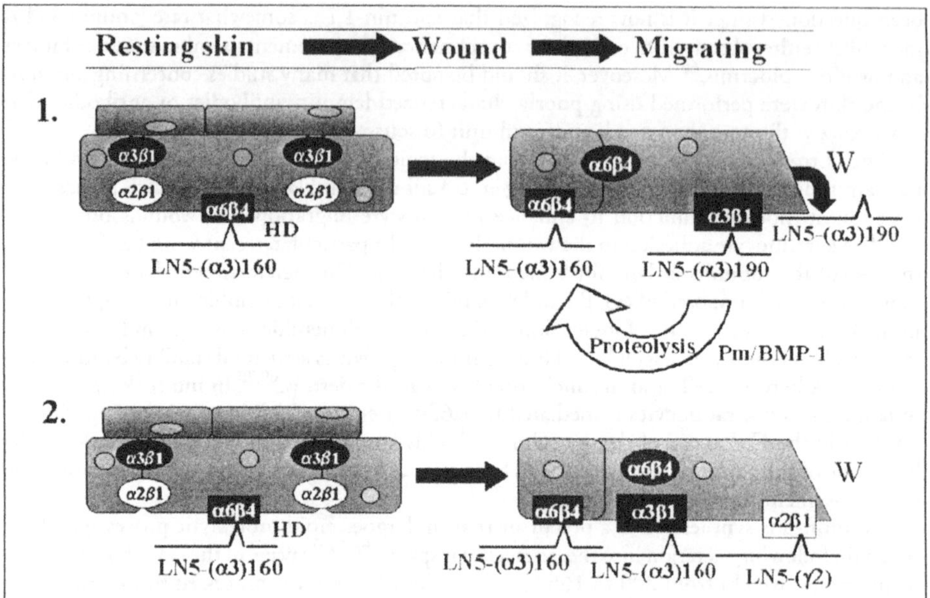

Fig. 5. Diagrammatic representation of the potential role of laminin-5 and its integrin receptors (α3β1, α6β4 and α2β1) in wound repair of the epidermis (adapted from 28). In resting skin the α6β4 integrin heterodimer is ligated to laminin-5 containing a processed α3 subunit (α3-160) in stable matrix anchors termed hemidesmosomes (HD). In this condition, α3β1 and α2β1 integrin complexes are located at sites of cell-cell contact. Model I is based on the studies of Goldfinger et al[26] and Nguyen et al[84] while model 2 is derived from the work of Decline and Rousselle.[27] In model 1, laminin-5 containing a 190 kD unprocessed α3 subunit (α3-190) is secreted from the wound cells (red arrow) and incorporates into the provisional matrix along the bed of the wound (W). The α3β1 integrin drives keratinocyte migration over this matrix. Alternatively, the keratinocytes lay down unprocessed laminin-5 into their matrix, and then this matrix supports the spreading or migration of the entire sheet of cells over the wound bed. Proteolysis of laminin-5 occurs via the action of plasmin (Pm) or BMP-1 and subsequent ligation to the α6β4 integrin nucleates hemidesmosome assembly and stable anchorage of the cells to the wound bed. In model 2, the α2β1 integrin mediates the migration of cells on laminin-5 via its interaction with the γ2 laminin subunit. In this Model, α3β1 integrin stabilizes adhesion of the migrating cells to the wound bed prior to α3 laminin subunit processing and hemidesmosome assembly triggered by processed laminin-5.

role in stable adhesion of keratinocytes to the newly deposited basement membrane by inducing reestablishment of hemidesmosomes and acting as a stop signal for migration.[26,78]

It is now well established that formation of mature hemidesmosomes in keratinocytes is induced by α6β4 integrin ligation to laminin-5 and that hemidesmosome assembly correlates with decreased cell motility.[78] Clusters of α6β4 integrin are observed along the substratum attached surfaces of a number of different cell types in the absence of laminin-5 ligand but these clusters most likely represent immature hemidesmosomes at best.[86] Alternatively, a number of workers have suggested that keratinocyte migration on laminin-5 matrix is mediated by the α3β1 integrin.[25,26,84,87] The evidence for this comes primarily from antibody inhibition studies. For example, antibodies against the α3 integrin subunit inhibit closure of wounds created in confluent monolayers of epithelial cells maintained in vitro.[26,84] However, Decline and Rousselle recently questioned this dogma as they showed that antibodies against the α_2 integrin subunit inhibit migration of keratinocytes over their own laminin-5 matrix towards a surface coated with fibronectin and type IV collagen.[27] Moreover, the α_2 integrin subunit codistributes with laminin-5 in the matrix of migrating keratinocytes. These workers identified a novel integrin binding site in the laminin-5 heterotrimer as they presented evidence that the α2β1 integrin complex shows direct interaction the γ2 subunit of laminin-5. Interestingly, the latter workers favor a hypothesis that, like α6β4 integrin, the α3β1 integrin heterodimer mediates adhesion to rather than migration over a laminin-5 matrix.

Concluding Remarks

Collectively, the data we discuss in this Chapter indicate that extracellular matrix proteins with which wound keratinocytes interact, namely type I collagen and laminin-5, profoundly influence gene expression and cell migration. In essence, we propose that ligation with preexisting dermal type I collagen initiates a wound phenotype, whereas the subsequent contact with newly deposited laminin-5 directs a program leading to a stationary, differentiating phenotype. At the onset of repair, a regulatory and functional interaction between collagenase-1 and collagen-mediated signaling is established as keratinocytes migrate from the basement membrane and onto the denuded dermis. Because in intact skin they are in contact with the dermis, ligation of the α2β1 integrin on keratinocytes with type I collagen provides an unambiguous spatial signal that the skin has been wounded. Interestingly, the integrin-collagen interaction that transduces this information is of a sufficiently high affinity as to hinder cell movement. Contact with collagen initiates a wound phenotype in keratinocytes characterized by expression of collagenase-1, which cleaves type I collagen, thereby loosening the affinity of the integrin-matrix interaction. Keratinocytes can then move over the wound bed, where they interact with more native collagen, which in turn, maintains the expression of collagenase-1 and allows the cycle to continue until the wound is closed.

Another cell-matrix interaction that affects keratinocyte behavior is the synthesis, processing, and deposition of laminin-5. Studies on the role of this ECM protein during healing have produced disparate results, leading to multiple, often incongruous interpretations of the role that laminin-5 plays during migration and re-formation of the basement membrane. Recent advances in conditional crelox knockout technology[88,89] would allow ablation of laminin-5 and its receptors in a localized area of the skin. Thus, one could wound knockout and nonknockout skin in the same animal and compare the rate, extent and efficiency of wound healing. To date, it has been impossible to critically address the role of laminin-5 in an in vivo model of wound repair as animals engineered lacking the α3 laminin subunit die are not viable to adulthood.[90] The data generated from laminin-5 subunit inducible knockouts may provide important new insights into the role of the ECM and ECM receptors in the skin repair process and critically determine the role of this protein during wound repair.

References

1. Clark EA, Brugge JS. Integrins and signal transduction pathways: The road taken. Science 1995; 268:233-239.
2. Burgeson RE, Christiano AM. The dermal-epidermal junction. Curr Opin Cell Biol 1997; 9:651-658.
3. Champliaud MF et al. Human amnion contains a novel laminin variant, laminin-7, which like laminin-6, covalently associates with laminin-5 to promote stable epithelial-stromal attachment. J Cell Biol 1996; 132:1189-1198.
4. Miner JH, Cunningham J, Sanes JR. Roles for laminin in embryogenesis: Exencephaly, syndactyly, placentopathy in mice lacking the laminin a5 chain. J Cell Biol 1998; 143:1713-1723.
5. Sorokin LM et al. Developmental regulation of the laminin alpha5 chain suggests a role in epithelial and endothelial cell maturation. Develop Biol 1997; 189:285-300.
6. Tunggal P, Smyth N, Paulsson M, Ott MC. Laminins: Structure and genetic regulation. Micr Res Tech 2000; 51:214-227.
7. Larjava H et al. Novel function for beta 1 integrins in keratinocyte cell-cell interactions. J Cell Biol 1990; 110:803-815.
8. Carter WG, Wayner EA, Bouchard TS, Kaur P. The role of integrins a2b1 and a3b1 in cell-cell and cell-substrate adhesion of human epidermal cells. J Cell Biol 1990; 110:1387-1404.
9. Tenchini ML et al. Evidence against a major role for integrins in calcium-dependent intercellular adhesion of epidermal keratinocytes. Cell Adhes Commun 1993; 1:55-66.

10. Coulombe PA. Towards a molecular definition of keratinocyte activation after acute injury to stratified epithelia. Biochem Biophys Res Commun 1997; 236:231-238.
11. Clark R. The Molecular and Cellular Biology of Wound Repair. New York: Plenum Press, 1996.
12. Martin P. Wound healing—Aiming for perfect skin regeneration. Science 1997; 276:75-81.
13. Woods A, Johansson S, Hook M. Fibronectin fibril formation involves cell interactions with two fibronectin domains. Exp Cell Res 1988; 177:272-83.
14. Mueller SC, Hasegawa T, Yamada SS, Yamada KM, Chen WT. Transmembrane orientation of the fibronectin receptor complex (integrin) demonstrated directly by a combination of immunocytochemical approaches. J Histochem Cytochem 1988; 36:297-306.
15. Vallen EA, Eldridge KA, Culp LA. Heparan sulfate proteoglycans in the substratum adhesion sites of human neuroblastoma cells: Modulation of affinity binding to fibronectin. J Cell Physiol 1988; 135:200-212.
16. Kurpakus MA, Quaranta V, Jones JCR. Surface relocation of alpha6 beta4 integrins and assembly of hemidesmosomes in an in vitro model of wound healing. J Cell Biol 1991; 115:1737-1750.
17. Kainulainen T et al. Laminin-5 expression is independent of the injury and microenvironment during the reepithelialization of wounds. J Histo Chem 1998; 46:353-60.
18. Stepp MA, Spurr-Michaud S, Gipson IK. Integrins in the wounded and unwounded stratified squamous epithelium of the cornea. Investig Ophthalmol Vis Sci 1993; 34:1829-1844.
19. Stepp MA, Zhu L, Cranfill R. Changes in beta 4 integrin expression and localization in vivo in response to corneal epithelial injury. Investig Ophthalmol Vis Sci 1996; 37:1593-1601.
20. Saarialho-Kere UK et al. Cell-matrix interactions modulate interstitial collagenase expression by human keratinocytes actively involved in wound healing. J Clin Invest 1993; 92:2858-2866.
21. Guo M et al. Altered processing of integrin receptors during keratinocyte activation. Exp Cell Res 1991; 195:315-322.
22. Cavani A et al. Distinctive integrin expression in the newly forming epidermis during wound repair. J Invest Dermatol 1993; 101:600-604.
23. Juhasz I, Murphey GF, Yan H-C, Herlyn M, Albelda SM. Regulation of extracellullar matrix proteins and integrin cell substratum adhesion receptors on epithelium during cutaneous wound healing in vivo. Am J Pathol 1993; 143:1458-1469.
24. Clark RAF et al. Fibronectin and fibrin provide a provisional matrix for epidermal cell migration during wound reepithelization. J Invest Dermatol 1982; 79:264-269.
25. Nguyen BP, Gil SG, Carter WG. Deposition of laminin 5 by keratinocytes regulates integrin adhesion and signaling. J Biol Chem 2000; 275:31896-31907.
26. Goldfinger LE et al. The a3 laminin subunit, a6b4 and a3b1 integrin coordinately regulate wound healing in cultured epithelial cells and in the skin. J Cell Sci 1999; 112:2615-2629.
27. Decline F, Rousselle P. Keratinocyte migration requires a2b1 integrin-mediated interaction with the laminin 5 g2 chain. J Cell Sci 2001; 114:811-823.
28. Woessner JF, Jr. The matrix metalloproteinase family. In: Parks WC, Mecham RP, eds. Matrix Metalloproteinases. San Diego: Academic Press, Inc., 1998:1-14.
29. Jeffrey JJ. Interstitial collagenases. In: Parks WC, Mecham RP, eds. Matrix Metalloproteinases. San Diego: Academic Press, Inc., 1998:15-42.
30. Lohi JL, Wilson CL, Roby JD, Parks WC. Epilysin: A novel human matrix metalloproteinase (MMP-28) expressed in testis and keratinocytes and in response to injury. J Biol Chem 2001; 276:10134-10144.
31. Pei D. Identification and characterization of the fifth membrane-type matrix metalloproteinase MT5-MMP. J Biol Chem 1999; 274:8925-8932.
32. Pei D. Leukolysin/MMP25/MT6-MMP: A novel matrix metalloproteinase specifically expressed in the leukocyte lineage. Cell Res 1999; 9:291-303.
33. Pei D. CA-MMP: A matrix metalloproteinase with a novel cysteine array, but without the classic cysteine switch. FEBS Lett 1999; 457:262-270.
34. Saarialho-Kere UK, Crouch EC, Parks WC. Matrix metalloproteinase matrilysin is constitutively expressed in adult human exocrine epithelium. J Invest Dermatol 1995; 105:190-6.
35. Wilson CL et al. Regulation of intestinal alpha-defensin activation by the metalloproteinase matrilysin in innate host defense. Science 1999; 286:113-117.
36. Grillo HC, Gross J. Collagenolytic activity during mammalian wound repair. Devel Biol 1967; 15:300-317.
37. Agren MS, Taplin CJ, Woessner JF, Eaglstein WH, Mertz PM. Collagenase in wound healing: Effect of wound age and type. J Invest Dermatol 1992; 99:709-714.
38. Buckley-Sturrock A. et al. Differential stimulation of collagenase and chemotatic activity in fibroblasts derived from rat wound repair tissue and human skin by growth factors. J Cell Physiol 1989; 138:70-78.

39. Saarialho-Kere UK, Chang ES, Welgus HG, Parks WC. Distinct localization of collagenase and TIMP expression in wound healing associated with ulcerative pyogenic granuloma. J Clin Invest 1992; 90:1952-1957.
40. Saarialho-Kere UK et al. Interstitial collagenase is expressed by keratinocytes which are actively involved in reepithelialization in blistering skin diseases. J Invest Dermatol 1995; 104:982-988.
41. Saarialho-Kere UK, Chang ES, Welgus HG, Parks WC. Expression of interstitial collagenase, 92 kDa gelatinase, and TIMP-1 in granuloma annulare and necrobiosis lipoidica diabeticorum. J Invest Dermatol 1993; 100:335-342.
42. Stricklin, GP, Nanney LB. Immunolocalization of collagenase and TIMP in healing human burn wounds. J Invest Dermatol 1994; 103:488-492.
43. Stricklin GP, Li L, Nanney LB. Localization of mRNAs representing interstitial collagenase, 72-kda gelatinase, and TIMP in healing porcine burn wounds. J Invest Dermatol 1994; 103:352-358.
44. Stricklin GP, Li L, Jancic V, Wenczak BA, Nanney LB. Localization of mRNAs representing collagenase and TIMP in sections of healing human burn wounds. Am J Pathol 1993; 143:1657-1666.
45. Vaalamo M et al. Patterns of matrix metalloproteinase and TIMP-1 expression in chronic and normally healing cutaneous wounds. Br J Dermatol 1996; 135:52-59.
46. Vaalamo M et al. Distinct populations of stromal cells express collagenase-3 (MMP-13) and collagenase-1 (MMP-1) in chronic ulcers but not in normally healing wounds. J Invest Dermatol 1997; 109:96-101.
47. Okada A et al. Expression of matrix metalloproteinases during rat skin wound healing: Evidence that membrane type-1 matrix metalloproteinase is a stromal activator of pro-gelatinase A. J Cell Biol 1997; 137:67-78.
48. Pilcher BK et al. The activity of collagenase-1 is required for keratinocyte migration on a type I collagen matrix. J Cell Biol 1997; 137:1445-1457.
49. Garlick JA, Parks WC, Welgus HG, Taichman LB. Re-epithelialization of human oral keratinocytes in vitro. J Dent Res 1996; 75:912-918.
50. Inoue M, Kratz G, Haegerstrand A, Ståhle-Bäckdahl M. Collagenase expression is rapidly induced in wound-edge keratinocytes after acute injury in human skin, persists during healing, and stops at reepithelialization. J Invest Dermatol 1995; 104:479-483.
51. Sudbeck BD, Pilcher BK, Pentland AP, Parks WC. Modulation of intracellular calcium levels inhibits secretion of collagenase-1 by migrating keratinocytes. Mol Biol Cell 1997; 8:811-824.
52. Sudbeck BD, Pilcher BK, Welgus HG, Parks WC. Induction and repression of collagenase-1 by keratinocytes is controlled by distinct components of different extracellular matrix compartments. J Biol Chem 1997; 272:22103-22110.
53. Librach CL et al. 92-kD type IV collagenase mediates invasion of human cytotrophoblasts. J Cell Biol 1991; 113:437-449.
54. Fisher C et al. Interstitial collagenase is required for angiogenesis in vitro. Dev Biol 1994; 162:499-510.
55. Shipley JM, Wesselschmidt RL, Kobayashi DK, Ley TJ, Shapiro SD. Metalloelastase is required for macrophage-mediated proteolysis and matrix invasion in mice. Proc Natl Acad Sci USA 1996; 93:3942-3946.
56. Cohen IK, Diegelmann RF, Linblad WJ. Epithelialization. In: Parks WC, Mecham RP, eds. Wound Healing: Biochemical and Clinical Aspects. Philadelphia: W.B. Saunders Co., 1992:115-127.
57. Moore WM, Spilburg CA. Purification of human collagenases with a hydroxamic acid affinity column. Biochemistry 1986; 25:5189-5195.
58. Wu H et al. Generation of collagenase-resistant collagen by site-directed mutagenesis of murine pro-a1(I) collagen gene. Proc Natl Acad Sci USA. 1990; 87:5888-5892.
59. Liu X et al. A targeted mutation at the known collagenase cleavage site in mouse type I collagen impairs tissue remodeling. J Cell Biol 1995; 130:227-237.
60. Carter WG, Wayner EA, Bouchard TS, Kaur P. The role of integrins alpha 2 beta 1 and alpha 3 beta 1 in cell-cell and cell-substrate adhesion of human epidermal cells. J Cell Bio. 1990; 110:1387-1404.
61. Symington BE, Takada Y, Carter WG. Interaction of integrins $\alpha3\beta1$ and $\alpha2\beta1$: Potential role in keratinocyte intercellular adhesion. J Cell Biol 1993; 120:523-535.
62. Staatz WD, Rajpara SM, Wayner EA, Carter WG, Santoro SA. The membrane glycoprotein Ia-IIa (VLA-2) complex mediates the Mg^{+2}-dependent adhesion of platelets to collagen. J Cell Biol 1989; 108:1917-1924.
63. Sudbeck BD, Parks WC, Welgus HG, Pentland AP. Collagen-mediated induction of keratinocyte collagenase is mediated by tyrosine kinase and protein kinase C activities. J Biol Chem 1994; 269:30022-30029.

64. Dumin JA, Dickeson SK, Stricker TP, Bhattacharyya-Pakrasi M, Roby JD, Santoro S et al. Procollagenase-1 (MMP-1) binds the integrin $\alpha_2\beta_1$ upon release from keratinocytes migrating on type I collagen. J Biol Chem 2001; 276:in press.
65. Woodley DT, Bachmann PM, O'Keefe EJ. Laminin inhibits human keratinocyte migration. J Cell Physiol 1988; 136:140-146.
66. Ferletta M, Ekblom P. Identification of laminin-10/11 as a strong cell adhesive complex for a normal and a malignant human epithelial cell line. J Cell Sci 1999; 112:1-10.
67. Chen M et al. Interactions of the amino-terminal noncollagenous (NC1) domain of type VII collagen with extracellular matrix components. A potential role in epidermal-dermal adherence in human skin. J Biologic Chem 1997; 272:14516-14522.
68. Rousselle P et al. Laminin 5 binds the NC-1 domain of type VII collagen. J Cell Biolog 1997; 138:719-728.
69. Borradori L, Sonnenberg A. Structure and function of hemidesmosomes: More than simple adhesion complexes. J Invest Dermatol 1999; 112:411-418.
70. Jones JCR, Hopkinson SB, Goldfinger LE. Structure and assembly of hemidesmosomes. BioEssays 1998; 20:488-494.
71. Baker SE et al. Laminin-5 and hemidesmosomes: Role of the alpha 3 chain subunit in hemidesmosome stability and assembly. J Cell Sci 1996; 109:2509-2520.
72. Giancotti FG. Signal transduction by the a6b4 integrin: Charting the path between laminin binding and nuclear events. J Cell Sci 1996; 109:1165-1172.
73. Mainiero F et al. The coupling of a6b4 integrin to the Ras-MAP kinase pathways mediated by Shc controls keratinocyte proliferation. EMBO J 1997; 16:2365-2375.
74. Clarke AS, Lotz MM, Cha, C, Mercurio AM. Activation of the p21 pathway of growth arrest and apoptosis by the b4 integrin cytoplasmic domain. J Biol Chem 1995; 270:22673-22676.
75. Shaw LM, Rabinovitz I, Wang HH-F, Toker A, Mercurio AM. Activation of phosphoinositide 3-OH kinase by the a6b4 integrin promotes carcinoma invasion. Cell 1997; 91:949-960.
76. Matsui C, Wang CK, Nelson CF, Hoeffler WK. The assembly of laminin-5 subunits. J Biol Chem 1995; 270:23496-23503.
77. Marinkovich MP, Lunstrum GP, Burgeson RE. The anchoring filament protein kalinin is synthesized and secreted as a high molecular weight precursor. J Biol Chem 1992; 267:17900-17906.
78. Goldfinger LE, Stack MS, Jones JCR. Processing of laminin-5 and its functional consequences: Role of plasmin and tissue-type plasminogen activator. J Cell Biol 1998; 141:255-265.
79. Amano S et al. Bone morphogenetic protein-1 (BMP-1) is an extracellular processing enzyme of the laminin 5 g2 chain. J Biol Chem 2000:in press.
80. Koshikawa N, Giannelli G, Cirulli V, Miyazaki K, Quaranta V. Role of cell surface metalloprotease MT1-MMP in epithelial cell migration over laminin-5. J Cell Biol 2000; 148:615-624.
81. Giannelli G, Falk-Marzillier J, Schiraldi O, Stetler-Stevenson WG, Quaranta V. Induction of cell migration by matrix metalloprotease-2 cleavage of laminin-5. Science 1997; 277:225-228.
82. Qin P, Kurpakus MA. The role of laminin-5 in TGFalpha/EGF-mediated corneal epithelial cell motility. Exp Eye Res 1998; 66:569-579.
83. Salo S et al. Laminin-5 promotes adhesion and migration of epithelial cells: Identification of a migration-related element in the g2 chain gene (LAMC2) with activity in transgenic mice. Matrix Biol 1999; 18:197-210.
84. Nguyen BP, Ryan MC, Gil SG, Carter WG. Deposition of laminin 5 in epidermal wounds regulates integrin signaling and adhesion. Curr Op Cell Biol 2000; 12:554-562.
85. Baker SE et al. Morphogenetic effects of soluble laminin-5 on cultured epithelial cells and tissue explants. Exp Cell Res 1996; 228:262-270.
86. Nievers MG et al. Ligand independent role of the beta4 integrin subunit in the formation of hemidesmosomes. J Cell Sci 1998; 111:1659-1672.
87. Mizushima H et al. Identification of integrin-dependent and -independent cell adhesion domains in COOH-terminal globular region of laminin-5 a3 chain. Cell Growth Diff 1997; 8:979-987.
88. Fleming RE et al. Mouse strain differences determine severity of iron accumulation in Hfe knockout model of hereditary hemochromatosis. Proc Natl Acad Sci USA 2001; 98:2707-2711.
89. Cao T, Longley M, Wang X, Roop D. An inducible mouse model for epidermolysis bullosa simplex. Implications for gene therapy. J Cell Biol 2001; 152:651-6.
90. Ryan MC, Lee K, Miyashita Y, Carter WG. Targeted disruption of the LAMA3 gene in mice reveals abnormalities in survival and late stage differentiation of epithelial cells. J Cell Biol 1999; 145:1309-1323.

CHAPTER 10

Matrix Metalloproteinases and their Inhibitors in Clinical Oncology

Taina Turpeenniemi-Hujanen

Abstract

The number of members in matrix metalloproteinase (MMP) family is rapidly increasing. It is evident that several members in this family contribute to cancer metastasis. Both invasion and angiogenesis are essential in tumor progression, and the members of the MMP-family are important in both of these cancer-related but not cancer-specific phenomena. Most clinical data published support the concept that in particular gelatinases A and B (MMP-2 and –9) are linked to the aggressive behavior of cancer. The aggressive clinical course has been linked to the expression of a MMP in a surprisingly large variety of cancer. The overexpression of gelatinases has especially turned out to be prognostic for survival or to correlate with grade or stage in several cancer types. MMP-2 has even been shown to predict relapse during antiestrogen adjuvant chemotherapy in breast carcinoma. The role of TIMPs (tissue inhibitors of metalloproteinases) is more complex in determining the clinical course of different types of cancers. In some tumors the relative amount of metalloproteinase and its inhibitor is associated with aggressive behavior of the disease. The main data, however, states for the time being that aggressive cancers also overexpress TIMPs, which may be associated with tumor progression rather than with an indolent clinical course of a tumor. There seems to be a larger variation between different cancer types in this context. Studies to control tumor invasion and angiogenesis by using synthetic inhibitors for MMPs have so far shown little clinical promise. Angiogenesis is controlled by many inhibitors and activators. It seems difficult to be able to affect it with very specific tools. This does not rule out that combinations of different antiangiogenic agents may be useful in the future as treatment options for cancer patients. MMPIs are now studied in many clinical trials, and the results of the ongoing studies will define their role in cancer treatment.

Metalloproteinase System in Cancer

Increased proteolysis is important in extracellular matrix (ECM) degradation in clinical conditions characterized by rapid matrix turnover as in tumor progression and metastasis. Matrix metalloproteinases (MMPs), a family of zinc-dependent endopeptidases, seem to act in concert to degrade various ECM proteins.[1] The metalloproteinases and their inhibitors in malignant tissues are produced by the tumor cells, stromal cells or both. The proteolysis in tumors is, at least in part, conducted by tumor-stroma interactions.[2,3] The growth of the tumor is also associated with an increase in the turnover of the tumor stroma which in some cases could clinically be reflected in serum levels of several markers associated with increased remodeling of the connective tissue matrix.[4]

MMPs have been implicated in tumor angiogenesis in several studies. Classical angiogenesis, the recruitment of new blood vessels from pre-existing ones, is a prerequisite for tumor clusters to grow beyond a few millimeters in size.[5] Several factors stimulating endothelial cells

Cell Invasion, edited by Jyrki Heino and Veli-Matti Kähäri. ©2002 Eurekah.com.

are known to induce the production of metalloproteinases.[6,7] The migration of endothelial cells to form a tube and later a vessel has some similarity with the spread of a tumor cell. MMP-2 and MMP-9, in particular, are upregulated in angiogenic lesions.[8] The attachment to endothelial cells might even induce MMP-9 from monocytes.[9] The role of gelatinases in angiogenesis is particularly addressed in studies of Itoh et al (1998), which demonstrated a reduction of angiogenesis and tumor growth in MMP-2 knockout mice.[10] The C-terminal, hemopexin-like domain of MMP-2 (=PEX) is able to block MMP-activity, and it disrupts tumor angiogenesis and growth.[11] Also the studies by Mohan et al showing that the angiogenic response stimulated by FGF-2 includes the activation of MMP-9[12] supports this concept. Endothelial synthesis of MMPs seems, however, to also have opposite effects on tumor angiogenesis. MMPs may have an antiangiogenic effect in producing angiostatin.[13,14] In spite of the theoretical controversies the MMP inhibitors have been shown to block tumor angiogenesis as well as tumor invasion[15] confirming the profound role of metalloproteinases in both of these prerequisites for the tumor progression.

The activities of MMPs are controlled in vivo by tissue inhibitors of metalloproteinases (TIMPs) that can form complexes either with latent or activated enzymes.[16] TIMPs are known to have an inhibitory effect on tumor growth, invasion, angiogenesis and metastasis in experimental models.[17-19] TIMPs can also take part in the activation of MMPs.[17] Disruption of the balance between the MMPs and their endogenous tissue inhibitors takes place in several diseases where increased proteolytic activity is believed to play a clinicopathological role such as in malignant invasion and metastasis. The picture is not, however, so clear clinically. This has become evident after the first clinical studies aiming to explore the roles of the different members of the metalloproteinase system in the clinical course of different types of cancers. In theory, it is a fascinating idea to be able to control matrix degradation and the progress of the disease by changing the balance between metalloproteinases and their inhibitors in favor of inhibitors. The genetic manipulation required for this has not, however, so far offered any clinical applications in oncology. MMPs are, therefore, still attractive targets for designing new synthetic inhibitors, which recently have entered into clinical trials.

The present clinical works studying metalloproteinases aim to characterize their role in the progression of different types of cancers, to use them as prognostic markers for metastasis and survival as well as to use them as predictive markers for clinical decision making concerning the choice of the cancer therapy and finally to use them as biological targets for new types of cancer therapy.

Measuring MMPs or TIMPs in Clinical Samples

Gelatinases and their inhibitors can be studied in clinical context with various methods. Avidin-biotin-immunohistochemical staining[20] using monoclonal or polyclonal antibodies is performed either on frozen or paraffin sections of the tumor depending on the antibody. Polyclonal antibodies for MMP-2 may have a cross-reactivity with fibronectin. Also monoclonal antibodies recognize different forms of MMPs (latent/activated) either in complex with TIMPs or free in tissues depending on the antigenic site they are designed to recognize. This has to be kept in mind when comparing the results from different studies. In situ hybridization, RT-PCR or Northern analysis have been used in some clinical studies to measure the mRNA for the enzymes or inhibitors. These techniques also are semiquantitative. The two former ones are more suitable for clinical use. In situ hybridization also shows localization of mRNA in the tissue. In zymography (substrate gel electrophoresis), enzymes can be evaluated according to their molecular weight and their activity.[21,22] This technique requires fresh tissue and a thorough pretreatment of the sample, but the advantage is that the relationship of the active and latent enzyme forms can be evaluated. Enzyme-linked immunoassays (ELISA) detect specific enzyme or inhibitor proteins from liquid samples (tissue homogenates, cerebrospinal fluid, ascites, pleural fluid, serum or plasma) with the help of the antibodies. In modified ELISAs tissue inhibitors are used together with the monoclonal antibodies in order to detect free and inhibitor-bound enzyme forms.[23,24] Immunohistochemistry and ELISA are applicable for the routine clinical use.

MMPs in Breast Carcinoma

Breast cancer is the most common cancer in women. Several biological factors associated with aggressive behavior have been identified. The trend in breast cancer treatment favors adjuvant treatment even in early stage breast carcinoma. Some breast cancers are still very indolent and the patients have a favorable survival. The key question is, therefore, which predict cure without adjuvant treatment. There are data showing that MMP-2 immunoreactive protein is linked to an aggressive clinical course in breast carcinoma. In early stage breast carcinoma the immunoreactive protein for MMP-2 does correlate with relapse.[25] This study failed to show any correlation between MMP-2 and survival. Recently, it was shown in studies from our group that high expression of MMP-2 protein in breast cancer cells of the primary tumor correlates with poor survival. This study consisted of unselected patients (164 patients) who were followed for 10 years.[26] MMP-2 was shown to be an independent prognostic factor in a multivariate analysis. Moreover, it was demonstrated in this study that the adverse influence of MMP-2 positivity on survival seems to be roughly dependent on the grading of the positive immunoreaction. Patients with strong positivity fell in the worst category whereas those only weakly positive had smaller differences in their survival rates when compared to MMP-2 negative patients.[26] Later, the predictive role of MMP-2 protein was further explored in pre- and postmenopausal patients suffering from breast carcinoma with axillary lymph node metastases. These patients received either antiestrogen or combination chemotherapy as an adjuvant therapy. It was shown that young, MMP-2 -positive, premenopausal patients are more likely to relapse during or after an adjuvant chemotherapy than other patients.[27] It is also notable, that the MMP-2 positivity can predict failure for the adjuvant antiestrogen therapy in postmenopausal patients with node-positive breast carcinoma.[28] On the other hand, the prognosis of patients with a MMP-2-negative tumor is significantly more favorable suggesting that MMP-2 negative patients do not require aggressive adjuvant treatment. Further studies are needed to confirm these results.

Metalloproteinases and their Inhibitors in Urological Cancers

Bladder cancer is one of the most extensively studied models to explore the function of metalloproteinases in cancer invasion. Several clinical studies have confirmed the preclinical findings showing a profound role of metalloproteinases in transitional cell cancer progression. Kanayama et al (1998) studied MMP-2 as well as its activator (MT1 MMP) and inhibitor (TIMP-2) expression in 41 patients with bladder carcinoma. They showed that the expression of TIMP-2 and MMP-2 is high in muscle invasive cancer and MT1-MMP, MMP-2 as well as TIMP-2 are all associated with decreased overall survival.[29] The inverse association of TIMP-2 with survival had been previously shown also by Grignon et al (1996) in bladder cancer.[30] The members of the metalloproteinase system can be studied in sera. S-MMP-2 and MMP-3 are correlated with a short disease-free survival in urothelial cancer.[31] Moreover, S-MMP-2/TIMP-2 is an independent prognostic factor in this cancer.[32] Further studies will show whether serum MMPs have value in monitoring patients treated for urothelial cancer.

The role of MMPs in renal carcinoma is less well known. The work done by Walther et al (1997) link MMP-2 to aggressive behavior in renal tumors. Renal cancer cells were cultured in this study from the primary tumors and they expressed mRNA for MMP-2. The higher expression was found to correlate with an unfavorable prognosis.[33] It is interesting that, in this study, the immunoreactive protein seen in primary tumor was linked to the expression of mRNA for MMP-2 in the tumor cells, and both findings related to the clinical course of the disease.

Preclinical studies have addressed the important role of MMPs in prostate cancer invasion and metastasis. Modification of MMP-9 in cultured prostate carcinoma cells correlates, for instance, with the ability of the cells to produce lung metastases in animal models.[34] Several studies link both MMP-2 and MMP-9 to prostate cancer cell invasion in vitro[35] or to prostate cancer-induced angiogenesis.[36] FGF or TGFβ induce and IL-10 decrease the expression of MMPs in prostate cancer cells.[35-38] IL-10 stimulates the expression of TIMP-1,[35,36] and it also inhibits the prostate cancer cell induced angiogenensis[36] as well as metastasis in an animal model.[35]

In clinical studies the MMPs generally are associated with an aggressive course of prostate carcinoma. MT1-MMP and MMP-2 immunolocalization changes during the progression of PIN (prostate intraepithelial neoplasia).[39] MMP-2 or TIMP-2 mRNA is localized to the malignant epithelial cells in both low and high grade tumors and in areas of extracapsular involvement in prostate carcinomas. MMP-2 to TIMP-2 -ratio is high in high-grade or high-stage tumors.[40] The immunoreactive protein for MMP-2 localizes on the prostate cancer cell plasma membrane and correlates to a comparable Gleason score (a strong prognostic variable in prostate cancer).[41] MMP-7 mRNA in primary tumor correlates with invasion, stage and serum PSA (prostate specific antigen, another important prognostic variable in prostate cancer).[42] Finally, MMP-2, -9 and TIMP-1 in primary tumor are prognostic in prostate carcinoma patients.[43] All the clinical data combined indicate that gelatinases have a profound role in the aggressive behavior of prostate carcinoma. Plasma levels of MMP-1, MMP-3, TIMP-1 and MMP-1/TIMP-1 -ratio have been studied in patients with prostate carcinoma by Lein et al (1998). MMP-1 or MMP-1/TIMP-1 –ratio did not correlate with stage; MMP-3 and TIMP-1 were, however, higher in patients with metastases.[44] Interestingly, serum MMP-2 does correlate with the extent of the disease in prostate carcinoma.[45] Future studies will show whether this marker will give any further information over using only the PSA-value to monitor prostate cancer patients.

MMPs in Melanoma

Preclinical studies address the role of MMPs in melanoma progression. Several lines of cultured melanoma cells have been shown to produce at least MMP-1, -2, -3 and –9. Further, the MMP-activity has been correlated to melanoma invasion in vitro. The expression of MMP-2 and MMP-9 as well as the invasion, attachment and type IV collagen-degrading activity has been shown to decrease in human melanoma cells after long term exposure to interferon alpha or gamma[46,21] indicating that the effect of interferons on melanoma progression could partly be due to the inhibition of metalloproteinases. Similarly Hendrix et al (1990) have demonstrated a simultaneous decrease in the propensity to invasion and the expression of MMP-2 and MMP-9 in melanoma cells after treatment with retinoic acid.[47] Further evidence supporting the important role of MMPs in melanoma invasion comes from the studies showing that TIMPs inhibit melanoma cell growth, invasion and attachment to basement membrane components,[48,49] that synthetic inhibitors for MMPs decrease melanoma metastasis in animal models,[50] and that monoclonal antibodies prepared for MMP-2 inhibit the invasive capacity of cultured human melanoma cells.[51] Moreover, the progression and angiogenesis of melanoma have been shown to be decreased in gelatinase A-deficient mice,[10] and the incorporation of TIMP-2cDNA into murine melanoma cells has resulted in a decrease in the growth of the cells in mice.[23] Concerning the bulk of preclinical data confirming that MMPs are important in melanoma progression it is surprising that only a few studies have been published about the clinical significance of MMPs in melanoma. The preclinical findings have been translated to clinical reality in three studies showing that the immunoreactive protein for MMP-2 is, indeed, a prognostic factor in primary cutaneous[52,53] as well as in uveal melanoma.[23] Although MMP-2 expression in tumor cells is correlated with prognosis in cutaneous melanoma serum, MMP-2 does not seem to offer any significant value in prognosis.[54] The expression of MMP-2 is occasionally increased also in the early stages of melanoma progression.[52] High MMP-2 immunoreactivity is, however, associated with malignant melanoma and with hematogeneous metastasis in melanoma, especially in males.[52,53] MMP-2 positive male melanoma patients have significantly higher risk for dying due to melanoma than other melanoma patients, 10 year-survival rates being 39% with MMP-2 positive males compared to 79% in other patients.[53] Since interferons are known to be antiangiogenic in melanoma and decrease the expression of gelatinases in melanoma cells after continuous exposure, it is important in further clinical studies to see whether adjuvant interferon therapy should be administered for these higher risk groups selected according to their MMP-expression.

Markers of Matrix Degradation in Lung Carcinoma

Lung carcinoma is a common and aggressive cancer comprising histologically different forms that all have an unfavorable prognosis. There are a few studies suggesting that MMPs may be involved in lung cancer progression. The mRNA for MMP-2, MT1-MMP and MMP-11 are highly expressed in bronchopulmonary carcinomas and are associated with malignant phenotype and TNM stage.[55] Further, the high MMP-2 or MMP-9 levels are related to increased tumor spread and a poor prognosis in lung cancer.[56,57] An association between tumor MMP-2 and MMP-9 with survival has been reported in lung adenocarcinoma.[57] In situ hybridization of MMP-2, MMP-9 and E-cadherin can predict the prognosis in non-small cell lung carcinoma (NSCLC) patients. Herbst and Fidler reported recently that (MMP-2 + MMP-9)/2: E-cadherin –rate is prognostic and shows an inverse correlation with overall survival and disease-free survival.[12] It is possible that the MMP-status can be determined from blood since plasma MMP-9 and serum TIMP-1 levels are prognostic for poor survival in lung carcinoma patients.[24] It is notable that the type I collagen degradation fragment, ICTP (type I carboxyterminal propeptide), is associated with collagen degradation in tissues. Recently S-ICTP was also shown to be associated with an unfavorable prognosis in lung cancer.[4] These studies suggest that the use of these markers of matrix degradation could be promising in the follow-up of the lung cancer patients also, but future studies addressing this question are on going.

MMPs and their Inhibitors in Gastrointestinal Cancers

Several studies have linked either MMP-2 or MMP-9 or both to aggressive behavior or advanced stage in different gastrointestinal tumors. There is a positive correlation between tumor stage and high expression of MMP-2 or TIMP-2 in colorectal cancer.[58,59] Also tissue mRNA levels of MMP-9 predict relapse and poor survival in this malignancy.[60] In gastric cancer, the high levels of MMP-2 and MMP-9 are also associated with poor survival.[61,62] Pancreatic carcinoma is a malignancy with a poor prognosis. Very few patients survive more than a year after diagnosis. It is surprising therefore, that overexpression of MMP-2 could indicate an aggressive phenotype even in pancreatic carcinoma.[63] The possibility of using assays of blood MMP-levels in patient follow-up or monitoring the treatment response has been studied in gastrointestinal tumors. Increased MMP-9 plasma levels are associated with poor prognosis in stage IV gastrointestinal cancers[64] but there is little data.

MMPs in Brain Neoplasias

Brain tumors include a wide range of neoplasms with different invasive behavior and varying prognosis. Brain tumors with low malignancy have a tendency to recur after radical excision, and with most malignant brain tumors the survival time is usually short. The tendency towards malignant progression may be related to angiogenic as well as to invasive potential of the tumor cells. The treatment of the most aggressive forms of brain tumors is still a challenge. In spite of the development of new drugs the prognosis of aggressive gliomas has not changed. New options for treatment are thus needed especially with glioblastomas. There are a few studies showing a possible role of metalloproteinases in brain tumor progression. High expression of metalloproteinases, particularly both MMP-9 and MMP-2 have been detected in malignant brain tumors.[65-70] The detection of the MMPs is possible also in cerebrospinal fluid. Friedberg et al showed a correlation between MMP-9 and malignity in brain tumors. The active MMP-2 correlated in this study with the malignant cytology in the cerebrospinal fluid.[71] The most extensive clinical material has so far been studied by Jäälinoja et al (2000) for the localization and grade of expression of immunocreative protein for MMP-2 in different brain tumors.[72] MMP-2 correlates with an aggressive histological grade of the tumors and with decreased overall survival.[72] Interestingly all metastases appeared positive for MMP-2 in this study,[72] as was shown with the hematogenous metastases of melanoma.[52] These findings further support the role of MMPs in brain tumor progression.

MMPs in Gynecological Malignancies

Metalloproteinases have been studied in gynecological malignancies, most extensively in ovarian carcinoma where several preclinical studies have previously associated the activity of these enzymes with malignant behavior (invasion and metastasis).[2,3,73] Metalloproteinases and their mRNA have been detected both in tumor cells and in stroma in ovarian and cervical carcinomas.[74-76] Some studies correlate gelatinases to lymph node metastasis. There are only a few clinical studies concerning of the value of MMPs in predicting the clinical course of these malignancies. In endometrial carcinoma such evidence is so far lacking, but preliminary studies in ovarian[74,77,78] and cervical carcinomas[79] also associate MMPs with a poor outcome. These studies are addressed in following chapter.

It is possible that ovarian cancer cells are dependent on stromal interactions in their MMP expression although they have been shown to produce MMPs in culture conditions in several studies. Ovarian cancer cells are able to maintain MMP-2 expression after culturing from primary tumors, whereas levels of MMP-9 decrease over time.[73] The expression of gelatinases is influenced by soluble factors from ovarian cancer or by contacts between ovarian cancer and fibroblasts in cell culture[2,3] usually showing an increased expression of MMPs in fibroblasts[2,3] but also in the ovarian cancer cells.[3] Autoregulation in controlling the expression of gelatinases and the propensity to invade is possible in this tumor model.[3] Tamakoshi et al (1994) showed that the active form of MMP-2 is linked to the ovarian cancer.[80] Similarly, MMP-7 mRNA and protein are overexpressed in ovarian tumors possibly contributing to its invasive nature.[81] Several studies have confirmed the association of the increased MMP-2 levels with ovarian cancer.[74,80,82-84] Moreover, a correlation between increased MMP-2 and poor survival has been shown.[74,78] It is notable, however, that the stromal negativity for MMP-2 protein could also be important in ovarian malignant behavior.[74] Ovarian cancer cells express low amounts of MMP-2 mRNA whereas the immunoreactive protein for MMP-2 is increased in ovarian cancer cells.[85] It is possible that ovarian cancer cells are able to utilize exogenous MMP-2 to mediate the proteolysis.[86] Finally, high levels of MMP-2, MMP-9, MT1-MMP and TIMP-2 mRNA correlate with unfavorable prognosis in ovarian cancer patients suggesting that MMPs do have an important role in progression of ovarian carcinoma.[78] This is also supported by the study of Garzetti et al (1995) showing higher S-MMP-2 levels in ovarian cancer patients when compared with patients with borderline tumors or cystadenomas[84] which address the need for further clinical studies.

Gelatinases have been associated with malignancy in two studies concerning cervical neoplasias. MMP-9 showed a trend to be overexpressed in cancers with poor differentiation, vessel invasion and lymph node metastasis.[75] There are a few studies suggesting a profound role of MMP-2 in the progression of cervical carcinoma. Cervical cancer cells modified by a transfection of antiapotosis clone 11 have been shown to express more MMP-2 and MT-1MMP but low TIMP-2.[75] MMP-2, TIMP-2 and MT-1MMP are detected both in stromal and tumor cells in lesions of cervical carcinoma; elevated MMP-2, however, occurs only in tumor cells. Precancerous CIN lesions failed to express these markers although also controversial findings are published.[75,76] Finally, a correlation between tumor MMP-2 and TIMP-2 mRNA and advanced stage or poor survival is reported in one clinical study in cervical carcinoma,[79] which needs to be confirmed.

Although the hormonal regulation of MMPs in nonmalignant endometrium is under investigation, there are only a few clinical studies addressing the role of these enzymes in endometrial malignancy. In a study by Iurlalro et al an enhanced expression of mRNA for

MMP-2 and MMP-9 was seen in malignant endometrium when compared with benign endometrium.[87] The expression of gelatinases in this study was elevated with the tumor grade and invasion depth. In addition, Määttä et al (2000) showed that enhanced expression of MT1-MMP, TIMP-1, -2 and –3 mRNA is seen in endometrial carcinoma.[88] MMP-1, MMP-2 and TIMP-2 immunoreactive protein can also be detected in most cases of endometrial carcinomas (80-95%) but none of these markers have been correlated with the overall survival.[89]

Hematological Malignancies

Most clinical studies exploring the role of MMPs in cancer progression have been carried out in solid tumors. Very few clinical studies exist on the hematological malignancies and MMPs. All leukemias are malignant neoplasias but myeloid leukemias usually have an especially poor prognosis. The knowledge of the biology of AML (acute myelous leukemia) has increased and cytogenetic and molecular markers have superseded morphology as the most important prognostic factors. The immunoreactive proteins of MMP-2 and MMP-9 have been studied in pretreatment bone marrows in two series of leukemia patients.[90,91] MMP-9 protein is a typical finding in the mature granulated neutrophils whereas lymphocytes are negative for both MMP-2 and -9. Some blast cells appear positive for MMP-2 in both ALL and AML but are associated with a favorable prognosis in AML.[90] Some of the adult patients with ALL have positive staining in bone marrow blast cells for MMP-2 but MMP-2 positivity is rare in bone marrow of the pediatric ALL patients.[91] Positive MMP-2 correlates, however, with the appearance of extramedullary infiltrates in adult ALL[91] suggesting that this metalloproteinase could contribute to the invasive behavior of blasts in adult ALL. Larger clinical studies are the role of metalloproteinases in hematological neoplasias.

The Role of MMPs in Bone Metastasis

Bone metastases are common in several types of cancers, particularly in breast and prostate cancer, which often present with bone metastases. Biochemical studies involving MMPs in bone metastasis have been carried out especially in prostate cancer models. Bone resorption correlates to MMP activity in prostate carcinoma cells.[92] TGFβ1 produced by osteoblasts is able to induce MMP-9 in prostate cancer cells.[93] MMPs are involved in osteoclastic bone resorption which is probably the key degradative machinery in a bone metastases. This is supported by findings showing inhibition of breast cancer cell-associated osteolysis in a mouse model by a synthetic metalloproteinase inhibitor and TIMP-2 transfection.[94] Bisphosphonates have been shown to inhibit MMP activity as well as the invasion of cancer cells.[93] It is interesting that alendronate is able to decrease MMP-2 expression in cultured prostate carcinoma cells[95] which suggests that the inhibitory effect of bisphosphonates on bone metastasis progression could also partly be due to a direct effect on cancer cells, not only on osteoclasts.

Clinical Value of TIMPs

Although TIMPs are important regulatory proteins in homeostasis of the ECM, some TIMPs also affect the proliferation as well as inhibit apoptosis and angiogenesis. TIMPs may act as organ-specific regulators in tissue development. For example TIMP-1 is known to have a specific role in bone development.[96,97] The diagnostic and therapeutic value of TIMP in human cancer has been studied extensively, usually including the detection of TIMPs in studies exploring the role of MMPs in cancer progression. The use of TIMPs as markers to predict clinical outcome has been widely examined, but the evidence generated in this field has been so far conflicting. In general the appearance of TIMPs in tumors has been thought to predict a favorable clinical outcome. Indeed, in some studies this has been the case. The serum MMP-2/TIMP-2 ratio turned out to be an independent prognostic factor for instance in urothelial carcinoma.[98] On the other hand, serum TIMP-2 correlates with advanced prostate carcinoma.[99] Similarly, high serum levels of TIMP-1 are associated with poor outcome in lung carcinoma.[24] Further, the tissue TIMP-2 in tumor correlates with invasion and decreased survival in the patients suffering from urothelial carcinoma.[100] Also positive stromal TIMP-2 immunostaining has been reported to correlate with tumor recurrence in breast cancer.[101] High tissue levels of TIMP-2 appear in cancer,[102,103] and it is not surprising that the levels of this inhibitor may also correlate with aggressive clinical behavior. It is difficult to interpret the data published so far on the existence of TIMPs in different cancers. It is evident, however, that cancer tissue produces both metalloproteinases and their inhibitors, the relative amount of which may differ in different cancer types.

Synthetic Matrix Metalloproteinase Inhibitors (MMPIs)

Regulation of metalloproteinase activity could theoretically offer an interesting tool to effect several biological MMP-related processes which are difficult to treat such as some chronic but disabling diseases or cancer.[104] Low toxicity treatment options that would have specificity against cancer would be valuable. Small molecular agents that could block the active site of the metalloproteinases have, therefore, been designed and tested with tumor bearing animals and later in clinical studies.[105,106] MMPIs are able to reduce invasion and intravasation of tumor cells in experimental models.[50,107] Several of these agents are now in early clinical trials, some of them even in phase III trials (Table 1). Nearly all of the inhibitors analysed contain a zinc ion chelating group such as a hydroxyamate, a carboxylate or a thiol group, a peptidic variant mimicking the peptide substrate which can bind to the substrate recognition site of the catalytic center of a metalloprotease.

The first synthetic of MMPIs inhibited a wide spectrum of MMPs. Theoretically, these compounds could also cause a variety of side effects due to the role of MMPs in connective tissue turnover. First MMPI in clinical trials was batimastat (BB-94), which has shown growth inhibition of tumors and hemangiomas in preclinical studies.[108] Its clinical use is restricted by its intraperitoneal or intrapleural use. Phase I/II trials in patients with malignant ascites have demonstrated a clinical benefit in 90% of patients.[109] The most advanced studies have been carried out so far with marimastat (BB-2516) which is administered orally. Dosing has varied in clinical studies up to 50 mg twice in a day. This dose causes side effects in >60 % of patients when used over a month period[110,111] including frozen shoulder or Dupuytren's disease after a long treatment.[111] Musculoskeletal pain, stiffness and tenderness are common. Side effects decreased on a dose of 25 mg twice a day,[110] and the most recent studies have been carried out with a lower dose.[112] The clinical effect of MMPI is usually a tumor marker response or a stabilization of the markers.[110] Recent studies have aimed at stabilization of the growth of the tumor. In pancreatic carcinoma, a stabilization of CA 19-9 levels was found in phase I/II studies leading to phase II/III trials with oral MMPIs.[113,114] A phase II randomized study in early breast cancer is being conducted in patients that receive placebo or marimastat 5-10 mg p.o. bid for 12 months given after adjuvant chemotherapy or tamoxifen.[112] These low doses have been well-tolerated,[112] but reports of the effect are still unpublished. First reports regarding MMPs as targets for therapy in prostate carcinoma showed some promise. In prostatic carcinoma patients, marimastat decreased serum PSA values in a dose-dependent manner.[105] This suggests that in the future this treatment should be studied at least in prostate carcinoma. A randomized, double-blind, placebo-controlled, multicenter study in 184 patients with inoperable gastric adenocarcinoma was the first study to show tumor effect in treatment of cancer patients with MMPI.[115] The patients were randomized to receive placebo or marimastat 10 mg two times in a day up to 18 months after chemotherapy if the patient's disease had stabilized. Survival improved with MMPI, and the difference was higher with patients that did not have metastases.[115] MM-1270 is a broad spectrum inhibitor that can be administered orally 1600 mg a day divided into[116] two doses. In phase I trials musculoskeletal toxicity typical for MMPIs appeared as well as skin rashes typical for inhibitors of angiogenesis. It can be used in (300 mg twice in a day) combination with 5-fluorouracil in patients with colorectal carcinoma.[116]

Due to side effects seen with nonspecific MMPIs, more specific inhibitors were designed for gelatinases. BAY 12-9566 is an inhibitor of MMP-2 and MMP-9.[124] It can be administered

Table 1. Synthetic metalloproteinase inhibitors in clinical trials

Compound	References	Range of inhibition	po.	Phase	Adjuvant	Response
BB-94 batimastate	109	wide	-	I/II, ascites	-	50 %
BB-2516 marimastate	112	wide	+	III, several, breast	+	ad 58% (marker)
BAY 12-9566	118, 117	MMP-2, -3 MMP-9, -13	+	III, lung, pancreatic	-	SD (I)
MM 1270	116	wide	+	I, colorectal	-	-
AG 3340 prinomastat	119	MMP-2, -3 MMP-9, -13	+	III, lung, prostatic	-	-
Mestat, COI3	120	wide	+	I/II, several	-	-

orally in two doses of 800 mg. In phase I studies the side effects included thrombocytopenia and elevation of transaminases; musculoskeletal toxicity was low. It could be combined with paclitaxel and carboplatin, and phase III trials were conducted in pancreatic and ovarian cancer.[117] BAY-12-9566 was compared in a randomized trial with gemcitabine in advanced pancreatic carcinoma patients that had no prior chemotherapy. The second interim analysis with 277 patients revealed a failure of MMPI to effect the course of this disease and the trial was closed.[118] AG 3340 is a selective inhibitor for MMP-2, -3, -9 and –13. It is administered orally and in phase I studies with 2-100 mg twice a day; doses >5 mg produced plasma concentrations exceeding the Ki-value for the inhibition of MMP-2.[119] Doses less than 25 mg bid are well tolerated with few joint complaints.[119] Phase III studies in non-small cell lung cancer and hormone-independent prostate cancer were designed. Col-3 was recently tested in a phase I study in patients with refractory solid tumors using doses starting from 36 mg/m^2/day. The cutaneous phototoxicity was dose limiting at 98 mg/m^2/day. Some stablization of disease was reported.[120] Some other new MMPIs are in clinical trials, and the results are expected in coming years.[104,121]

References

1. Curran S, Murray GI. Matrix metalloproteinases in tumour invasion and metastasis. J Pathol 1999; 189:300-308.
2. Miyagi E, Gasumitsu H, Hirahara F et al. Marked induction of gelatinases, especially type B, in host fibroblasts by human ovarian cancer cell in anthymic mice. Clin Exp Metastasis 1995; 13:89-96.
3. Westerlund A, Hujanen E, Puistola U et al. Fibroblasts stimulate human ovarian cancer cell invasion and expression of 72-kDa gelatinase A (MMP-2). Gyn Oncology 1997; 67:76-82.
4. Ylisirniö S, Sassi ML, Risteli J et al. Serum type I collagen degradation markers, ICTP and crosslaps, are factors for poor survival in lung cancer. Cancer Res 1999; 19:5577-5582.
5. Hanahan D, Folkman J. Patterns and emerging mechanisms of the angiogenic switch during tumorigenesis. Cell 1996; 86:353-364.
6. Mach F, Schonbeck U, Fabunmi RP et al. T -lymphocytes induce endothelial cell matrix metalloproteinase expression by a CD40L-dependent mechanism: implications for tubule formation. Am J Pathol 1999; 154:229-238.
7. Lamoreaux WJ, Fitzgerald ME, Reiner A et al. Vascular endothelial growth factor increases release of gelatinase A and decreases release of tissue inhibitor of metalloproteinases by microvascular endothelial cells in vitro. Microvascular Res 1998; 55:29-42.
8. Kurizaki T, Toi M, Tominaga T. Relationship between matrix metalloproteinase expression and tumor angiogenesis in human breast carcinoma. Oncology Rep 1998; 5:673-677.

9. Amorino GP, Hoover RL. Interactions of monocytic cells with human endothelial cells stimulate monocytic metalloproteinase production. Am J Pathol 1998; 152:199-207.
10. Itoh T, Tanioka M, Yoshida H et al. Reduced angiogenesis and tumour progression in gelatinase A-deficient mice. Cancer Res 1998; 58:1048-1051.
11. Brooks PC, Silletti S, von Schalscha TL et al. Disruption of angiogenesis by PEX, a noncatalytic metalloproteinase fragment with integrin binding activity. Cell 1998; 92:391-400.
12. Mohan R, Sivak J, Ashton P et al. Curcuminoids inhibit the angiogenestic response stimulated by fibroblast growth factor-2, including expression of matrix metalloproteinase gelatinase B. J Biol Chem 2000; 275:10405-12.
13. O'Reilly MS, Wiederschain D, Stetler-Stevenson WG et al. Regluation of angiostatin production by matrix metalloproteinase-2 in a model of concomitant resistance. J Biol Chem 1999; 274:29568-71.
14. Patterson BC, Sang QA. Angiostatin-converting enzyme activities of human matrilysin (MMP-7) and gelatinase B/type IV collagenase (MMP-9). J Biol Chem 1997; 272:28823-28825.
15. Hibner B, Cad A, Flynn C et al. BAY 12-9566, a noverl, biphenyl matrix metalloproteinase inhibitor, demonstrates anti-invasive and anti-angiogenic properties. Proc Am Assoc Cancer Res 1998; 39:A2063.
16. Fujimoto N, Ward RV, Shinya T et al. Interaction between tissue inhibitor of metalloproteinases-2 and progelatinase A: immunoreactivity analysis. Biochem J 1996; 313:827-833.
17. Chambers AF, Matrisian LM. Changing views of the role of matrix metalloproteinases in metastasis. J Natl Canc Inst 1997; 89:1260-1270.
18. Stetler-Stevenson WG, Krutzsch HC, Liotta LA. Tissue inhibitor of metalloproteinase (TIMP-2). J Biol Chem 1989; 264:17374-78.
19. Kähäri V-M, Saarialho-Kere U. Matrix metalloproteinases and their inhibitors in tumour growth and invasion. Ann Med 1999; 31:34-45.
20. Hsu SM, Raine L, Fanger H. The use of avidin-biotin peroxidase comples (ABC) in immunoperoxidase technique: A comparison between ABC and unlabeled antibody (PAP) procedures. J Histochem Cytochem 1981; 29:577-580.
21. Hujanen ES, Väisänen A, Zheng A et al. Modulation of Mr 72.000 and Mr 92.000 type IV collagenase (gelatinase A and B) gene expression by interferons alpha and gamma in human melanoma. Int J Cancer 1994; 58:582-586.
22. Heussen C, Dowdle EB. Electrophoretic analysis of plasminogen activators in polyacrylamide gels containing sodium dodecylsulfate and copolymerized substrates. Ann Biochem 1980; 102:192-202.
23. Väisänen A, Kallioinen M, von Dickhoff K et al. Matrix metalloproteinase-2 (MMP-2) immunocreative protein – a new prognostic marker in ureal melanoma? J Pathol 1999; 188:56-62.
24. Ylisirniö S, Höyhtyä M, Turpeenniemi-Hujanen T. Serum matrix metalloproteinases-2, -9 and tissue inhibitors of metalloproteinases-1, -2 in lung cancer – TIMP-1 as a prognostic marker. Anticancer Res 2000; 20:1311-1316.
25. Daidone MG, Silvestrini R, D'Errico A et al. Laminin receptors, collagenase IV and prognosis in node negative breast cancers. Int J Cancer 1991; 48:529-532.
26. Talvensaari-Mattila A, Pääkkö P, Höyhtyä M et al. MMP-2 immunocreative protein, a marker of aggressiveness in breast carcinoma. Cancer 1998; 83:1153-1162.
27. Talvensaari-Mattila A, Pääkkö P, Turpeenniemi-Hujanen T. MMP-2 positivity and age less than 40 years increases the risk for recurrence in premenopausal patients with node-positive breast carcinoma. Breast Cancer Res Treatm 1999; 58:287-293.
28. Talvensaari-Mattila A, Pääkkö P, Blanco-Sequeiros G et al. Matrix metalloproteinase-2 (MMP-2) is associated with the risk for a relapse in postmenopausal patients with node-positive breast cardinoma treated with antiestrogen adjuvant therapy. Breast Cancer Res Treatm 2001; 65:55-61.
29. Kanayama H, Yokota K, Kurokawa Y et al. Prognostic values of matrix metalloproteinase-2 and tissure inhibitor of metalloproteinase-2 expression in bladder cancer. Cancer 1998; 82:1359-1366.
30. Grignon DJ, Saler W, Toth M et al. High levels of tissue inhibitor of metalloproteinase-2 (TIMP-2) expression are associated with poor outcome in invasine bladder cancer. Cancer Res 1996; 56:1654-1659.
31. Gohji K, Fujimoto N, Komiyama T et al. Elevation of serum levels of matrix metalloproteinase-2 and –3 as new predictors of recurrence in patients with urothelial carcinoma. Cancer 1996; 78:2379-2387.
32. Gohji K, Fujimoto N, Ohkawa J et al. Imbalance between serum matrix metalloproteinase-2 and its inhibitor as a predictor of recurrence of urothelial cancer. Br J Cancer 1998; 77:650-655.
33. Walther MM, Kleiner DE, Lubesnky IA et al. Progelatinase A mRNA expression in cell lines derived from tumors in patients with metastatic renal cell carcinoma correlates inversely with survival. Urology 1997; 50:295-301.

34. Sehgal G, Hua J, Bernhard EJ et al. Requirement for matrix metalloproteinase-9 (gelatinase B) expression in metastasis by murine prostate carcinoma. Am J Pathol 1998; 152:591-596.
35. Stearns ME, Fudge K, Garcia F et al. IL-10 inhibition of human prostate PC-3ML cell metastases in SCID mice: IL-10 stimulation of TIMP-1 and inhibition of MMP-2/MMP-9 expression. Invasion Metastasis 1997; 17:62-74.
36. Stearns ME, Rhim J, Wang M. Interleukin 10 (IL-10) inhibition of primary human prostate cell-induced angiogenesis: IL-10 stimulation of tissue inhibitor of metalloproteinase-1 and inhibition of matrix metalloproteinase (MMP)-2/MMP-9 secretion. Clin Cancer Res 1999; 5:189-196.
37. Klein RD, Maliner-Jongewaard MS, Udayakusmar TS et al. Promatrilysin expression is induced by fibroblast growth factors in the prostatic carcinoma cell line LnlcaP but not in normal primary prostate epithelial cells. Prostate 1999; 41:215-223.
38. Festuccia C, Angelucci A, Gravina GL et al. Osteoblastderined T6F-beta 1 modulates matrix degrading protease expression and activity in prostate cancer cells. Int J Cancer 2000; 85:407-415.
39. Upadhyay J, Shekarrig B, Nemeth JA et al. Membrane type-1 matrix metalloproteinase (MT1-MMP) and MMP-2 immunolocalization associated with high-grade prostatic intraepithelial neoplasia. Clinical Cacner Res 1999; 5:4105-10.
40. Still K, Robson CN, Autzen P et al. Localization and quantification of mRNA for matrix metalloproteinase-2 (MMP-2) and tissue inhibitor of matrix metalloproteinase-2 (TIMP-2) in human bening and malignant prostatic tissue. Prostate 2000; 42:18-25.
41. Stearns ME, Stearns M: Immunohistochemical studies of activated matrix metalloproteinase-2 (MMP2a) expression in human prostate cancer. Onc Res 1996; 8:63-7.
42. Hashimoto K, Kihira Y, Motuo Y et al. Expression of matrix metalloproteinase-7 and tissue inhibitor of metalloproteine se-1 in human prostate. J Urology 1998; 160:1872-1876.
43. Wood M, Fudge K, Mohler JL et al. In situ hybridization studies of metalloproteinases 2 and 9 and TIMP-1 and TIMP-2 expression in human prostate cancer. Clin Exp Met 1997; 15:246-258.
44. Lein M, Nowak L, Jung K et al. Metalloproteinases (MMP-1, MMP-3) and their inhibitors (TIMP) in blood plasma of patients with prostate carcinoma. Urologe Ausgabe A 1998; 37:377-81.
45. Gohji K, Fujimoto N, Hara I et al. Serum matrix metalloproteinase-2 and its density in men with prostate cancer as a new predictor of disease extension. Int J Cancer 1998; 79:96-101.
46. Hujanen ES, Turpeennieim-Hujanen T. Recombinant interferon alpha and gamma modulate the invasive potential of human melanoma in vitro. Int J Cancer 1991; 47:576-581.
47. Hendrix MJ, Wood WR, Seftor EA et al. Retinoic acid inhibitor of human melanoma cell invasion through a reconstituted basement membrane and its relation to decreases in the expression of proteolytic enzymes and mortility factor receptor. Cancer Res 1990; 50:4121-4130.
48. Valente P, Fassina G, Melchiori A et al. TIMP-2 overexpression reduces invasion and angiogenesis and protects B16F10 melanoma cells from apoptosis. Int J Cancer 1998; 75:246-253.
49. Montgomery AMP, Mueller BM, Resifeld RA et al. Effect of tissue inhibitor of the matrix metalloproteinase-2 expression on the growth and spontaneous metastasis of a human melanoma cell line. Cancer Res 1994; 54:5467-73.
50. Chirivi RG, Garofalo A, Crissamin MJ et al. Inhibition of the metastatic spread and growth of BI6-BL6 murine melanoma by a synthetic matrix metalloproteinase inhibitor. Int J Cancer 1994; 58:460-464.
51. Höyhtyä M, Hujanen E, Turpeenniemi-Hujanen T et al. Modulation of type IV collagenase activity and invasive behavior of metastaic human melanoma (A2058) cells in vitro by monoclonal antibodies to type IV collagenase. Int J Cancer 1990; 46:282-286.
52. Väisänen A, Tuominen H, Kallioinen M et al. Matrix metalloproteinase-2 (72 kD type IV collagenase) expression occurs in the early stage of human melanocytic tumour progression and may have prognostic value. J Pathol 1996; 180:283-289.
53. Väisänen A, Kallioinen M, Taskinen PJ et al. Prognostic value of MMP-2 immunocreative protein (72 kD type IV collagenase) in primary skin melanoma. J Pathol 1998; 186:51-58.
54. Vuoristo MS, Kellokumpu-Lehtinen P, Parvinen LM et al. Serum matrix metalloproteinase-2 as a prognostic marker in advanced cutaneous melanoma. Acta Oncol 2000; 39:877-9.
55. Narvrocki B, Polette M, Marchand V et al. Expression of matrix metalloproteinases and their inhibitors in human bronchopulmonary carcinomas: quantitative and morphological analysis. Int J Cancer 1977; 72:556-64.
56. Brown PD, Bloxidge RE, Stuart NSA et al. Association between expression of activated 72-kD gelatinase and tumor spread in non-small-cell lung carcinoma. J Natl Cancer Inst 1993; 85:574-578.
57. Kodate M, Kasai T, Hashimoto H et al. Expression of MMP (gelatinase) in T1 adenocarcinoma of the lung. Pathol Int 1997; 47:461-469.
58. Lesey AT, Cioce V, Sobel ME et al. Increased expression of the mr. 72000 type IV collagenase in human colonic adenocarcinoma. Cancer Res 1991; 51:439-444.

59. Ring P, Johansson K, Höyhtyä M et al. Expression of tissue inhibitor of metalloproteinase TIMP-2 in human colorectal cancer – a predictor of tumor stage. Br J Cancer 1997; 76:805-811.
60. Zeng ZS, Huang Y, Cohen AM et al. Prediction of colorectal cancer relapse and survival via tissue RNA levels of matrix metalloproteinase-9. J Clin Oncol 1996; 14:3133-40.
61. Grigioni WF, D'Errico A, Fortunate C et al. Prognosis of gastric carcinoma revealed by interactions between tumor cells and basement membrane. Mol Pathol 1994; 7:220-225.
62. Sier CF, Kubben FJ, Ganesh S et al. Tissue levels of matrix metalloproteinases MMP-2 and MMP-9 are related to the overall survival of patients with gastric carcinoma. Br J Cancer 1996; 74:413-417.
63. Bramhall SR, Neoptolemis JP, Stamp GWH et al. Imbalance of expression of matrix metalloproteinases (MMPs) and tissue inhibitors of the matrix metalloproteinases (TIMPs) in human pancreatic cancer. J Pathol 1997; 182:347-355.
64. Zucker S, Lysik RM, DiMassimo et al. Plasma assay of gelatinase B: tissue inhibitor of metalloproteinase complexes in cancer. Cancer 1995; 76:700-8.
65. Rao JS, Steck PA, Mohanam S et al. Elevated levels of Mr 92,000 type IV collagenase in human brain tumors. Cancer Res 1993; 53:2208-2211.
66. Nakagawa T, Kubota T, Kabuto M et al. Production of matrix metalloproteinases and tissue inhibitor of metalloproteinases-1 by human brain tumors. J Neurosurg 1994; 81:69-77.
67. Nakano A, Tani E, Miyazaki K et al. Matrix metalloproteinases and tissue inhibitors of metalloproteinases in human gliomas. J Neurosurg 1995; 83:298-307.
68. Saxena A, Robertson JT, Kufta C et al. Increased expression of gelatinase A and TIMP-2 in primary human glioblastomas. Int J Oncol 1995; 7:469-473.
69. Rab JS, Yamamoto M, Mohanam S et al. Expression and localization of 92 kDa type IV collagenase/gelatinaseB (MMP-9) in human gliomas. Clin Exp Metastasis 1996; 14:12-18.
70. Sawaya RE, Yamamoto M, Yokaslan ZL et al. Expression and localization of 72 kDa type IV collagenase (MMP-2) in human malignant gliomas in vivo. Clin Exp Metastasis 1996; 14:35-42.
71. Friedberg MH, Glantz MJ, Klempner MS et al. Specific matrix metalloproteinase profiles in the cerebrospinal fluid correlated with the presence of malignant astrocytomas, brain metastases, and carcinomatous meningitis. Cancer 1998; 82: 923-930.
72. Jäälinoja J, Herva R, Korpela M et al. Matrix metalloproteinase 2 (MMP-2) immunoreactive protein is associated with poor grade and survival in brain neoplasms. J Neuro-Ongolocy 2000; 46:81-90.
73. Fishman DA, Bafetti LM, Banionis S et al. Production of extracellular matrix-degrading proteinases by primary cultures of human epithelial ovarian carcinoma cells. Cancer 1997; 80:1457-1463.
74. Westerlund A, Apaja-Sarkkinen M, Höyhtyä M et al. Gelatinase A –immunoreactive protein in ovarian lesions – prognostic value in epithelial ovarian cancer. Gynecologic Oncology 1999; 79:91-98.
75. Inoue Y, Abe K, Obata K et al. Immunohistochemical studies on matrix metalloproteinase-9 (MMP-9) and type-IV collagen in endometrial carcinoma. J Obst Gynaec Res 1997; 23(2):139-145.
76. Talvensaari-Mattila Anne, Apaja-Sarkkinen M, Höyhytä M et al. Matrix metalloproteinase 2 immunocreative protein appears early in cervical epithelial dedifferentiation. Gynecol Oncol 1999; 72:306-311.
77. Davidson B, Goldberg I, Golieb WH et al. High levls of MMP-2, MMP-9 MT1-MMP and TIMP-2 mRNA correlate with poor survival in ovarian carcinoma. Clin Exp Metastas 1999; 17:799-808.
78. Davidson B, Goldberg I, Gotlieb WH et al. High levels of MMP-2, MMP-9, MT1-MMP and TIMP-2 mRNA correlate with poor survival in ovarian carcinoma. Clin Exp Metastasis 1999; 17:799-808.
79. Davidson B, Goldberg I, Kopolovic J et al. MMP-2 and TIMP-2 expression correlates with poor prognosis in cervical carcinoma – a clinicopathologic study using immunohistochemistry and mRNA in situ hybridization. Gynecologic Oncology 1999; 73:372-382.
80. Tamakoshi K, Kikkawa F, Nawa A et al. Different pattern of zymography between human gynecologic normal and malignant tissues. Am J Obstet Gynecol 1994; 2:478-484.
81. Tanimoto H, Underwood LJ, Shigemasa K et al. The matrix metalloproteinase pump-1 (MMP-7, Matrilysin): a candidate marker/target for ovarian cancer detection and treatment. Tumor Biology 1999; 20:88-98.
82. Naylor MS, Stamp GWH, Davies B et al. Expression and activity of MMPs and their regulators in ovarian cancer. In J Cancer 1994; 58:50-56.
83. Takemura M, Azuma C, Kimura T et al. Type IV collagenase and tissue inhibtor of metalloproteinase in ovarian cancer tissue. Int J Gyn Obstet 1994; 46:303-309.
84. Garzetti GG, Ciavattini A, Lucarini G et al. Tissue and serum metalloproteinase (MMP-2) expression in advanced ovarian serous cystadenocarcinomas: clinical and prognostic implications. Anticancer Res 1995; 15(6B):2799-2804.
85. Autio-Harmainen H, Karttunen T, Hurskainen T. Expression of 72 kDa type IV collagenase (gelatinase A) in benign and malignant ovarian tumors. Lab Invest 1993; 63:312-321.

86. Young TN, Pizzo SV, Stack MS. A plasma membrane-associated component of ovarian adenocarcinoma cells enhances the catalytic efficiency of matrix metalloproteinase-2. J Biol Chem 1995; 270:999-1002.
87. Iurlaro M, Loverro G, Vacca A et al. Angiogenesis extent and expression of matrix metalloproteinase-2 and -9 correlate with upgrading and myometrial invasion in endometrial carcinoma. Eur J Clin Invest 1999; 29:793-801.
88. Määttä M, Soini Y, Liakka A et al. Localization of MT1-MMP, TIMP-1, TIMP-2 and TIMP-3 messenger RNA in normal, hyperplastic, and neoplastic endometrium. Enhanced expression by endometrial adenocarcinomas is associated with low differentation. Am J Clin Pathol 2000; 114:402-411.
89. Moser PL, Hefler L, Tempfer C et al. Immunohistochemical detection of matrix metalloproteinases (MMP) 1 and 2, and tissue inhibitor of metalloproteinase 2 (TIMP 2) in stage I and II endometrial cancer. Anticancer Res 1999; 19(3B):2365-7.
90. Kuittinen O, Savolainen E-R, Koistinen P et al. T. Gelatinase A and B (MMP-2 , MMP-9) in leukaemia MMP-2 may indicate a good prognosis in AML. Anticancer Res 1999; 19:4395-400.
91. Kuittinen O, Savolainen E-R, Koistinen et al. MMP-2 and MMP-9 expression in adult and childhood acute lymphatic leukemia (ALL). Leukemia Res 2001; 25:125-32.
92. Sanchez-Sweatman OH, Orr FW, Singh G. Human metastatic prostate PC3 cell lines degrade bone using matrix metalloproteinases. Invasion Metastasis 1999; 18:297-305.
93. Boissier S, Ferreras M, Peyruchaud O et al. Bisphosphonates inhibit breast and prostate carcinoma cell invasion, an early event in the formation of bone metastases1. Cancer Research 2000; 60:2949-2954.
94. Yoneda T, Sasaki A, Dunstan C et al. Inhibition of osteolytic bone metastasis of breast cancer of combined treatment with the bisphosphonate ibandronate and tissue inhibitor of the matrix metalloproteinase-2. J Clin Invest 1997; 99:2509-2517.
95. Stearns ME. Alendronate blocks TGF-beta 1 stimulated collagen 1 degradation by human prostate PC-3 ML cells. Clin Exp Met 1998; 16:332-9.
96. Nomura S, Hogan BL, Wills AJ et al. Developmental expression of tissue inhibitor of metalloproteinase (TIMP) RNA. Development 1989; 105:575-583.
97. Flenniken AM, Williams BR. Developmental expression of the endogenous TIMP gene and a TIMP-lacZ fusion gene in transgenic mice. Genes Dev 1990; 4:1094-1106.
98. Gohji K, Fujimoto N, Ohkawa J et al. Imbalance between serum matrix metalloproteinase-2 and its inhibitor as a predictor of recurrence of urothelial cancer. Br J Cancer 1998; 77:650-5.
99. Jung K, Nowak L, Lein M et al. Matrix metalloproteinases -1 and -3, tissue inhibitor of metalloproteinase-1 and the complex of metalloproteinase-1/tissue inhibitor in plasma of patients with prostate cancer. Int J Cancer 1997; 74:220-223.
100. Kanayama H, Yokota K, Kurokawa Y et al. Prognostic values of matrix metalloproteinase-2 and tissue inhibitor of metalloproteinase-2 expression in bladder cancer. Cancer 1998; 82:1359-1366.
101. Visscher DW, Höyhtyä M, Ottosen SK et al. Enhanced expression of tissue inhibitor of MMP-2 (TIMP-2) in the stroma of breast cancer correlates with tumor reccurence. Int J Cancer 1994; 59:339-344.
102. Visscher DW, Höyhtyä M, Ottosen et al. Enhanced expression of tissue inhibitor of metalloproteinase-2 (TIMP-2) in the stroma of breast carcinomas correlates with tumor reccurence. Int J Cancer 1994; 59:339-344.
103. Grignon DJ, Sater W, Toth M et al. High levels of tissue inhibitor of metalloproteinase-2 (TIMP-2) expression are associated with poor outcome in invasive bladder cancer. Cancer Res 1996; 56:1654-1659.
104. Brown PD. Ongoing trials with matrix metalloproteinase inhibitors. Expert Opinion on Investigational Drugs 2000; 9:2167-2177.
105. Wojtowicz-Praga SM, Dickson RB, Hawkins MJ. Matrix metalloproteinase inhibitors. Invest New Drugs 1997; 15:61-75.
106. Denis LJ, Verweij J. Matrix metalloproteinase inhibitors: present achievements and future prospects. Invest New Drugs 1997; 15:175-185.
107. Kim J, Yu W, Kovalski K, Ossowski L: Requirement for specific proteases in cancer cell intravasation as revealed by a novel semiquantitative PCR-based assay. Cell 1998; 94:353-62.
108. Taraboletti G, Garofalo A, Belotti D et al. Inhibitor of angiogenesis and murine hemangioma growth by batimastat, a synthetic inhibitor of matrix metalloproteinase. J Natl Cancer Inst 1995; 87:293-298.
109. Parsons SL, Watson SA, Steele RJ. Phase I/II trial of batimastat, a matrix metalloproteinase inhibitor, in patients with malignant ascites. European Journal of Surgical Oncology. 1997; 23(6):526-31.
110. Nemunaitis J, Poole C, Primrose J et al. Combined analysis of studies of the effects of the matrix metalloproteinase inhibitor marimastat on serum tumor markers in advanced cancer: selection of a biologically active and tolerable dose for longer-term studies. Clin Cancer Res 1998; 4(5):1101-9.

111. Hutchinson JW, Tierney GM, Parsons SL et al. Dupuytren's disease and frozen shoulder induced by treatment with a matrix metalloproteinase inhibitor. Journal of Bone & Joint Surgery—British Volume. 1998; 80(5):907-8.
112. Miller KD, Gradishar WJ, Schuchter LM et al. A randomized phase II pilot trial of adjuvant marimastat (MAR) in patients (Pts) with early breast cancer. Proceedings of Am Soc Clin Oncol 2000; 19:96a.
113. Evans JD, Ghaneh P, Kawesha A, et al. Role of matrix metalloproteinases and their inhibitors in pancreatic cancer. Digestion 1997; 58(6):520-8.
114. Rosemurgy A, Harris J, Langleben A et al. Marimastat in patients with advanced pancreatic cancer: a dose-finding study. Am J Clin Oncol. 1999; 22(3):247-52.
115. Fielding J, Scholefield J, Stuart R et al. A randomized double-blind placebo-controlled study of marimastat in patients with inoperable gastric adenocarcinoma. Proceedings of Am Soc Clin Oncol 2000; 19:240a.
116. Eatock M, Cassidy J, Johnson J et al. A phase I study of the matrix metalloproteinase inhibitor MM1270 (previously termed CYS 27023A) with 5FU and folinic acid. Proceedings of Am Soc Clin Oncol 1999; 18:209a.
117. Goel R, Hirte H, Major P et al. Clinical pharmacology of the metalloproteinase (MMP) and angiogenesis inhibitor Bayer 12-9566 in cancer patients. Proceedings of Am Soc Clin Oncol 1999; 18:160a.
118. Moore MJ, Hamm J, Eisenberg P et al. A comparison between gemicitabine (GEM) and the matrix metalloproteinase (MMP) inhibitor BAY12-9566 (9566) in patients (Pts) with advanced pancreatic cancer. Proceedings of American Society of Clinical Oncology 2000; 19:240a.
119. Pithavala Y, Schalinsky D, Wilding G et al. Comparison of preclinical efficacy and associated plasma concentrations of AG 3340, a matrix metalloproteinase (MMP) inhibitor, with plasma concentrations achieved clinically. Proceedings of Am Soc Clin Oncol 1999; 18:223a.
120. Rudek MA, Figg WD, Dyer V et al. Phase I clinical trial of oral col-3, a matrix metalloproteinase inhibitor, in patients with refractory metastatic cancer. J Clin Oncol 2001; 19:584.
121. Heath EI, Grochow LB. Clinical potential of matrix metalloproteinase inhibitors in cancer therapy. Drugs 2000; 59:1043-1055.

Index

A

B

C

E

F

G

H

I

K

L

M

N

O

P

R

S

T

U

V

W